$26.75

Illustrated Guide to the NATIONAL ELECTRICAL CODE 1993

by John E. Traister

Craftsman Book Company
6058 Corte del Cedro / P.O. Box 6500 / Carlsbad, CA / 92018

Acknowledgements

A deep and grateful bow must be made in the direction of several individuals and organizations who helped in the preparation and development of this book. Without their help, there would be no book.

National Fire Protection Association, Inc.

Square D Company

John Karns, Art Work

Linda Karns, Typist

Ruby Updike, Typist and Organizer

Library of Congress Cataloging in Publication Data

Traister, John E.
Illustrated guide to the National electrical code, 1993 / by John E. Traister.
p. cm.
Includes index.
ISBN 0-934041-80-6
1. Electric engineering--United States--Insurance requirements.
2. Electric wiring--United States--Insurance requirements.
3. National Fire Protection Association. National Electrical Code (1993) I. Title.
TK260.T73 1992
621.319'24'021873--dc20 92-36229
CIP

Contents

INTRODUCTION TO THE NATIONAL ELECTRICAL CODE®

Whether you are installing a new electrical system or repairing an existing one, all electrical work must comply with the current National Electrical Code (NEC) and all local ordinances. Like most laws, the NEC is easier to work with once you understand the language and know where to look for the information you need.

In this introduction, you will learn the key terms and basic layout of the NEC. A brief review of the individual NEC sections that apply to electrical systems will be covered. Sample installations will be discussed in chapters to follow throughout this book.

This book, however, is not a substitute for the NEC. You need a copy of the most recent edition and it should be kept handy at all times. The more you know about the code, the more you are likely to refer to it.

NEC Terminology

There are two basic types of rules in the NEC: mandatory rules and advisory rules. Here is how to recognize the two types of rules and how they relate to all types of electrical systems.

Mandatory rules—All mandatory rules have the word *shall* in them. The word "*shall*" means *must*. If a rule is mandatory, you must comply with it.

Advisory rules—All advisory rules have the word *should* in them. The word "*should*" in this case means *recommended but not necessarily required*. If a rule is advisory, compliance is discretionary. If you want to comply with it, do so. But you don't have to if you don't want to.

Be alert to local amendments to the NEC. Local ordinances may amend the language of the NEC, changing it from *should* to *shall*. This means that you must do in that county or city what may only be recommended to some other area. The office that issues building permits will either sell you a copy of the code that's enforced in that county or tell you where the code is sold. In rare instances where none are available locally, order a copy from the National Fire Protection Association, Batterymarch Park, Quincy, MA 02269.

Learning the Layout of the NEC

Begin your study of the NEC with Articles 100 and 110. These two articles have the basic information that will make the rest of the NEC easier to understand. Article 100 defines terms you will need to understand the code. Article 110 gives the general requirements for electrical installations. Read these two articles over several times until you are thoroughly familiar with all the information they contain. It's time well spent.

Once you are familiar with Articles 100 and 110 you can move on to the rest of the code. There are several key sections you will use often in servicing electrical systems. Let's discuss each of these important sections.

Wiring Design and Protection: Chapter 2 of the NEC discusses wiring design and protection, the information electrical technicians need most often. It covers the use and identification of grounded conductors, branch circuits, feeders, calculations, services, overcurrent protection and grounding. This is essential information for *any* type of electrical system, regardless of the type.

Chapter 2 is also a "how-to" chapter. It explains how to provide proper spacing for conductor supports, how to provide temporary wiring and how to size the proper grounding conductor or electrode. If you run into a problem related to the design or installation of a conventional electrical system, you can probably find a solution for it in this chapter.

Wiring Methods and Materials: Chapter 3 has the rules on wiring methods and materials. The materials and procedures to use on a particular system depend on the type of building construction, the type of occupancy, the location of the wiring in the building, the type of atmosphere in the building or in the area surrounding the building, mechanical factors and the relative costs of different wiring methods.

The provisions of this article apply to all wiring installations except remote control switching (Article 725), low-energy power circuits (Article 725), signal systems (Article 725), communication systems and conductors (Article 800) when these items form an integral part of equipment such as motors and motor controllers.

There are three basic wiring methods used in most modern electrical systems. Nearly all wiring methods are a variation of one of these three basic methods:

- Sheathed cables of two or more conductors, such as NM cable and BX armored cable (Articles 330 through 339)
- Raceway wiring systems, such as rigid and EMT conduit (Articles 342 to 358)
- Busways (Article 364)

Article 310 in Chapter 3 gives a complete description of all types of electrical conductors. Electrical conductors come in a wide range of sizes and forms. Be sure to check the working drawings and specifications to see what sizes and types of conductors are required for a specific job. If conductor type and size are not specified, choose the most appropriate type and size meeting standard NEC requirements.

Articles 318 through 384 give rules for raceways, boxes, cabinets and raceway fittings. Outlet boxes vary in size and shape, depending on their use, the size of the raceway, the number of conductors entering the box, the type of building construction and atmospheric conditions of the areas. Chapter 3 should answer most questions on the selection and use of these items.

The NEC does not describe in detail all types and sizes of outlet boxes. But manufacturers of outlet boxes have excellent catalogs showing all of their products. Collect these catalogs. They are essential to your work.

Article 380 covers the switches, push buttons, pilot lamps, receptacles and convenience outlets you will use to control electrical circuits or to connect portable equipment to electric circuits. Again, get the manufacturers' catalogs on these items. They will provide you with detailed descriptions of each of the wiring devices.

Article 384 covers switchboards and panelboards, including their location, installation methods, clearances, grounding and overcurrent protection.

Equipment for General Use

Chapter 4 of the NEC begins with the use and installation of flexible cords and cables, including the trade name, type letter, wire size, number of conductors, conductor insulation, outer covering and use of each. The chapter also includes fixture wires, again giving the trade name, type letter and other important details.

Article 410 on lighting fixtures is especially important. It gives installation procedures for fixtures in specific locations. For example, it covers fixtures near combustible material and fixtures in closets. The NEC does not describe how many fixtures will be needed in a given area to provide a certain amount of illumination.

Article 430 covers electric motors, including mounting the motor and making electrical connections to it. Motor controls and overload protection are also covered.

Articles 440 through 460 cover air conditioning and heating equipment, transformers and capacitors.

Article 480 gives most requirements related to battery-operated electrical systems. Storage batteries are seldom thought of as part of a conventional electrical system, but they often provide standby emergency lighting service. They may also supply power to security systems that are separate from the main AC electrical system.

Special Occupancies

Chapter 5 of the NEC covers *special occupancy* areas. These are areas where the sparks generated by electrical equipment may cause an explosion or fire. The hazard may be due to the atmosphere of the area or just the presence of a volatile material in the area. Such areas will seldom, if ever, be encountered in residential construction, but to review the entire NEC, it is briefly covered. Commercial garages, aircraft hangers and service stations are typical special occupancy locations.

Articles 500 through 501 cover the different types of special occupancy atmospheres where an explosion is possible. The atmospheric groups were established to make it easy to test and approve equipment for various types of uses.

Articles 501-4, 502-4 and 503-3 cover the installation of explosion-proof wiring. An explosion-proof system is designed to prevent the ignition of a surrounding explosive atmosphere when arcing occurs within the electrical system.

There are three classes of special occupancy locations:

- Class I (Article 501): Areas containing flammable gases or vapors in the air. Class I areas include paint spray booths, dyeing plants where hazardous liquids are used and gas generator rooms.
- Class II (Article 502): Areas where combustible dust is present, such as grain-handling and

storage plants, dust and stock collector areas and sugar-pulverizing plants. These are areas where, under normal operating conditions, there may be enough combustible dust in the air to produce explosive or ignitable mixtures.

- Class III (Article 503): Areas that are hazardous because of the presence of easily ignitable fibers or flyings in the air, although not in large enough quantity to produce ignitable mixtures. Class III locations include cotton mills, rayon mills and clothing manufacturing plants.

Article 511 and 514 regulate garages and similar locations where volatile or flammable liquids are used. While these areas are not always considered critically hazardous locations, there may be enough danger to require special precautions in the electrical installation. In these areas, the NEC requires that volatile gases be confined to an area not more than 4 feet above the floor. So in most cases, conventional raceway systems are permitted above this level. If the area is judged critically hazardous, explosion-proof wiring (including seal-offs) may be required.

Article 520 regulates theaters and similar occupancies where fire and panic can cause hazards to life and property. Drive-in theaters do not present the same hazards as enclosed auditoriums. But the projection rooms and adjacent areas must be properly ventilated and wired for the protection of operating personnel and others using the area.

Chapter 5 also covers residential storage garages, aircraft hangars, service stations, bulk storage plants, health care facilities, mobile homes and parks, and recreation vehicles and parks.

Special Equipment

Residential electrical workers will seldom need to refer to the Articles in Chapter 6 of the NEC, but to cover the NEC completely these Articles will again be discussed briefly.

Article 600 covers electric signs and outline lighting. Article 610 applies to cranes and hoists. Article 620 covers the majority of the electrical work involved in the installation and operation of elevators, dumbwaiters, escalators and moving walks. The manufacturer is responsible for most of this work. The electrician usually just furnishes a feeder terminating in a disconnect means in the bottom of the elevator shaft. The electrician may also be responsible for a lighting circuit to a junction box midway in the elevator shaft for connecting the elevator cage lighting cable and exhaust fans. Articles in Chapter 6 of the NEC give most of the requirements for these installations.

Article 630 regulates electric welding equipment. It is normally treated as a piece of industrial power equipment requiring a special power outlet. But there are special conditions that apply to the circuits supplying welding equipment. These are outlined in detail in Chapter 6 of the NEC.

Article 640 covers wiring for sound-recording and similar equipment. This type of equipment normally requires low-voltage wiring. Special outlet boxes or cabinets are usually provided with the equipment. But some items may be mounted in or on standard outlet boxes. Some sound-recording electrical systems require direct current, supplied from rectifying equipment, batteries or motor generators. Low-voltage alternating current comes from relatively small transformers connected on the primary side to a 120-volt circuit within the building.

Other items covered in Chapter 6 of the NEC include: X-ray equipment (Article 660), induction and dielectric heat-generating equipment (Article 665) and machine tools (Article 670).

If you ever have work that involves Chapter 6, study the chapter *before work begins*. That can save a lot of installation time. Here is another way to cut down on labor hours and prevent installation errors. Get a set of rough-in drawings of the equipment being installed. It is easy to install the wrong outlet box or to install the right box in the wrong place. Having a set of rough-in drawings can prevent those simple but costly errors.

Special Conditions

In most commercial buildings, the NEC and local ordinances require a means of lighting public rooms, halls, stairways and entrances. There must be enough light to allow the occupants to exit from the building if the general building lighting is interrupted. Exit doors must be clearly indicated by illuminated exit signs.

Chapter 7 of the NEC covers the installation of emergency lighting systems. These circuits should be arranged so that they can automatically transfer to an alternate source of current, usually storage batteries or gasoline-driven generators. As an alternative, you can connect them to the supply side of the main service so disconnecting the main service switch would not disconnect the emergency circuits. See Article 700.

Notes on using the NEC

Mandatory rules are characterized by the use of the word:

SHALL

A recommendation or that which is advised but not required is characterized by the use of the word:

SHOULD

Explanatory material in the form of Fine Print Notes is designated:

(FPN)

| A change bar in the margins indicates that a change in the NEC has been made since the last edition

• A bullet indicates that something has been deleted from the last edition of the NEC

chapter 1

ROUGH WIRING

A completely roughed-in electrical system for any type of building includes the following:

- All outlet boxes should be properly secured to the building structure—that is, outlet boxes for receptacles, wall switches, lighting, fixtures, junction boxes, and the like.

- All concealed wiring feeding the outlet boxes should be in place and properly secured, and all splices made in the outlet boxes. Wiring that will be partially exposed and partially concealed should also be installed during the rough-in wiring stage of the electrical installation.

- Flush-mounted panelboards, electric fans, electric heaters, and other flush-mounted equipment should be mounted and all wires connected to the housing of the equipment. The wires do not necessarily need to be connected to the equipment terminals as long as they are accessible for connection later on. For example, where a flush-mounted panelboard is used, only the housing will be installed and all cables and conduits connected to this housing. The loose wires are then left inside the empty housing until the wall covering is applied and finished. Once finished, the panel interiors (circuit breakers, fuseblocks, cable terminals, etc.) are installed, and the wiring is connected to their respective terminals. Unless the panel is energized before the final inspection, the panel cover is also left off until the final inspection.

- Service conductors from flush-mounted panels should be installed as well as service conductors running through a concealed area. If a conduit raceway system is used, only the raceway or conduit needs to be installed during the rough-in inspection; the conductors may be pulled in later.

Outlet Locations

NEC regulations must be prevalent at all times during the outlet layouts, and then double-checked by the designer or electricians before the layout is completed. As an example, the sketch in Fig. 1-1 shows a rough (and simple) layout of convenience outlets (120-volt receptacles). With no room designations it is difficult to determine if the layout in this drawing meets NEC regulations or not. If, for example, these rooms were for office use, Article 210 of the NEC would be the general reference. If these rooms were in a dwelling, additional Articles would include 210-52. Therefore, it is absolutely necessary to determine the usage of each

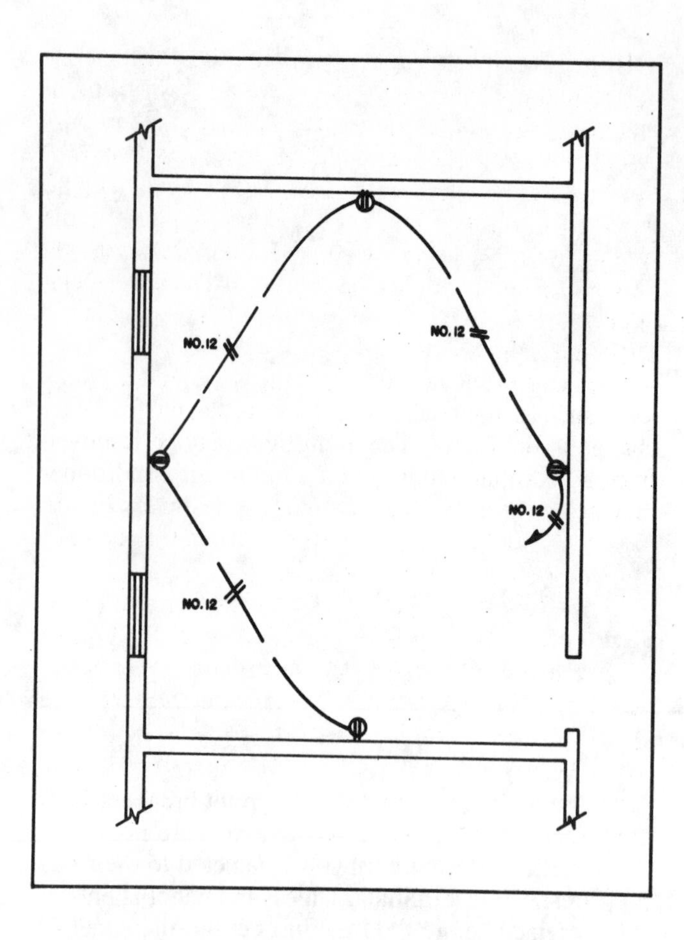

Fig. 1-1: Duplex receptacles laid out in a room. What type of room is it? The designer or electrician must know before laying out the receptacles.

room or area, and the type of occupancy before the design is started. To do otherwise would be a waste of time and money.

Article 210-52 of the NEC specifically states the minimum requirement for the location of receptacles in residential buildings:

- *"In every kitchen, family room, dining room, breakfast room, living room, parlor, library, den, sun room, bedroom, recreation room, or similar rooms, receptacle outlets shall be installed so that no point along the floor line in any wall space is more than 6 ft, measured horizontally, from an outlet in that space, including any wall space 2 ft or more in width and the wall space occupied by sliding panels in exterior walls. The wall space afforded by fixed room dividers, such as free-standing bar-type counters or railings, shall be included in the 6-ft measurement."*

In kitchen and dining areas a receptacle outlet must be installed at each counter space wider than 12 inches. Countertop spaces separated by range tops, refrigerators, or sinks shall be considered as separate countertop spaces. Receptacles rendered inaccessible by appliances fastened in place or appliances occupying dedicated space shall not be considered as these required outlets.

Receptacle outlets in the floor shall *not* be counted as part of the required number of receptacle outlets unless they are located close to the wall. "Close" has not been specified exactly in the NEC, but most inspection authorities defines "close" to be within 12 inches of the wall or baseboard. But don't take this as *gospel*. Some areas may require that floor outlets be located within 6 inches of the wall or baseboard to be counted as the minimum required.

At least one wall receptacle outlet shall be installed in the bathroom adjacent to the basin location. This receptacle shall be a 120-volt (15- or 20-ampere) receptacle and have a ground-fault circuit interrupter protection for personnel.

At least one duplex receptacle must be installed outside the building. Also, at least one receptacle outlet in addition to any provided for laundry equipment shall be installed in each basement and in each attached garage.

At least one wall-switch controlled lighting outlet must be installed in every habitable room, and in bathrooms, hallways, stairways, and attached garages as well as outdoor garages. One lighting outlet must also be installed in the following spaces:

- attic
- crawl space
- utility room
- basement

It is recommended that one convenience outlet be placed in hallways, regardless of their size. In larger hallways, one outlet should be provided for every 15 feet (linear) of hallway. Such outlets are normally used for vacuum cleaners, floor polishers, table lamps, etc.

The NEC does not specify the maximum number of receptacle outlets to be connected to each circuit in residential applications. However, it is a good practice never to load a circuit to more than 80 percent of its current-carrying capacity. In the case of a 15-ampere circuit at 120 volts, this would be 15 x 120 x 0.8 = 1.44 kVA. Still it is difficult to predetermine just what will be connected to these outlets in actual use. In residential applications, most designers allow 300 watts per duplex receptacle unless it is provided for a known

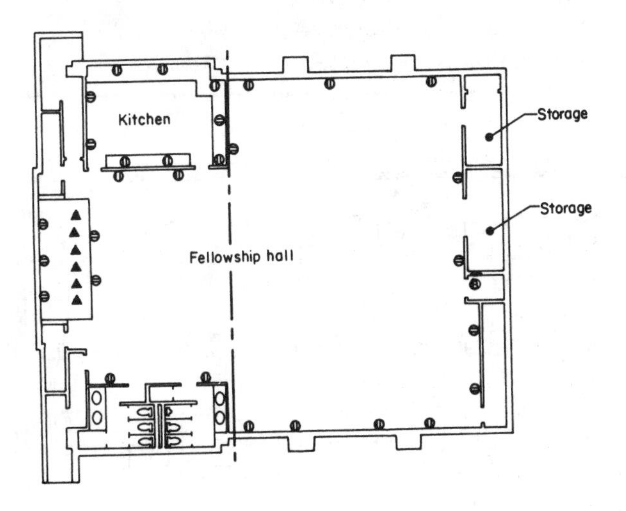

Fig. 1-2: Floor plan of building with receptacle outlet boxes located at various intervals.

usage. This means four or five outlets per circuit if it is provided with a 15-ampere overcurrent protective device. If this figure is used in designing residential electrical systems, the end result will be a well-designed branch circuit with minimal voltage drop.

Other electricians and designers figure 1.5 amperes per duplex receptacle; this method allows a maximum of ten outlets per circuit. Even by this method, common sense indicates that certain outlets will not use 1.5 amperes. These would include outlets for clocks, night lights, etc.

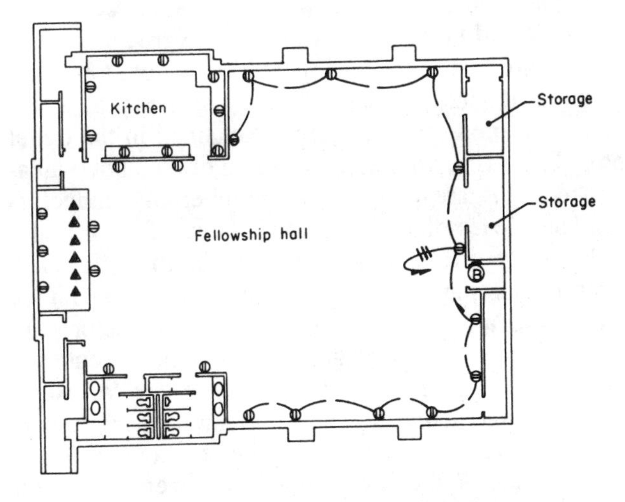

Fig. 1-3: Circuiting of the receptacles shown above.

If such low-wattage items are known to be connected to the circuit, the designer would be justified in increasing the maximum number of ten outlets to possibly twelve outlets per circuit. The reverse is also true. For example, if outlets were to be installed over a home workbench where it is known that heavy power tools will be used, the designer will most certainly provide outlets and circuits to handle this load. The outlets and circuits will probably be similar to those installed in the kitchen or laundry area.

The only problem with allowing for small wattage loads such as night lights and clocks is that the load may change in the future. The night light may be removed from a bedroom outlet, and a window air conditioner may be connected. This is why most designers like to play it safe and allow 300 watts minimum for each duplex receptacle.

In laying out receptacle outlets for residential applications, the first step is to provide outlets for known usages such as washer, dryer, dishwasher, workbench outlets, bedside outlets, etc. The information can usually be obtained from the architect, owners, and/or interior decorator. The outlets are then located so that no point along the floor line, in any wall space, will be more than 6 feet from an outlet in that space.

Circuiting of the outlets should be such that all home runs are as short as possible, and all looping between outlets should also be made as short as possible. But on any given electrical installation, there are many different combinations or groupings of outlets into circuits—all of which will be technically correct; the problem is to determine quickly the route which will conserve the most wire and in turn keep voltage drop to a minimum.

In commercial buildings, the plug receptacles or convenience outlets are usually placed on separate circuits and are not made a part of the general lighting circuit as is sometimes done in a residence.

The NEC states that a minimum of 180 watts must be allowed for this type of outlet. Therefore, since a duplex receptacle has two plugs, a minimum of 350 watts must be allowed for each duplex receptacle. Most electrical workers round this figure off and allow 400 watts for each duplex receptacle when sizing the circuit conductors and overcurrent protection.

The floor plan in Fig. 1-2 shows a room with several duplex receptacles located at various intervals within the room. Assuming a 20-ampere branch circuit, the number of circuits required to feed all the receptacles may be found by first calculating the number of receptacles that can be connected to a 20-ampere branch circuit.

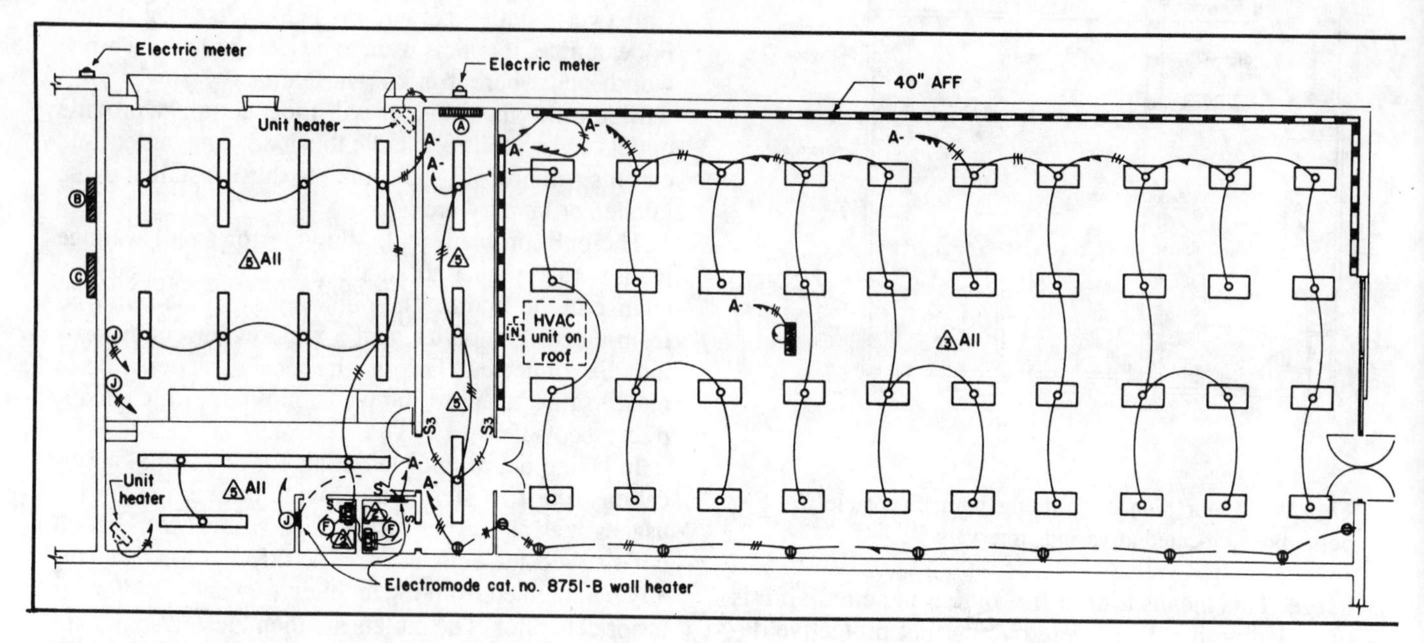

Fig. 1-4: Floor plan of a commercial appliance store showing multioutlet assemblies.

120 (V) x 20 (A)/400 (watts per receptacle) = 6 (number allowed on each circuit)

Next the number of duplex receptacles are counted. Since there are twelve duplex receptacles in the area, they will require two branch circuits ($12 \div 6 = 2$) and may be circuited as shown in Fig. 1-3.

The NEC defines a multioutlet assembly as a type of surface or flush raceway designed to hold conductors and plug receptacles, assembled in the field or at the factory. "Plugmold" is one of several brand names of multioutlet assemblies available, and is a product of The Wiremold Co. The NEC further states that where fixed multioutlet assemblies are employed, each 5 feet or fraction thereof of each separate and continuous length shall be considered as one outlet of not less than 180 volt-amperes (VA) capacity, except in locations where a number of appliances are likely to be used simultaneously, when each foot or fraction thereof shall be considered as an outlet of not less than 180 VA.

The floor plan shown in Fig. 1-4 gives an example of multioutlet assemblies used in a commercial store area. Multioutlet assemblies were used because several television sets were to be demonstrated in these areas. However, multioutlet assemblies are not limited to such locations only, but are also practical anywhere that receptacles are desired fairly close together.

Outlet Boxes

Electrical workers installing electrical systems in residential occupancies must be familiar with outlet box capacities, the spacing of duplex receptacles, and other requirements of the NEC. In general, the maximum numbers of conductors permitted in standard outlet boxes are listed in Table 370-16(a) of the NEC. These figures apply where no fittings or devices such as fixture studs, cable clamps, switches, or receptacles are contained in the box and where no grounding conductors are part of the wiring within the box. Obviously, in all modern residential wiring systems there will be one or more of these items contained in the outlet box. Therefore, where one or more of the above mentioned items are present, the number of conductors shall be one less than shown in the tables. For example, a deduction of one conductor must be made for each strap containing a device such as a switch or duplex receptacle; a further deduction of one conductor shall be made for one or more grounded conductors entering the box. A 3-inch x 2-inch x 2¾-inch box for example, is listed in the table as containing a maximum number of six No. 12 wires. If the box contains cable clamps and a duplex receptacle, three wires will have to be deducted from the total of six–providing for only three No. 12 wires. If a ground wire is

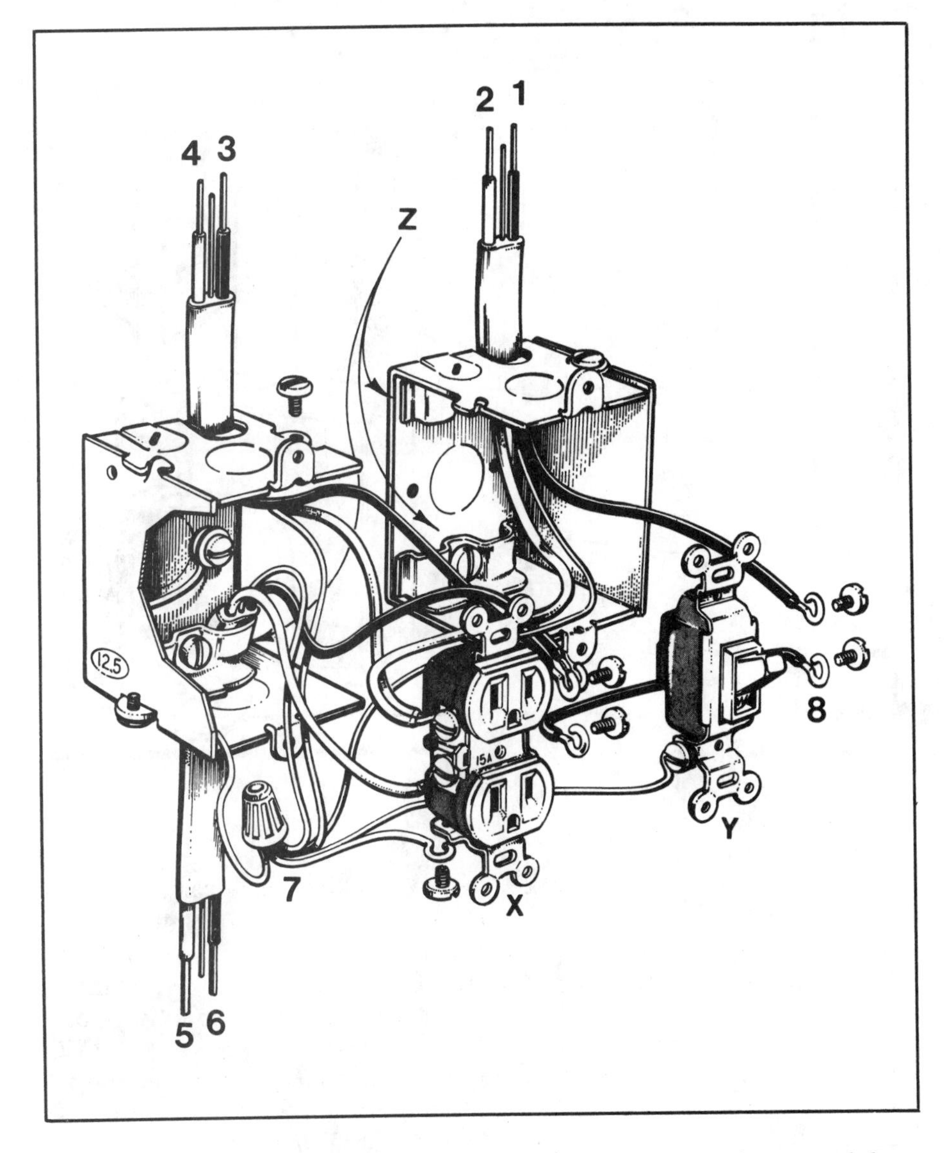

Fig. 1-5: Two ganged switch boxes with six No. 12 AWG conductors and three ground wires entering. Does this arrangement comply with the NE Code?

used, only two No. 12 wires may be used, which might be the case when this is the last outlet on the circuit.

Fig. 1-5 illustrates one possible wiring configuration for outlet boxes and the maximum number of conductors permitted in them as governed by Section 370-16 of the NEC. This example shows two single-gang switch boxes joined or "ganged" together to hold a single-pole toggle switch and a duplex receptacle. This type of arrangement is likely to be found above a workbench whereas the duplex receptacle is provided for power tools and the single-pole switch could be used to control lighting above the bench.

Since Table 370-16(a) gives the capacity of one 3 x 2 x 2¼-inch device box as 12.5 cubic inches, the total capacity of both boxes in Fig. 1-5 is 25 cubic inches. These two boxes have a capacity to allow 10 No. 12 AWG conductors, or 12 No. 14 AWG conductors, less the deductions as listed below.

- Two conductors must be deducted for each strap-mounted device. Since there is one duplex receptacle (X) and one single-pole toggle switch (Y), four conductors must be deducted from the total number stated in the above paragraph.
- Since the combined boxes contain one or more cable clamps (Z), another conductor must be deducted. Note that only one deduction is made for similar clamps, regardless of the number. However, any unused clamps may be removed to facilitate the electrical worker's job; that is, allowing for more work space.

Therefore, to comply with the NEC, and considering the combined deduction of five conductors, only five No. 12 AWG conductors (seven No. 14 AWG conductors) may be installed in the outlet-box configuration in Fig. 1-5.

Fig. 1-5 shows three type nonmetallic-sheathed (NM) cables, designated 12/2 with ground, entering the ganged outlet boxes. This is a total of six current-carrying conductors and three ground wires, for a total of nine. Is this arrangement in violation of the NEC?

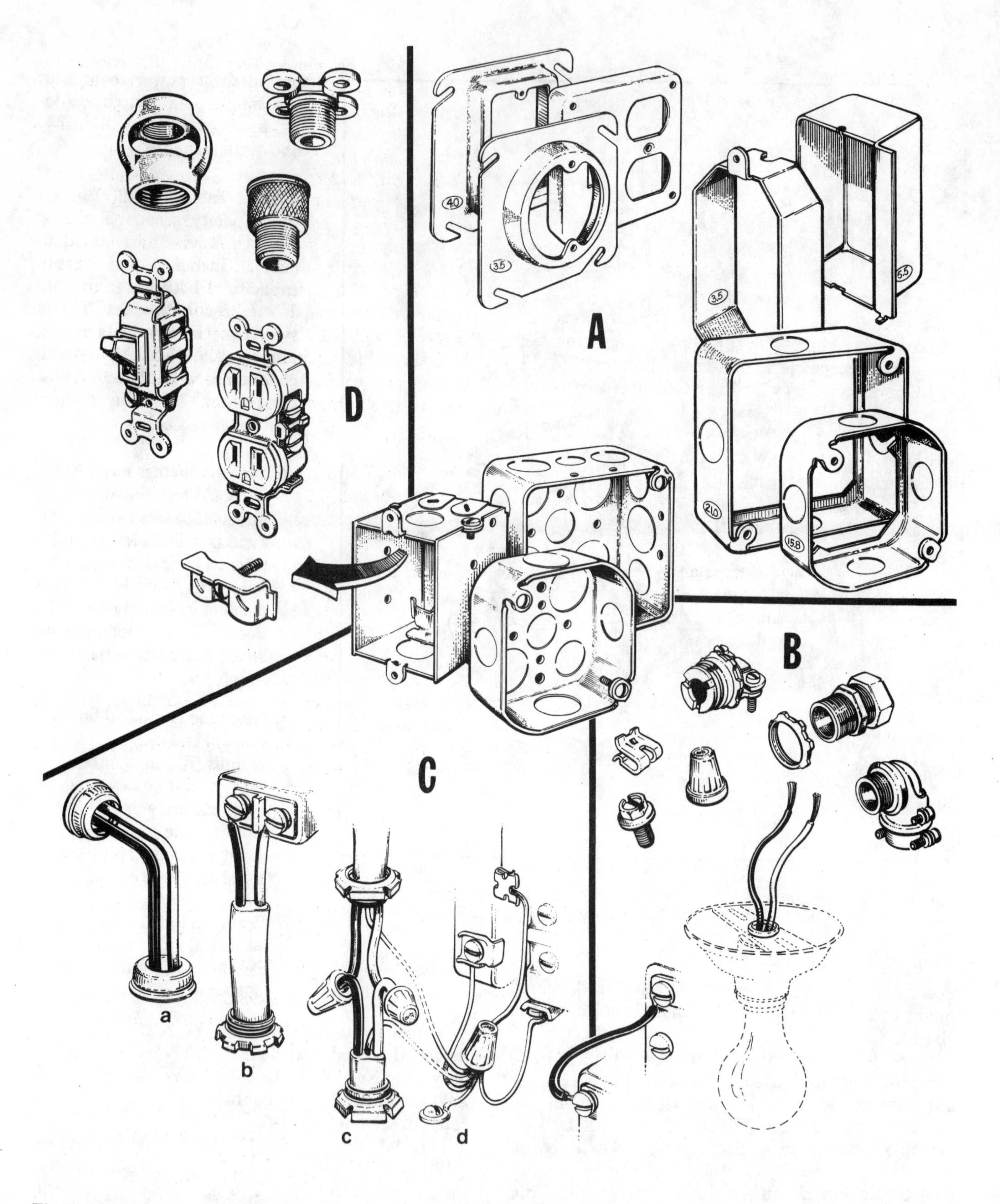

Fig. 1-6: Summary of NEC rules governing the number of conductors allowed in outlet boxes.

Yes, even though the ground wires count as only one conductor, regardless of the number, there are still 7 conductors entering the box, and with the previous deductions, only five were allowed. However, if the wire size is reduced to No. 14 AWG, this configuration will comply with the 1993 NEC.

Also note the jumper wire in Fig. 1-5; this is numbered "8" in the drawing. Conductors that both originate and end in the same outlet box are exempt from being counted against the allowable capacity of an outlet box. This jumper wire (8) taps off one terminal of the duplex receptacle to furnish a "hot wire" to the single-pole toggle switch. Therefore, this wire originates and terminates in the same set of ganged boxes and is not counted against the total number of conductors. By the same token, the three grounding conductors extending from the wire nut to the individual grounding screws on the devices originate and terminate in the same set of boxes. These conductors are also exempt from being counted with the total.

A pictorial definition of stipulated conditions as they apply to Section 370-16 of the NEC is shown in Fig. 1-6. Section "A" illustrates an assortment of raised covers and outlet box extensions. These components, when combined with the appropriate outlet boxes, serve to increase the usable work space. Each type is marked with their cubic-inch capacity which may be added to the figures in Table 370-16(a) of the NEC to calculate the increased number of conductors allowed.

Section "B" shows components which may be used in outlet boxes without affecting the total number of conductors. Such items include grounding clips and screws, wire nuts and box connectors when the latter is inserted through knock-out holes in the outlet box and secured with locknuts. Pre-wired fixture wires are not counted against the total number of allowable conductors in an outlet box; neither are conductors originating and ending in the box.

Section "C" shows typical wiring configurations which must be counted as conductors when calculating the total capacity of outlet boxes. A wire passing through the box without a splice or tap (a) is counted as one conductor. Therefore, a cable containing two wires that passes in and out of an outlet box without a splice or tap is counted as two conductors. However, wires which enter a box and are either spliced or connected to a terminal, and then exit again (c) are counted as two conductors. In the case of two 2-wire cables, the total conductors charged will be four. Wires that enter and terminate in the same box (b) are charged as individual conductors and in this case, the total charge will be two conductors. Remember, when one or more grounding wires enter the box and are joined (d), a deduction of only one is required, regardless of their number.

Section "D" in Fig. 1-6 illustrates components that require deduction adjustment from those specified in Table 370-16(a). Such items include fixture studs, hickeys, and fixture-stud extensions. One conductor must be deducted from the total for *each* type of fitting used. Deductions are also required for strap-mounted devices, like duplex receptacles and wall switches; two conductors must be deducted for each one used. An additional deduction of one conductor is made when one or more internally mounted cable clamps are used.

Conductors

Conductor Size: The American Wire Gauge (AWG) is currently used in the United States to identify the sizes of wire and cable up to and including No. 4/0, which is commonly pronounced in the electrical trade as "four-aught" or "four-naught." These numbers run in reverse order as to size; that is, No. 14 AWG is smaller than No. 12 AWG and so on up to size No. 1 AWG. To this size wire (No. 1 AWG), the larger the gauge number, the smaller the size of the conductor. However, the next larger size, after No. 1 AWG, is No. 1/0 AWG, then 2/0 AWG, 3/0 AWG, and 4/0 AWG. At this point, the AWG designations end and the larger sizes of conductors are identified by circular mils (CM). From this point, the larger the size of wire, the larger the number of circular mils. For example, 300,000 CM is larger than 250,000 CM. In writing these sizes, the "thousand" numerals (,000) were replaced by the letter M (the Roman numeral for thousand) in older code versions. Instead of 500,000 CM, it was written 500 MCM. Today it has been changed from "MCM" to "kcmil." So the code says 500 kcmil.

Branch Circuits and Feeders

The conductors that extend from the panelboard to the various outlets are called circuits and are defined by the NEC as *"that point of a wiring system extending beyond the final overcurrent device protecting the circuit...."*

In general, the size of the branch-circuit conductors varies depending on the current requirements of the electrical equipment connected to the outlet. Most, however, will consist of either No. 14, 12, 10, or 8 AWG. Conductors larger than No. 8 AWG are usually considered to be feeders rather than branch circuits.

A simple branch circuit requires two wires or conductors to provide a continuous path for the flow of electric current. The usual branch circuit for receptacles, as might be used to plug in a window unit (air conditioner) operates at 120 volts. The most common wire size will be No. 12 AWG, although size No. 14 AWG is allowed for some residential circuits.

Fractional-horsepower motors and small electric heaters usually operate at 120 volts also and are connected to a simple 120-volt branch circuit either by means of a receptacle, a junction box, or a direct connection. Larger electric motors, air conditioning, duct and unit heaters, and other large current-consuming equipment operate on a two- or three-wire circuit at 240 or 480 volts.

The NEC specifically states that the total load on a branch circuit, other than motor loads, must not exceed 80 percent of the circuit rating when the load will constitute a continuous load, that is, a load that is in continuous operation for 3 hours or longer.

To size the total load permitted on a circuit of, say, 20 amperes consisting of No. 12 AWG copper wire and fused at 20 amperes that will be in operation for 3 hours or longer, multiply the amperes (20) by 0.80 (percent) to obtain 16 amperes. This figure multiplied by the voltage (120) gives a maximum of 1920 VA that may be connected to the circuit. However, on new wiring systems it is best in most cases to limit the load on 20-ampere circuits to a maximum of 1200 to 1600 VA. This permits some loads to be added in the future, and the practice also keeps the temperature of the conductors lower for better performance and less voltage drop. However, most motors, oil burners, duct heaters, and the like should be fed with one circuit only.

Wiring Methods

Several types of wiring methods are currently in use for wiring installations. The methods used for a given application are determined by several factors:

- The requirements set forth in the NEC.
- Local codes and ordinances.
- Engineer's specifications.
- Type of building construction.
- Location of the wiring in the building.
- Importance of the wiring system's appearance.
- Costs and budgets.

In general, electrical wiring is used for two types of applications in electrical wiring systems: wiring to provide power to operate the electrical components and equipment, and control wiring to regulate the equipment. Wiring may be further subdivided into either open or concealed wiring.

In open wiring systems, the cable and/or raceways are installed on the surface of the walls, ceilings, columns, and the like where they are in view and readily accessible. Such wiring is often used in areas where appearance is not important.

Concealed wiring systems have all cable and raceway runs concealed inside of walls, partitions, ceilings, columns, and behind baseboards or molding where they are out of view and not readily accessible.

This type of wiring system is generally used in all new construction with finished interior walls, ceiling, and floors and is the preferred type where good appearance is important.

Cable systems: Several types of cable systems are used in wiring systems for building construction to feed or supply power to equipment, and include the following:

Nonmetallic sheathed cable: Type NM cables are manufactured in two or three wires, and with varying sizes of conductors. The jacket or covering consists of rubber, plastic, or fiber. This type of cable may be concealed in the framework of buildings, or in some instances, may be run exposed on the building surfaces.

Underground feeder cable: Type UF cable may be used underground, including direct burial in the earth, as a feeder or branch-circuit cable when provided with overcurrent protection at the rated ampacity as required by the NEC. When type UF cable is used above grade where it will come in direct contact with the rays of the sun, its outer covering must be sun resistant.

Service-entrance cable: Type SE cable, when used for an electrical service, must be installed as required by the NEC, and may be used in interior wiring systems provided all the circuit conductors of the cable are insulated with rubber or thermoplastic insulation.

Armored cable: BX cable is manufactured in two-, three-, and four-wire assemblies, with varying sizes of conductors, and is used in locations similar to those where NM cable is used. The metallic spiral covering on BX cable offers a greater degree of mechanical protection than NM cable, and the metal jacket also provides for a continuously grounded system without the need for additional grounding conductors. This type of cable may be used for under-plaster extensions, as provided in the NEC, and embedded in plaster finish, brick, or other masonry, except in damp or wet locations. It also may be run or "fished" in the air voids of masonry block or tile walls, except where such walls are exposed or subject to excessive moisture or dampness or are below grade.

Mineral-insulated metal-sheathed cable: Type MI cable is a factory assembly of one or more conductors insulated with a highly compressed refractory mineral insulation and enclosed in a liquid-tight and gas-tight continuous copper sheath. It may be used for services, feeder, and branch circuits in dry, wet, or continuously moist locations. Furthermore, it may be used indoors or outdoors, embedded in plaster, concrete, fill, or other masonry, whether above or below grade. This type of cable may also be used in hazardous locations, where exposed to oil or gasoline, where exposed to corrosive conditions not deteriorating to the cable's sheath, and in underground runs where suitably protected against physical damage and corrosive conditions. In other words, MI cable may be used in practically any electrical installation.

Raceways

A raceway is any channel used for holding wires, cables, or busbars, which is designed and used solely for this purpose. Types of raceways include rigid metal conduit, intermediate metal conduit (IMD), rigid nonmetallic conduit, flexible metal conduit, liquid-tight flexible metal conduit, electrical metallic tubing (EMT), underfloor raceways, cellular metal floor raceways, cellular concrete floor raceways, surface metal raceways, wireways, and auxiliary gutters. Raceways are constructed of either metal or insulating material such as PVC (plastic).

Raceways provide mechanical protection for the conductors that run in them and also prevent accidental damage to insulation and the conducting metal. They also protect conductors from the harmful chemical attack of corrosive atmospheres and prevent fire hazards to life and property by confining arcs and flame due to faults in the wiring system.

Another function of metal raceways is to provide a path for the flow of fault current to ground, thereby preventing voltage buildup on conductor and equipment enclosures. This feature, of course, helps to minimize shock hazards to personnel. To maintain this feature, it is extremely important that all raceway systems be securely bonded together into a continuous conductive path and properly connected to a grounding electrode such as a water pipe or a ground rod.

The NEC provides rules on the maximum number of conductors permitted in raceways. In conduits, for either new work or rewiring of existing raceways, the maximum fill must not exceed 40 percent of the conduit cross-sectional area. In all such cases, fill is based on using the actual cross-sectional areas of the particular types of conductors used. Other derating rules specified by the NEC may be found in Article 310. For example, if more than three conductors are used in a single conduit, a reduction in current carrying capacity is required. Ambient temperature is another consideration that may call for derating of wires below the values given in NEC tables.

Residential Rough Wiring

In nearly all instances, the major portion of wiring for residential branch circuits will consist of either type NM cable (nonmetallic sheathed) or type AC cable (armored cable). Type UF (underground feeder) cable is used extensively for underground wiring for outside lights and for feeding buildings that are not attached to the home itself. Rigid conduit and EMT (electrical metallic tubing) are sometimes used for service-entrance masts and for a few other limited applications in the home, but such wiring methods are insignificant compared to most of the wiring.

Since type NM cable is the least expensive, this wiring method will be the type most often used in residential wiring systems. NM cable may be used in all types of dwelling units not exceeding three floors above grade. However, it may not be exposed to corrosive fumes or vapors (found in a shop, for example), or embedded in masonry, concrete, or plaster.

NM cable must be secured in place at intervals not exceeding 4½ feet and within 12 inches from every cabinet, box, or fitting. The staples, straps, or other fasteners must be designed and installed so as not to injure the cable. Where the cables pass through strips, or other approved means, this protection should extend at least 6 inches above the floor.

Where the cable is run at right angles with joists in unfinished basements, cables containing two No. 6 or three No. 8 conductors may be secured directly to the lower edges of the joists. Smaller cables, however, must be run either through bored holes in joists or on running boards. Of course, where cables are run parallel to the joists, cables of any size must be secured to the sides or faces of the joists.

Type AC cable is permitted for use in dry locations only. This type of cable is not approved for direct burial in the earth. The required supports for this cable are the same as those for type NM cable. Make sure that all bends in the cable have a radius of the curve of the inner edge not less than 5 times the diameter of the cable.

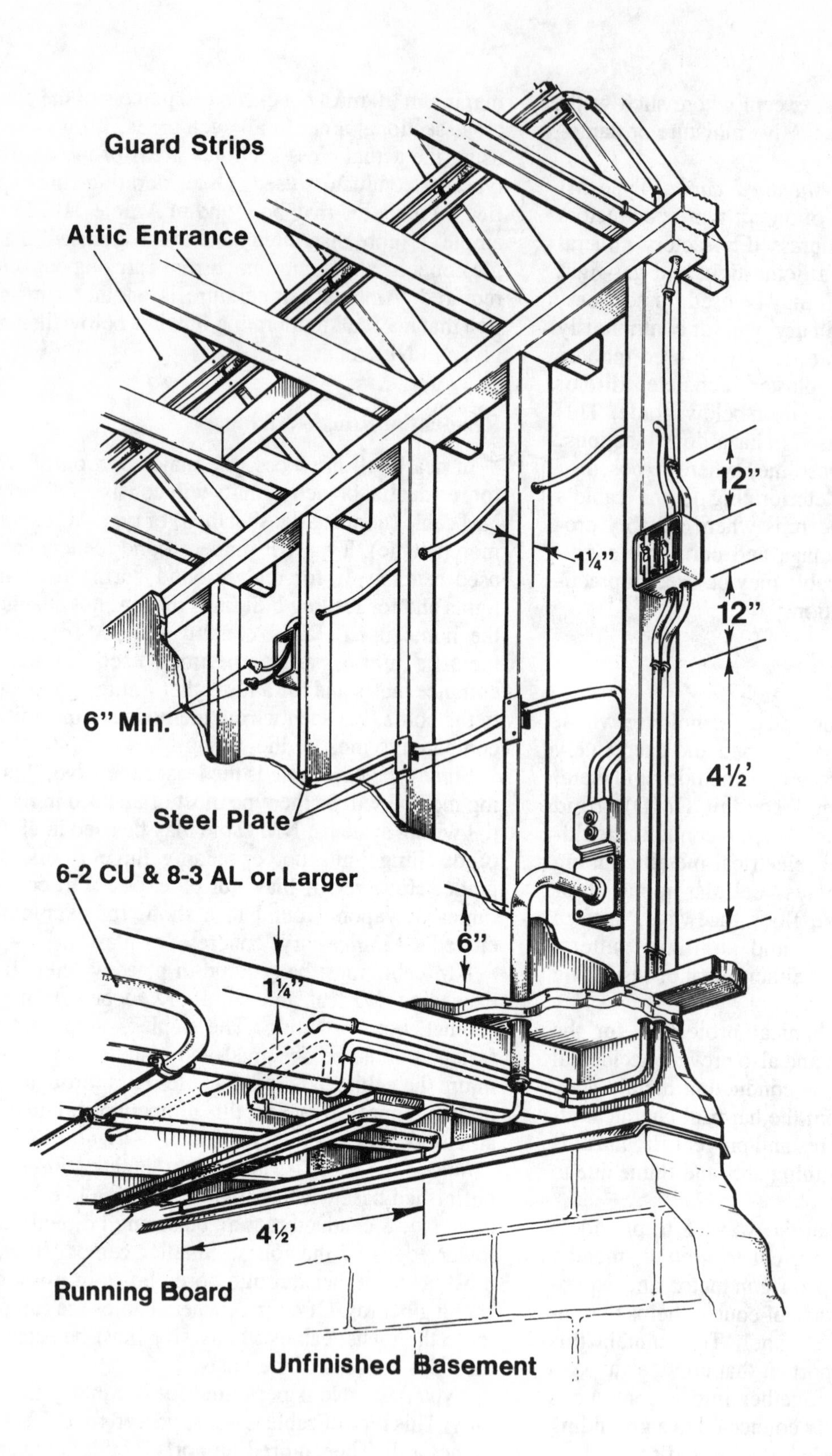

Fig. 1-7: NEC rules summarizing the installation of type NM cable.

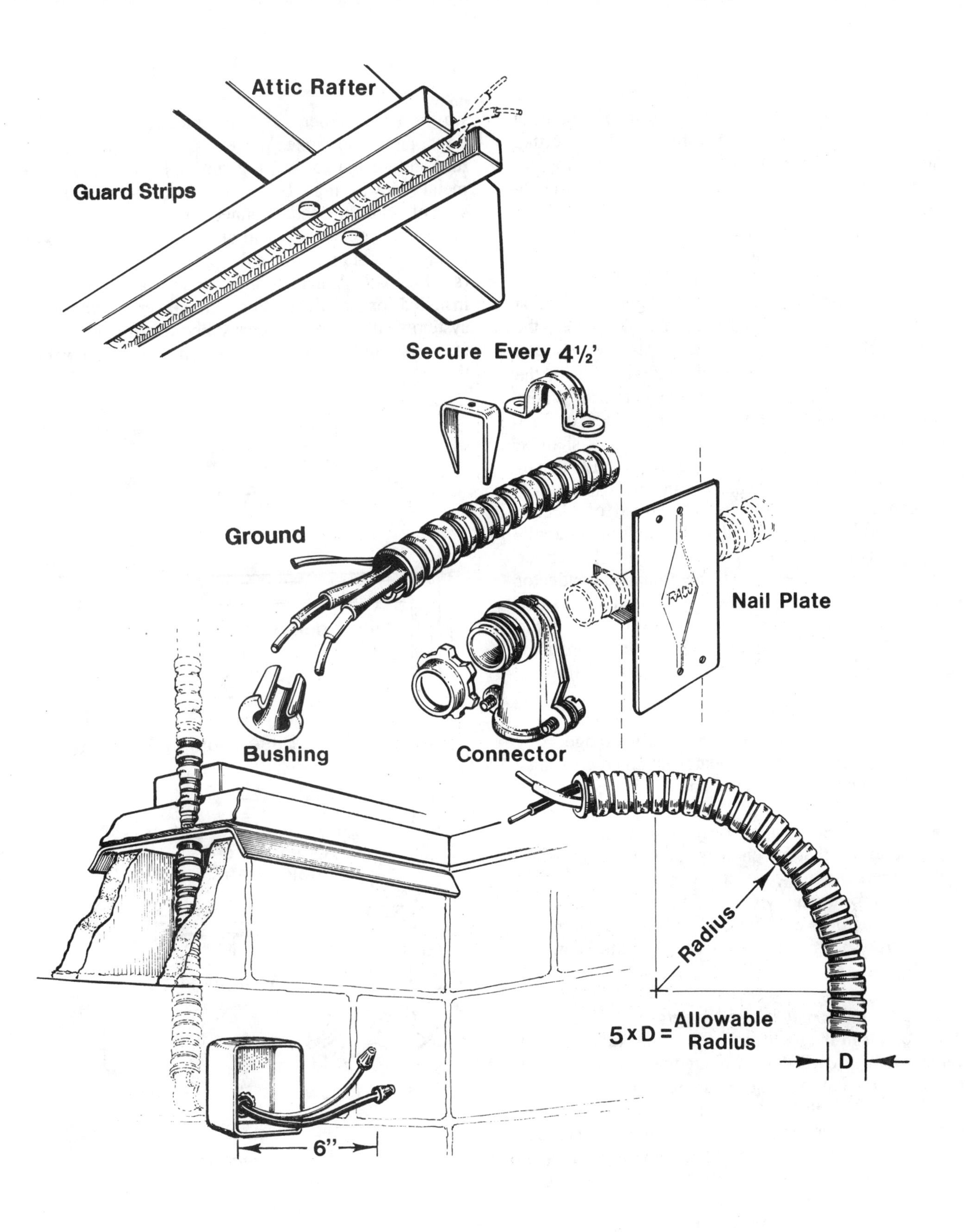

Fig. 1-8: NEC rules summarizing installation requirements for type AC (BX) cable.

Furthermore, an anti-short bushing must be installed at every termination point of this cable. The NEC further requires that such bushings will be visible for inspection when the cable is connected to outlet boxes or cabinets.

Where run across the top of floor joists, or within 7 feet of floor or floor joists across the face of rafters or studding, or in accessible attics and roof spaces, the cable shall be protected by substantial guard strips that are at least as high (or thick) as the cable. Where the space is not accessible by permanent stairs or ladders, protection shall be required only within 6 feet of the nearest edge of the scuttle hole or attic entrance. Where cable is carried along the sides of rafters, studs, or floor joists, neither guard strips nor running boards shall be required. Figs. 1-7 and 1-8 summarize the NEC installation requirements for Type NM and AC cables.

Because of the relatively heavy load of electric ranges, clothes dryers, and other household appliances, many electrical contractors use type SE (service entrance) cable for feeders. If the grounded conductor of the SE cable is insulated, it may be used in the same way that type NM cable is used. However, if the grounded conductor is not insulated, the cable shall not be used within a building except for the following ways.

- As a branch circuit to supply only a range, wall-mounted oven, counter-mounted cooking unit, or clothes dryer as covered in Section 250-60 of the NEC.
- As a feeder to supply only those other buildings that are on the same premises.
- Where the fully insulated conductors are used for circuit wiring and the uninsulated conductor is used for equipment grounding purposes—not as a grounded or neutral conductor.

Type UF cable may be used underground, including direct burial in the earth, as a feeder or branch circuit cable where provided with overcurrent protection of the rated ampacity as required in Section 339-4 of the NEC. Type UF cable is also permitted for interior wiring in wet, dry, or corrosive locations under the recognized wiring methods of the NEC. This type of cable should not be used as service entrance cables or embedded in poured cement, concrete, or aggregate, or where exposed to direct rays of the sun unless the covering is approved for the purpose.

In some areas of the home, especially for modernization work in existing homes, surface-metal raceway is used. When used, the raceway, elbows, fittings, and outlet boxes must be of the same manufacture and design for use together. The size of surface metal raceway must be approved for the number and size of wires needed. Connections shall be made to other types of raceways in an approved manner with fittings manufactured for the purpose and application (see Fig. 1-9). Where combination metal raceways are installed for signal, lighting, and power circuits, each system shall be run in separate compartments clearly identified and maintaining the same relative position throughout the system.

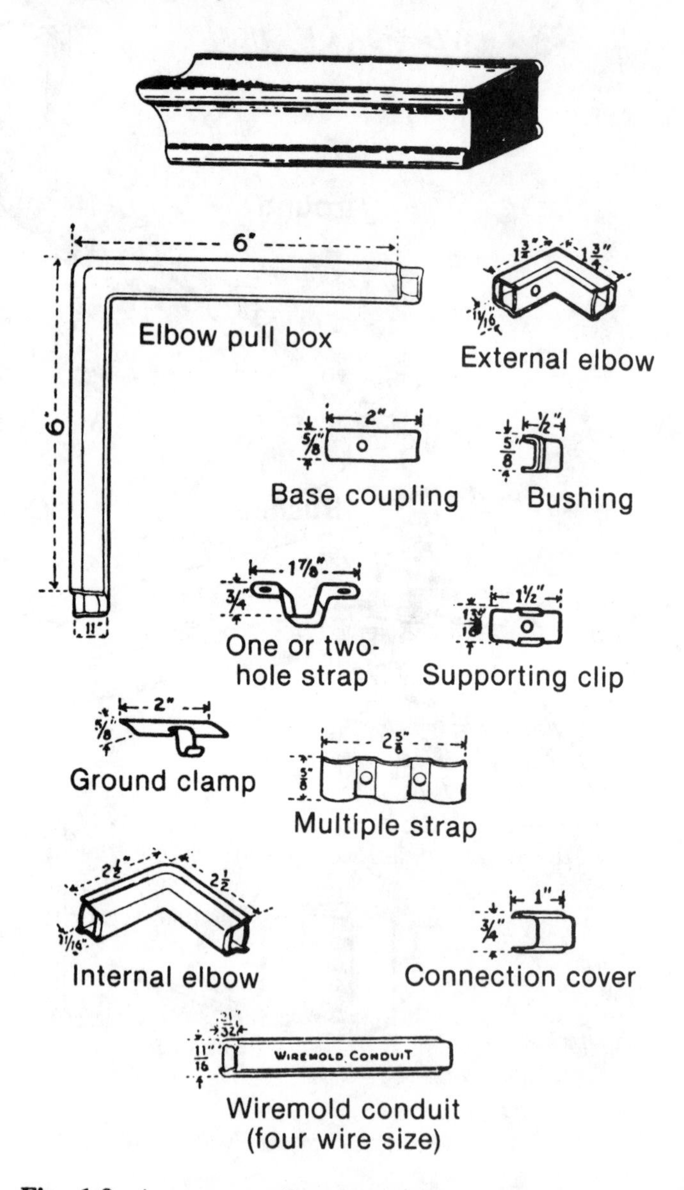

Fig. 1-9: Assortment of surface-metal raceway manufactured by The Wiremold Co.

Sizing Conductors

In all electrical systems, the conductors should be sized so that the voltage drop never exceeds 3 percent for power, heating, and lighting loads or combinations of these. Furthermore, the maximum total voltage drop for conductors for feeders and branch circuits combined should never exceed 5 percent.

The voltage drop in any two-wire, single-phase circuit consisting of mostly resistance-type loads, with the inductance negligible, may be found by the following equation:

$$VD = 2K \times L \times I/CM$$

where

VD =	drop in circuit voltage
L =	length of conductor
I =	current in the circuit
CM =	area of conductor in circular mils
K =	resistivity of conductor metal, that is, 11 for copper and 18 for aluminum.

With this equation, the voltage drop in a circuit consisting of No. 10 AWG copper wire, 50 feet in length, and carrying a load of 20 amperes would be

$$VD = 2(11) \times 50 \times 20/10{,}380$$
(area in CM of No. 10 wire)

which equals 2.12 volts. Divided by 120, this is .01 or 1 percent, well within the limits of the allowed voltage drop.

It is also quite common to deal with the problem of sizing conductors for special electronic and electrical equipment on which conventional wiring sizing equations and calculations cannot be used. For example, computerized HVAC control equipment is highly sensitive to voltage variations, and the circuits must be sized to obtain the very minimum of voltage drop.

To illustrate, assume that a solid-state computer control device is located 460 feet from the main distribution panelboard and it is necessary to keep the voltage drop less than the 3 percent allowable. The service consists of a three-phase, four-wire, 120/208-volt circuit, and the allowable voltage drop is 2 percent. Copper wire will be utilized.

To find the wire size that will carry the load with less than 2 percent voltage drop, use the equation

$$\text{Circular mils} = \text{length} \times \text{amperes} \times 2K/\text{volts lost}$$

AWG Wire Size	Resistance Copper	Resistance Aluminum	Area Circular Mils
18	7.95	13.1	1620
16	4.99	8.21	2580
14	3.14	5.17	4110
12	1.98	3.25	6530
10	1.24	2.04	10380
8	0.778	1.28	16510
6	0.491	0.808	26240
4	0.308	0.508	41740
3	0.245	0.403	52620
2	0.194	0.319	66360
1	0.154	0.253	83690
1/0	0.122	0.201	105600
2/0	0.0967	0.159	133100
3/0	0.0766	0.126	167800
4/0	0.0608	0.100	211600
250	0.0515	0.0847	—
300	0.0429	0.0707	—
350	0.0367	0.0605	—
400	0.0321	0.0529	—
500	0.0258	0.0424	—
600	0.0214	0.0353	—
700	0.0184	0.0303	—
750	0.0171	0.0282	—
800	0.0161	0.0265	—
900	0.0143	0.0235	—
1000	0.0129	0.0212	—
1250	0.0103	0.0169	—
1500	0.00858	0.0141	—
1750	0.00735	0.0121	—
2000	0.00643	0.0106	—

Fig. 1-10: A wire-size table must be used in voltage-drop calculations.

In this equation the number of feet must be measured or scaled one way, not both sides of the circuit; volts lost should be taken as the drop allowed in volts, not the percentage. Circular mils show the size of wire in AWG to use, while K = 11 for copper.

AWG Wire Size	Length of circuit, one way (ft) 25	50	75	100	125	150	175	200	225	250	275	300
20	29	14	10	7.2	5.8	4.8	4.1	3.8	3.2	2.9	2.8	2.4
18	58	29	19	14	11	9.6	8.2	7.2	6.4	5.8	5.2	4.8
16	86	43	29	22	17	14	12	11	9.6	8.7	7.8	7.2
14	133	67	44	33	27	22	19	17	15	13	12	11

Fig. 1-11: Table for calculating voltage-drop in low-voltage wiring—the most common circuits for signaling and alarm systems.

There are only two unknowns in this situation: the circular mils and the volts lost. To solve for the voltage drop or volts lost, multiply 208 (the circuit voltage between phases) by 0.02 (their percentage of voltage drop permitted for the particular piece of equipment). The answer will be 4.16 volts. The nameplate on the piece of equipment gives a full-load ampere rating of 87 amperes at 208 volts. Substituting all known values in the equation, we have

Circular mils = 460 x 87 x 22/4.16 = 211,644 CM

Referring to a wire-size table, such as the one in Fig. 1-10, 250,000 CM (250 kcmil) is the closest wire size normally available and will therefore be the size to use for the circuit feeding the piece of equipment.

Even when sizing wire for low-voltage (24-volt systems) control circuits, the voltage drop should be limited to 3 percent because excessive voltage drop causes:

- Failure of control coil to activate
- Control contact chatter
- Erratic operation of controls
- Control coil burnout
- Contact burnout

The voltage-drop calculations described previously may also be used for low-voltage wiring, but tables are quite common.

To use the table in Fig. 1-11 for example, assume a load of 35 VA with a 50-foot run for a 24-V control circuit. Referring to the table, scan the 50-foot column. Note that No. 18 AWG wire will carry 29 VA and No. 16 wire will carry 43 VA while still maintaining maximum of 3 percent voltage drop. In this case, No. 16 wire would be the size to use. When the length of wire is other than listed in the table in Fig. 1-11, the capacity may be determined by the following equation:

VA capacity = length of circuit (from table) x VA (from table) length of circuit (actual)

The 3 percent voltage-drop limitation is imposed to assure proper operation when the power supply is below the rated voltage. For example, if the rated 240-V supply is 10 percent low (216V), the transformer does not produce 24V; only 21.6V. When normal voltage drop is taken from this 21.6V, it approaches the lower operating limit of most controls. If it is assured that the primary voltage to the transformer will always be at rated value or above, the control circuit will operate satisfactorily with more than 3 percent voltage drop. However, during high power usage, power companies frequently set their transformer taps to reduce the voltage. When this occurs, the transformer voltage will be lower than normal.

In most installations, several lines connect the transformer to the control circuit. One line usually carries the full load of the control circuit from the hot side of the transformer to one control, with the return perhaps through several lines of the various other controls. Therefore, the line from the hot side of the transformer is the most critical regarding voltage drop and VA capacity and must be properly sized.

When low-voltage lines are installed, it is suggested that one extra line be run for emergency purposes. This can be substituted for any one of the existing lines that may be defective. Also, it is possible to parallel this extra line with the existing line carrying the full load of the control circuit if the length of run affects control operation because of a voltage drop. In many cases this will reduce the voltage drop and permit satisfactory operation.

Installing the Rough Wiring

Residential: Workers who install electrical wiring systems in residences usually have their own time-proven techniques. The following describes one way a residence might be wired.

The electrician begins by using a tape measure and marker to locate the duplex receptacles—always following the requirements of Section 210-52 of the NEC. During this layout work, the height of the outlets are also measured and marked. The height varies with each electrician or contractor, but most duplex receptacles are installed 15 inches above the floor to the bottom of the outlet box. Receptacles above countertops are usually installed 48 inches above the floor to the bottom of the box; wall switches are normally 50 inches above the finished floor. See. Fig. 1-12.

For simple residential outlet layouts most experienced electricians will make mental notes of the general wiring arrangement and mentally select outlet boxes to comply with the number of conductors required. For more complex layouts, a sketch should be made. The outlet boxes are then secured to the framing of the house. Where wood studs are used, the outlet boxes are usually secured with nails driven through the nail holes provided in the box and then directly into the wood studs. Some electricians prefer boxes with mounting brackets. For use on concrete blocks—as in a residential basement—masonry fasteners are used. Once all the outlet boxes are secured, the final layout should

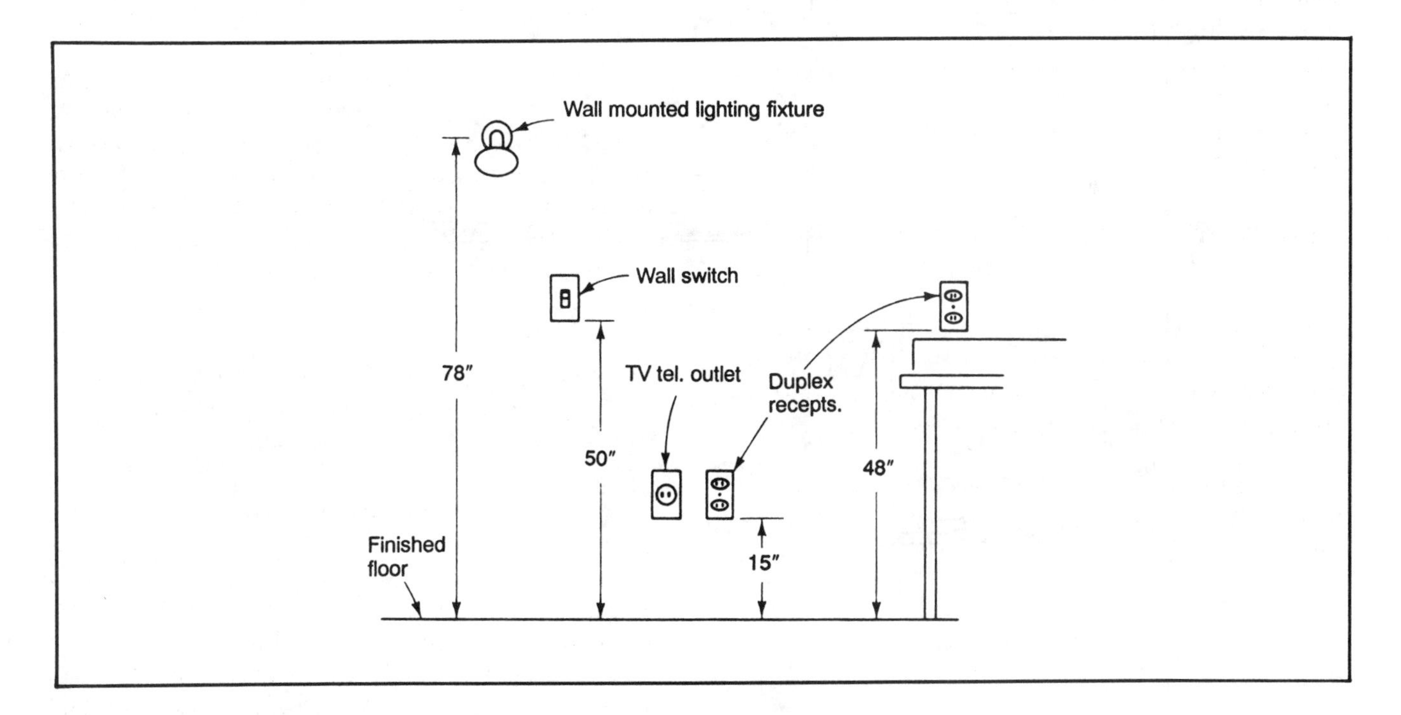

Fig. 1-12: Recommended mounting heights for various types of outlet boxes.

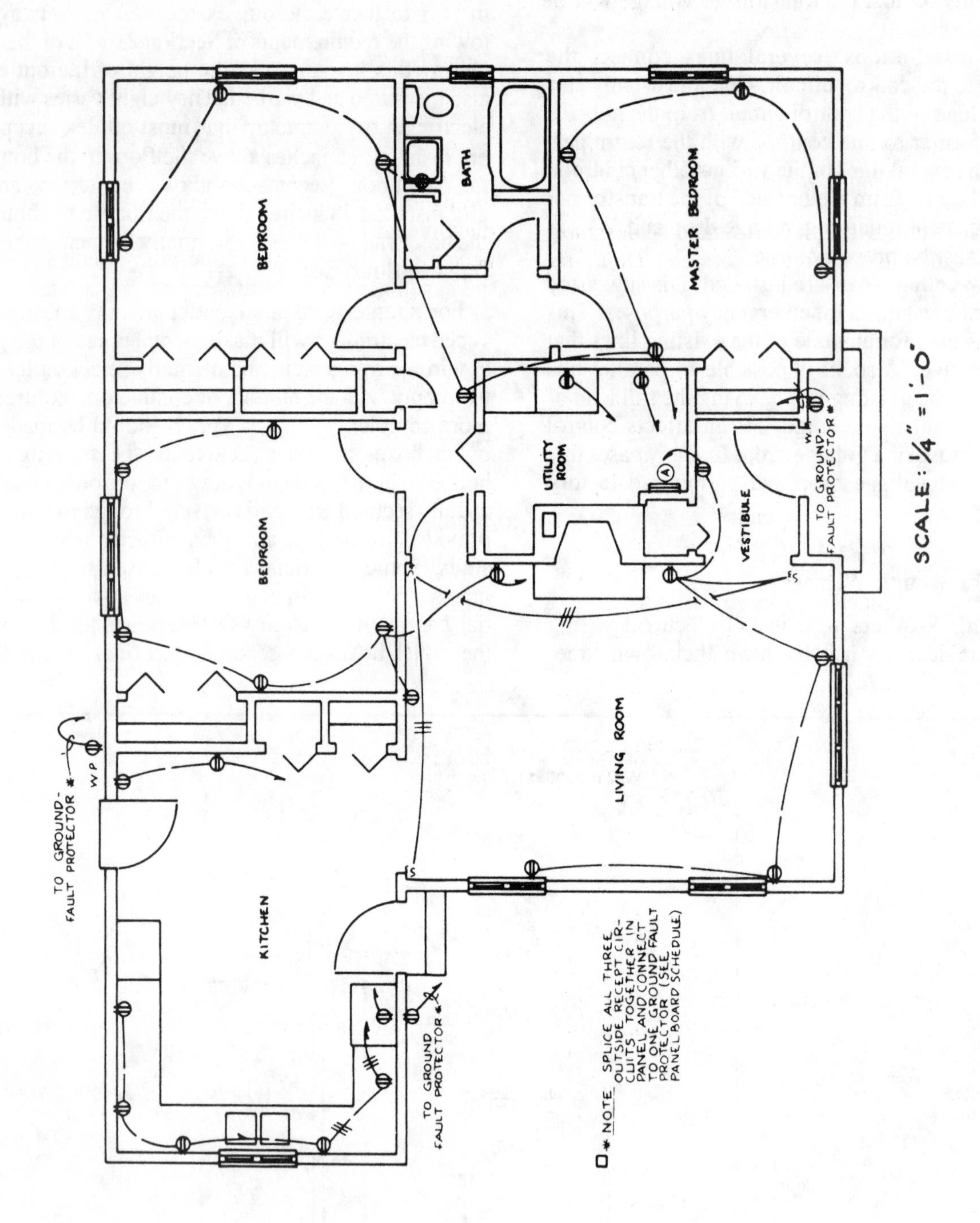

Fig. 1-13: Floor-plan layout of receptacles as described in the text.

appear similar to the one in Fig. 1-13. The lighting layout for this residence is shown in Chapter 6.

Before the branch circuits are installed, the location of the panelboard must be selected (see Chapter 2). The residence in Fig. 1-13 shows the panelboard located in the utility room, designated by the letter "A" enclosed in a circle. A small residential building will usually have only one panelboard to service the entire electrical circuits. In larger buildings, there may be dozens of subpanels throughout the building. When more than one panelboard is used the next panelboard will be designated "B" and then "C", etc.

The branch-circuit wiring layout should be designed to keep the homeruns as short as possible. Doing so will reduce the cost of the installation and also keep voltage drop to a minimum.

The layout in Fig. 1-13 shows the duplex receptacles serving the master bedroom and another bedroom (adjacent to the bathroom) connected to one circuit. All but two outlet boxes will have to be sized to contain four conductors. The outlet box on the bottom wall of the top bedroom will contain six conductors (the one feeding the bathroom receptacle). The outlet box on the left wall of the master bedroom will also have to be sized for six conductors because of the homerun to the panelboard (designated by the half arrowhead).

You will note that two outlets, serviced by individual circuits, are used in the utility room. Actually, both of these outlets could be installed on one circuit to comply with the NEC. However, the owners of this house wanted to use an automatic ironer next to the washing machine in this room. There is a possibility that when the washer and ironer were operating at the same time, one circuit would be overloaded. Thus, the reason for two circuits here. Remember that the NEC states the minimum requirements. It is often necessary to install an electrical system that surpasses these minimum requirements.

Continuing with our analogy, four of the receptacles in the living room are switched by two 3-way wall switches (S3) and one 4-way wall switch. These four receptacles will have table or floor lamps plugged into them and may be turned off and on at any of the three locations; that is, from the vestibule, the hallway, or kitchen entrance. In this case, when the receptacles are switched off, all four receptacles are off. Where the receptacles are used to also supply an electric clock, television, etc., it would be best to split-wire these four receptacles; that is, the bottom receptacle of each duplex receptacle is wired to be energized at all times. The top receptacle is switched at the three locations mentioned before. This allows the television, stereo, electric clock, etc. to be plugged into the lower outlet while the table lamps are connected to the top outlets for switching purposes. The other two receptacles in the living room are connected to the same circuit supplying the five duplex receptacles in the remaining bedroom.

The wiring in the kitchen again surpasses the NEC requirements. Actually, two 20-ampere circuits would suffice for all the kitchen outlets. However, due to the many small appliances now being used in the home, two additional circuits were supplied in the kitchen/dining area. Each circuit contains two #12 AWG conductors and a ground wire. Each circuit will be connected to a 20-ampere, single-pole circuit breaker in the panelboard in the utility room.

The floor plan of the residence in Fig. 1-13 shows three outside outlets; one would be sufficient to comply to NEC regulations. However, for convenience, one outlet is shown at the front entrance. This may be used for Christmas decorations, to plug in electric hedge or grass trimmers, and similar devices. The outside receptacle located at the kitchen door and under the car port is necessary to plug in a vacuum cleaner to vacuum the car, or perhaps for use with an electric car polisher. It may also be used for an electric rotisserie on, say, a barbecue grill, and many other applications. The remaining outside receptacle on the back of the house may be used for the same applications as mentioned previously.

Ground-Fault Circuit-Interrupters

The NEC states in Section 210-8 that certain residential outlets must be provided with ground-fault circuit-interrupters. This includes receptacle outlets installed in the following locations:

- Bathrooms, garages, outdoors, crawl spaces, unfinished basements and kitchen receptacles within 6 feet of the kitchen sink.

Referring again to the floor plan in Fig. 1-13 the two circuits supplying the countertop receptacles in the kitchen must have ground-fault circuit-interrupters installed—either at the panelboard where the circuits originate, or by installing ground-fault circuit-interrupters as a part of the receptacles at the outlet location. It is usually best to use the panelboard interrupters whenever possible.

The three outside receptacles also fall under this category. There is a note on the working drawings that states, "Splice all three outside receptacle circuits together in panelboard and connect to one ground-fault circuit-interrupter as indicated in the panelboard schedule."

The one bathroom outlet, next to the sink, is connected with the circuit feeding the bedroom outlets. Therefore, a ground-fault circuit-interrupter is provided here with one that is an integral part of the receptacle itself. See Fig. 1-14.

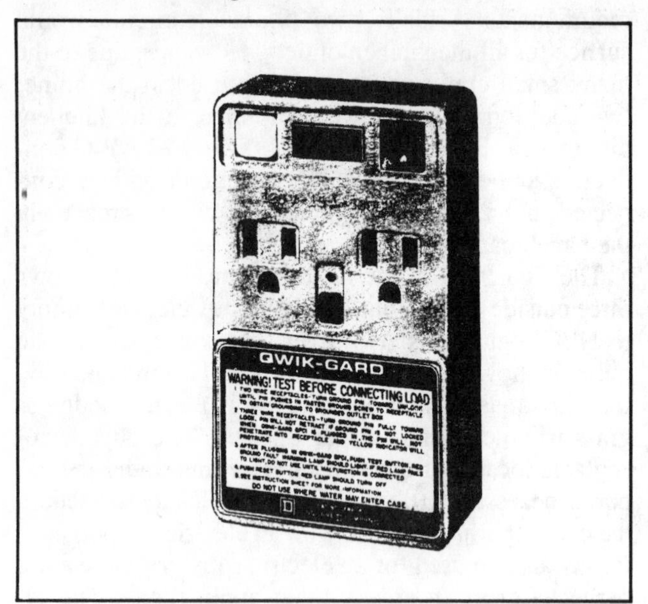

Fig. 1-14: Receptacle with integral ground-fault circuit-interrupter.

Special Outlets

Electrical workers must also consider wiring for special appliances during the rough-in stage. Such items include garbage disposals, exhaust fans, water heaters, electric ranges or cooktops and ovens, clothes dryers, and similar pieces of equipment.

Electric Ranges: Single electric ranges are supplied with a single circuit (usually rated at 40-50 amperes) from the panelboard to a 250-volt, 50-ampere range receptacle. The electric range is then connected by means of a range cord, although the circuit can be connected directly to a junction box on the range provided for such a purpose. Fig. 1-15 shows how a direct-connection circuit may be indicated on a working drawing (left) along with a range receptacle (right).

Cooking Tops and Ovens: The NEC permits both the cooking top and the wall-mounted oven—found in many modern homes—to be connected to one circuit provided the circuit has ample capacity to serve both appliances. The diagram in Fig. 1-16 shows how this might be applied to a built-in cooking top and oven in a residence.

In general, the No. 10 AWG tap conductors from the No. 6 AWG conductors must be as short as possible; that is, "no longer than necessary for servicing." Therefore, two appliances of approximately 30-ampere rating each may be fed by one 50-ampere circuit.

However, since tapping or splicing the smaller-size conductors from the 50-ampere circuit will require extra junction boxes, cable or conduit connectors, and wire connectors, as well as additional labor for making the splices and connections, it is advisable to use two separate 30-ampere circuits if the distance to the panelboard were of reasonable length. The two separate circuits would also offer greater protection in the event of a ground fault because of the lower-rated individual overcurrent protection.

Clothes Dryers: The clothes-dryer circuit usually will be a 120/240-volt, single-phase, three-wire circuit of No. 10 AWG wire, since most of the heating elements are 4.5 kW. A clothes-dryer outlet and feeder is shown in Fig. 1-17. It utilizes a 30-ampere, 250-volt dryer receptacle. This is the most common method of providing electric power to clothes dryers. The dryer is then connected to the outlet with a power cord similar to those used with electric ranges.

Water Heaters: Many water heaters used for residential and commercial applications require electric current for operation. The residential size may range from 1.2 to 4.5 kW, the smaller size at 120 volts and all others at 240 volts. Most utilize a 52-gallon tank with a maximum wattage of 4.5 kW. Section 220-32 (3) of the NEC requires that the nameplate rating be used when sizing conductors for water heaters. However, if the water heating elements are so interlocked that all elements cannot be energized at the same time, the maximum possible load is considered the nameplate rating. For example, a 52-gallon water heater with two 4500-watt heating elements might seem to have a total load of 9000 watts. However, most residential water heaters are controlled so that only one element will operate at a time. That is, when the bottom thermostat calls for heat, the contacts close and allow current to flow through the heating element. Whether or not the upper themostat calls for heat, the contacts will not close as long as current flows in the lower element.

When the desired temperature is reached at the bottom thermostat, the contacts open and stop the flow of current. At this time, if the upper themostat still calls for heat, the upper contacts will close and allow current to flow through this element. This method of control prevents the existence of a load of more than 4.5 kW. A wiring diagram showing the connections of the circuit

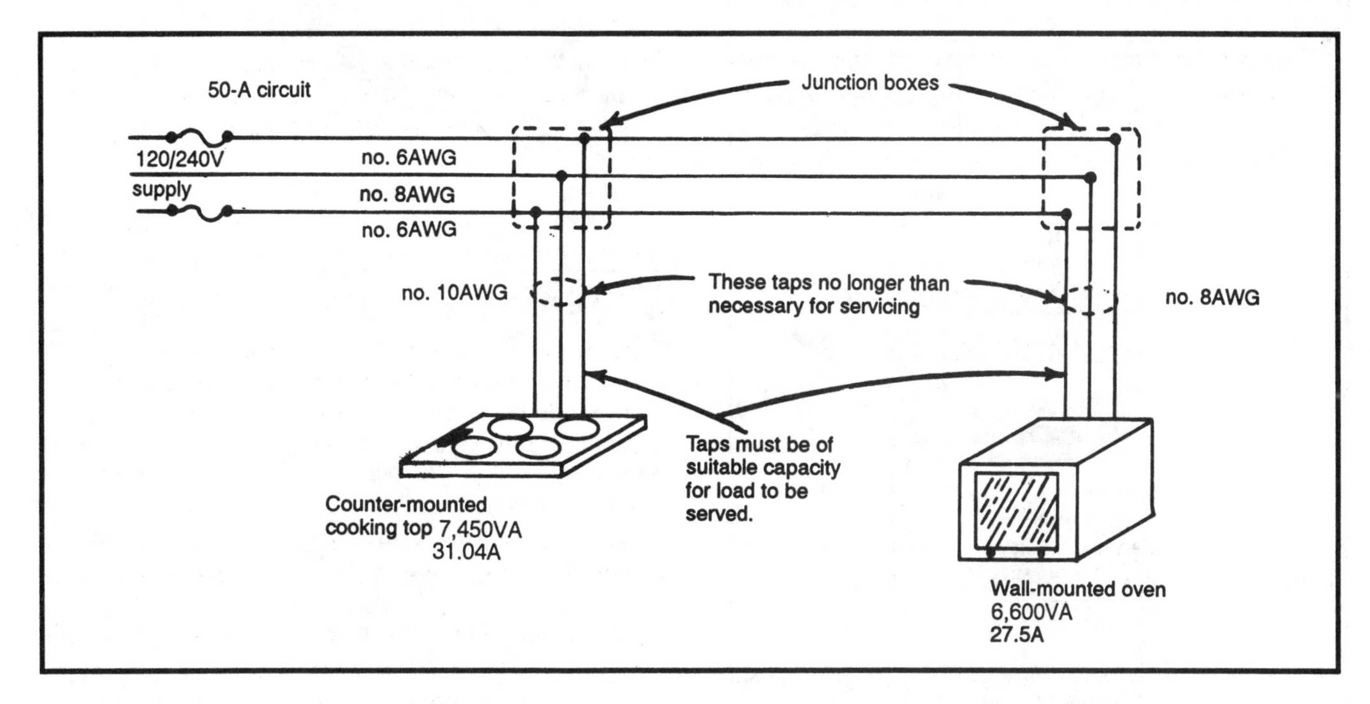

Fig. 1-15: Method of showing direct-connection circuit on a working drawing (left). Method of showing an outlet connection on a working drawing (right).

just described is shown in Fig. 1-18. Therefore, according to Section 220-32(3) of the NEC, the nameplate rating for this water heater will be 4500 watts; not 9000 watts. Depending upon the length of the circuit, the minimum wire size should be No. 10 AWG.

The water heater must also be equipped with a temperature-limiting means in addition to the conventional thermostat to disconnect all "hot" conductors when maximum water temperature is reached. See Section 422-14 of the NEC.

Garbage Disposals: The garbage disposal shown by symbol and notation on the floor plan in Fig. 1-19 is rated at 7.5 amperes at 120 volts. A separate 15-A, 120-volt branch circuit is shown going to panelboard "A" from the junction box provided for connection

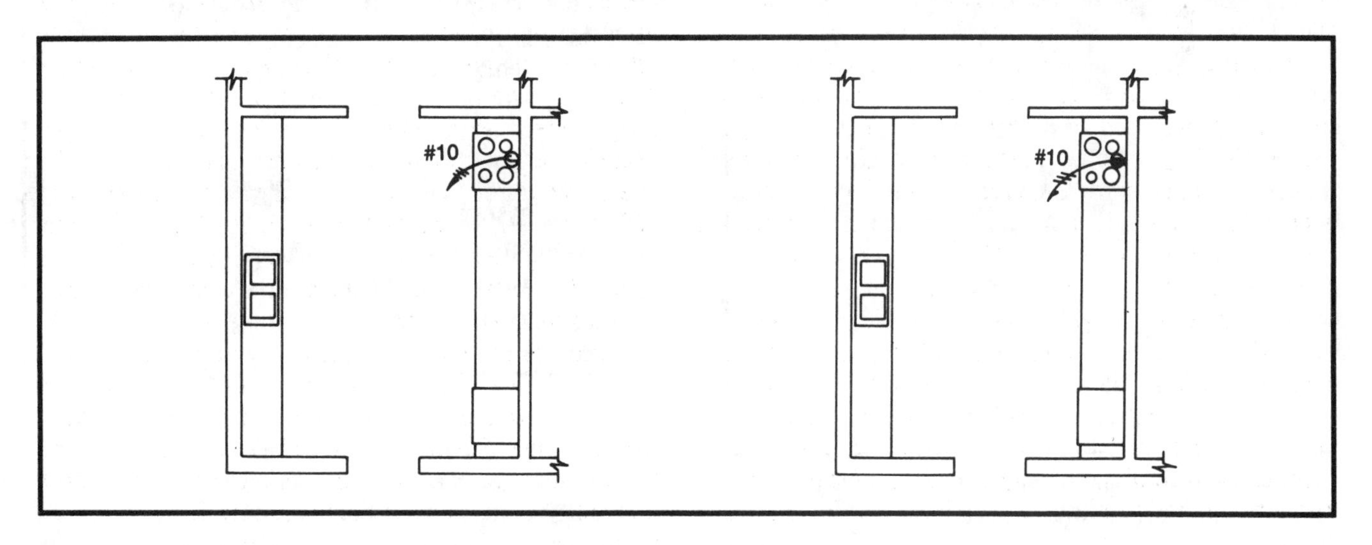

Fig. 1-16: Wiring diagram for cooking top and wall-mounted oven.

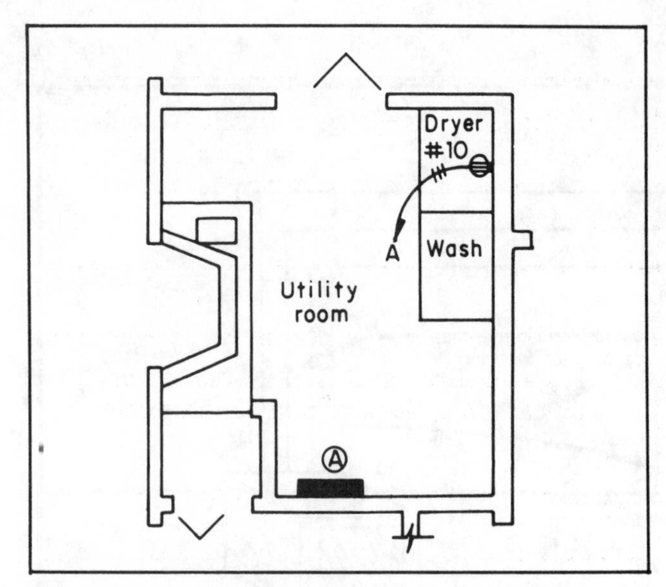

Fig. 1-17: Floor plan of a utility room showing a clothes-dryer outlet with three No. 10 wires feeding it.

purposes on the disposal unit. Most garbage-disposal units are powered by split-phase, 120-volt motors of between ⅓- and ½-hp rating; and according to the NEC, running-overcurrent protection is required and

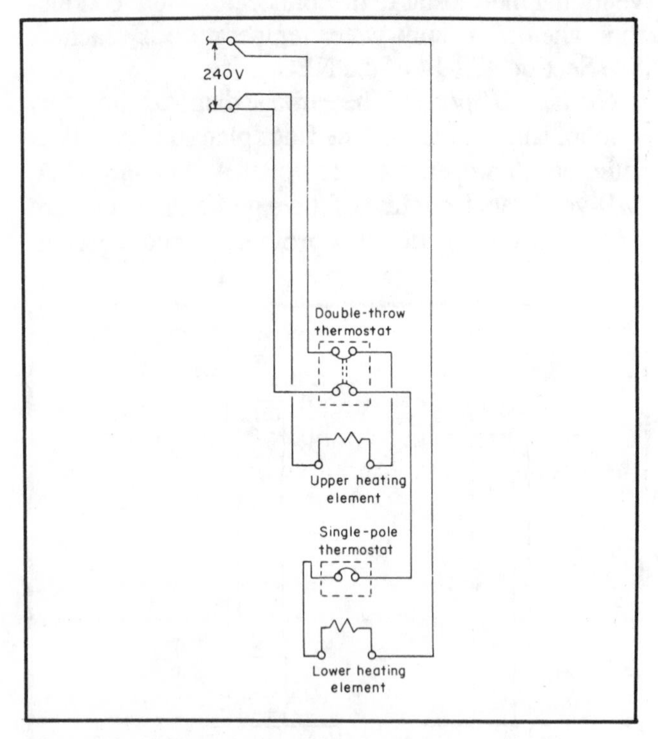

Fig. 1-18: Wiring diagram showing the connection for a typical residential water heater.

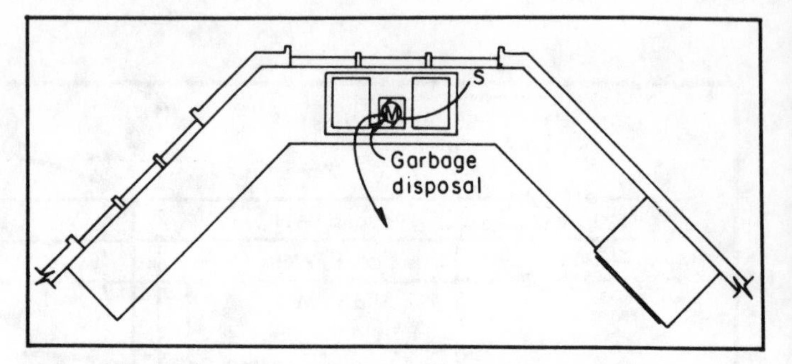

Fig. 1-19: Wiring diagram of residential garbage disposal.

must not exceed 125 percent of the full-load current rating of the motor. A conventional toggle switch is also used to turn the disposal off and on. Running-overcurrent protection is built into most modern garbage disposals so that only a branch circuit is needed to feed the unit. The table in Fig. 1-20 summarizes the 1993 NEC rough-wiring changes.

1993 NEC Section	Category
Section 300-1(a)	Scope
Section 300-4(d)	Protection
Section 300-5(d)	Protection
Section 300-8	Installation
Section 300-10	Metallic continuity
Section 310-11	Supports
Section 300-13	Electrical continuity
Section 300-15(c)	Fittings & Connectors
Section 300-18	Raceway installations
Table 300-19(a)	Conductor spacing
Section 300-22	Wiring in ducts
Section 300-35	Protection
Section 305-4(b)	Feeders
Section 305-4(c)	Branch circuits
Section 305(g)	Splices
Section 310-4	Conductors
Section 310-7	Direct burial conductors
Section 310-8	Wet locations
Section 310-11(a)(1)	Surface markings

Fig. 1-20: List of NEC Sections concerning rough-in wiring that have been changed with the 1993 Edition of the NE Code.

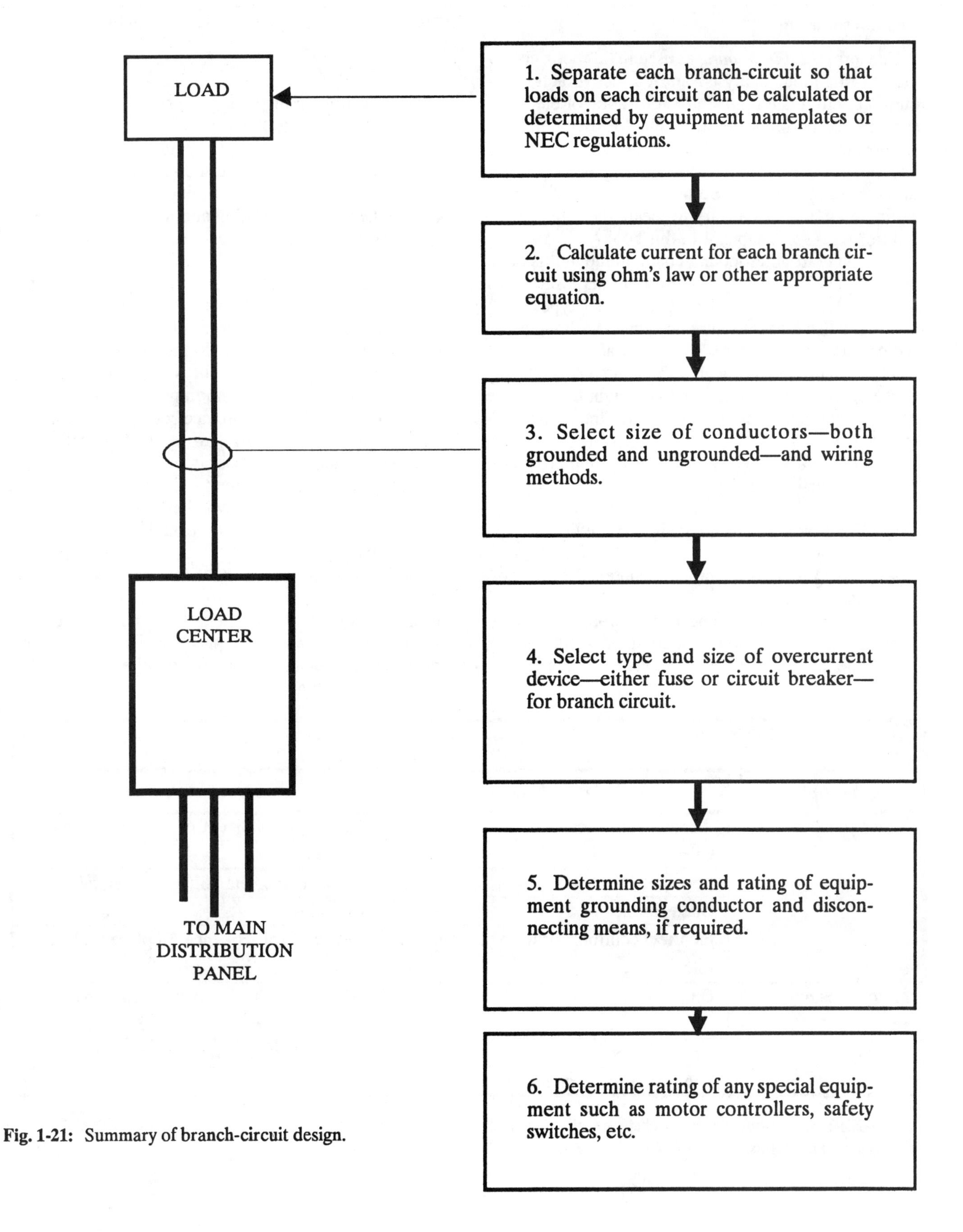

LOAD
1. Separate each branch-circuit so that loads on each circuit can be calculated or determined by equipment nameplates or NEC regulations.
2. Calculate current for each branch circuit using ohm's law or other appropriate equation.
3. Select size of conductors—both grounded and ungrounded—and wiring methods.
LOAD CENTER
4. Select type and size of overcurrent device—either fuse or circuit breaker—for branch circuit.
5. Determine sizes and rating of equipment grounding conductor and disconnecting means, if required.
TO MAIN DISTRIBUTION PANEL
6. Determine rating of any special equipment such as motor controllers, safety switches, etc.

Fig. 1-21: Summary of branch-circuit design.

Troubleshooting

The term *troubleshooting* as used in this book covers the investigation, analysis, and corrective action required to eliminate faults in the operation of electrical systems for building construction. Most troubles are simple and easily corrected; examples are a blown fuse or tripped circuit breaker—indicating a possible danger in the electrical system. Before replacing a fuse or resetting a circuit breaker, try to locate the cause of the failure. Also see Chapter 4 in this book.

The many troubleshooting charts found in this book give steps for correcting the most common problems that develop in electrical systems of all types. The charts are also meant to stimulate a train of thought and indicate a work procedure directed toward the source of trouble. To use them, find the complaint on the left side of the charts, then read across to the right for probable causes; continue across to the right for the proper corrective action.

Think before acting: Study the problem thoroughly, then ask yourself these questions:

- What were the warning signs preceding the trouble?
- What previous repair and maintenance work has been done?
- Has similar trouble occurred before?
- If the circuit or piece of equipment still operates, is it safe to continue operating it before further testing?

The answers to these questions can usually be obtained by:

- Questioning the system owners.
- Taking time to think the problem through.
- Looking for additional symptoms.
- Consulting the troubleshooting charts.
- Checking the simplest things first.
- Checking with testing and measuring instruments.

Furnace shutoff: Most modern buildings are equipped with one of three types of furnaces: oil, gas, or electric. No matter which type is used, electricity plays a key role in the unit's operation. In an emergency the furnace can be turned off by the switch controlling its electrical power. In most buildings the furnace control switch is located in the furnace room or on a nearby wall. Some switches are a part of an electrical supply box that contains the fuse or circuit breaker controlling the furnace's electrical circuit; others are simple *on/off* switches resembling wall switches.

To become familiar with troubleshooting charts, look at the one in Fig. 1-22. This chart is typical of those found throughout the book to help the reader locate and correct electrical problems. Remember, however, that troubleshooting charts do not have the answer to every conceivable problem. Even if this were possible, the amount of material required would fill an entire public library.

■ TROUBLESHOOTING
AC Motors

Trouble	Probable Cause	Correction Action
Motor stalls	Wrong application	Change type or size. Consult manufacturer.
	Overloaded motor	Reduce load.
	Low motor voltage	See that nameplate voltage is maintained.
	Open circuit	Fuses blown
	Incorrect control resistance of wound rotor	Check control sequence. Replace broken resistors. Repair open circuits.
Motor does not start	One phase open	See that no phase is open. Reduce load.
	Defective rotor	Look for broken bars or rings.
	Poor starter coil connection	Remove end bells

Fig. 1-22: Typical troubleshooting chart to facilitate investigation and corrective actions on electrical equipment and circuits.

Safety Reminders

Although electricity is one of our most useful energy sources, it is also potentially dangerous to life and property. Therefore, anyone who works with electricity should take precautions to guard against its dangers. Never touch any electrical apparatus without knowing its voltage and use characteristics. Never touch any uninsulated parts of an electrical apparatus until you are certain that power has been disconnected; that is, the purpose of the apparatus and how it relates to the overall system. Then proceed with caution, using appropriate testing instruments to verify that the circuit is "dead."

Each chapter in this book will contain a certain amount of safety reminders to enable you to work safer with electricity. Digest each of these sections thoroughly before continuing.

First Aid

If breathing has stopped, begin mouth-to-mouth resuscitation as soon as possible. If stoppage is due to poisonous gas or lack of oxygen, first move victim to fresh air. If it is due to electric shock, do not touch victim until he or she is separated from the electrical current. Dry your hands and disconnect electricity from the circuit. If this is not possible, use a length of dry wood to push the victim clear.

Administer mouth-to-mouth resuscitation as shown in Fig. 1-23.

Minor burns from electric shock: Immerse in clean, cold water or apply clean ice if skin is not broken. Do *not* remove stuck clothing. Place pad over burn and bandage loosely. Expose surrounding area to the air. If burns are extensive, cover burned body area with a sterile dressing. If victim is conscious and can swallow, give him plenty of non-alcoholic liquids.

Minor cuts: Many times those working on electrical panels systems will accidentally cut themselves on edges of sheet-metal equipment housings or ductwork. If this happens, wash wound and surround the cut area with soap and warm water, wiping away from the wound.

Hold a sterile pad firmly over the wound until the bleeding stops. Then change the pad and bandage loosely. Replace the sterile pad and bandage as necessary to keep them clean and dry. Wearing leather gloves while working will keep such cuts to a minimum.

Eye injuries: Do *not* rub. If injury is from a foreign body, give natural watering a chance to wash it away. If this fails to remove it, try flushing with lukewarm or cold water. If the foreign body still remains, do not try to take it out with a handkerchief or tissue—this risks corneal damage. Only a doctor should probe the eye.

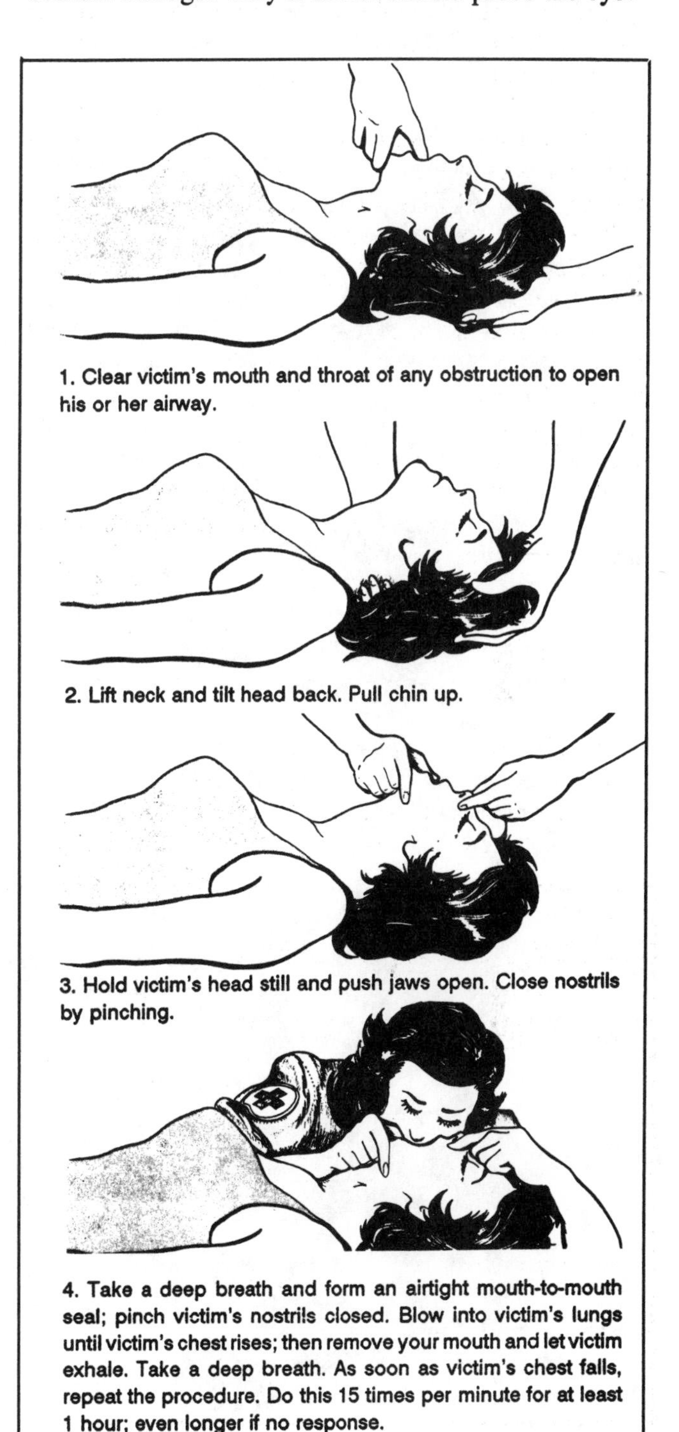

1. Clear victim's mouth and throat of any obstruction to open his or her airway.

2. Lift neck and tilt head back. Pull chin up.

3. Hold victim's head still and push jaws open. Close nostrils by pinching.

4. Take a deep breath and form an airtight mouth-to-mouth seal; pinch victim's nostrils closed. Blow into victim's lungs until victim's chest rises; then remove your mouth and let victim exhale. Take a deep breath. As soon as victim's chest falls, repeat the procedure. Do this 15 times per minute for at least 1 hour; even longer if no response.

Fig. 1-23: The four basic steps for mouth-to-mouth resuscitation.

chapter 2

ELECTRIC SERVICES

All buildings containing equipment that utilizes electricity require an electric service. An electric service will enable the passage of electrical energy from the power company's lines to points of use within the buildings. Figure 2-1 shows the basic sections of a residential electric service. In this illustration, note that the high-voltage lines terminate on a power pole near the building that is being served. A transformer is mounted on the pole to reduce the voltage to a usable level (120/240 volts in this case). The remaining sections are described as follows:

- *Service drop*: The overhead conductors, through which electrical service is supplied, between the last power company pole and the point of their connection to the service facilities located at the building or other support used for the purpose.

- *Service entrance*: All components between the point of termination of the overhead service drop or underground service lateral and the building main disconnecting device, except for the power company's metering equipment.

- *Service-entrance conductors*: The conductors between the point of termination of the overhead service drop or underground service lateral and the main disconnecting device in the building.

- *Service-entrance equipment*: Provides overcurrent protection to the feeder and service conductors, a means of disconnecting the feeders from energized service conductors, and a means of measuring the energy used by the use of metering equipment.

When the service conductors to the building are routed underground, as shown in Fig. 2-2, these conductors are known as the service lateral, defined as follows:

- *Service lateral*: The underground conductors through which service is supplied between the power company's distribution facilities and the first point of their connection to the building or area service facilities located at the building or other support used for the purpose.

Sizing the Service

Sometimes it is confusing just which comes first: the layout of the outlets, or the sizing of the service. In many cases, the service size (size of main disconnect, panelboard, service conductors, etc.) can be sized using NEC procedures before the outlets are actually laid out. In other cases, the outlets will have to be laid out first. However, in either case, the service-entrance and panelboard size will have to be calculated and located before the circuits may be designed or installed. Let's take an actual residence and size the service-entrance

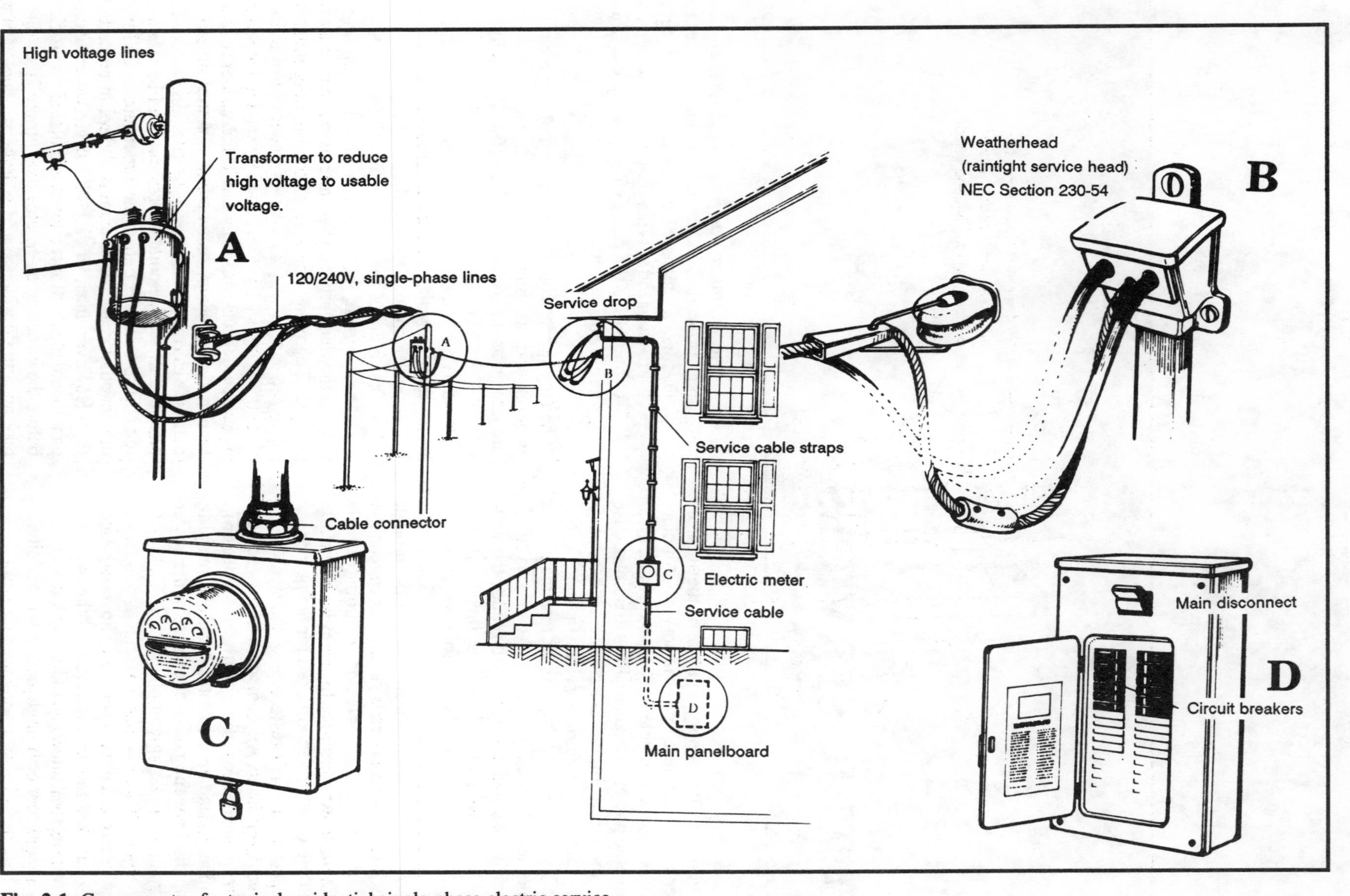

Fig. 2-1: Components of a typical residential single-phase electric service.

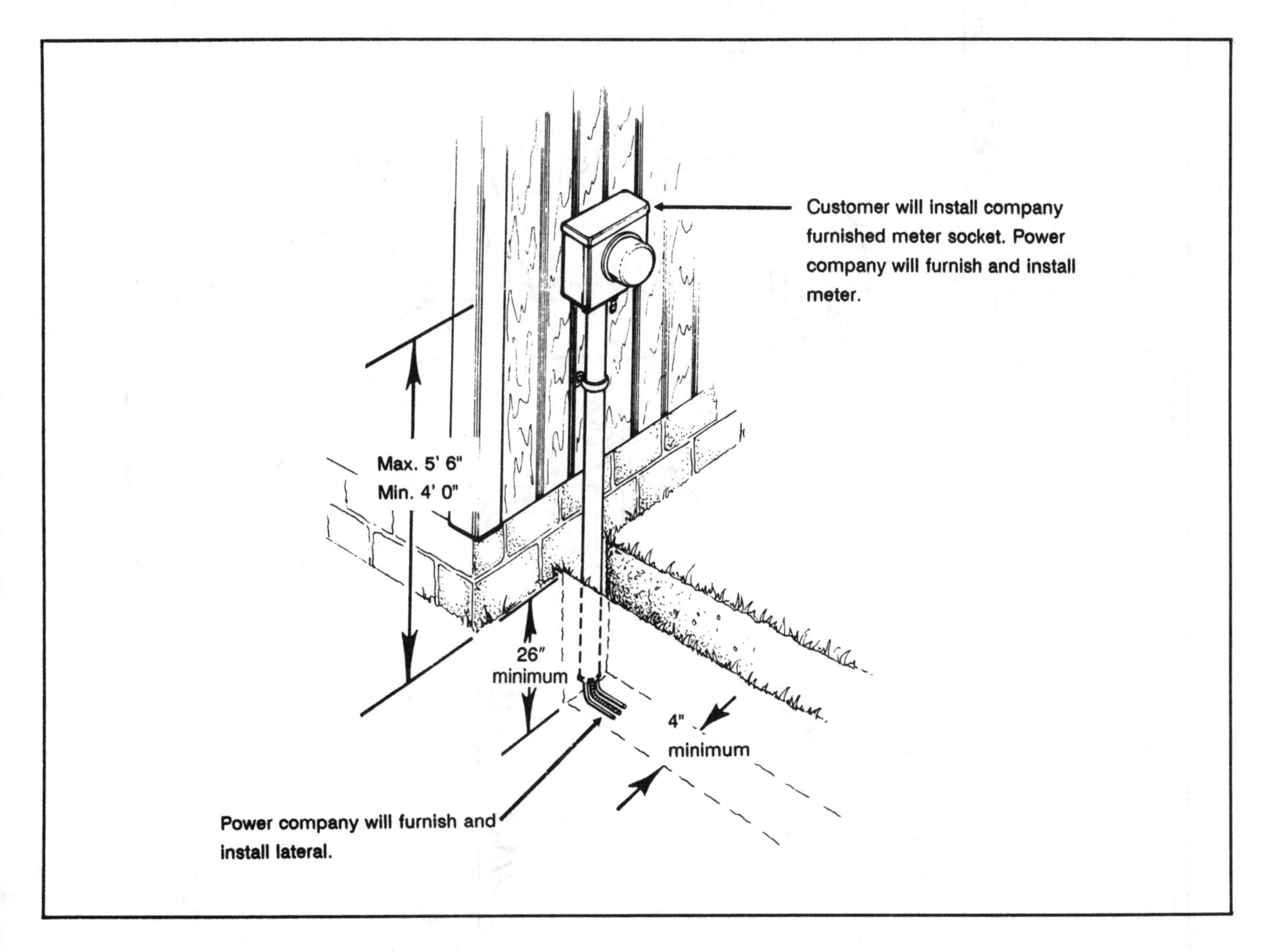

Customer will install company furnished meter socket. Power company will furnish and install meter.
Max. 5' 6"
Min. 4' 0"
26" minimum
4" minimum
Power company will furnish and install lateral.

Fig. 2-2: An underground electric service is known as service lateral.

according to the latest edition of the National Electrical Code.

A floor plan of a small residence is shown in Fig. 2-3. This building is constructed on a concrete slab with no basement or craw space. There is an unfinished attic above the living area, and an open carport just outside the kitchen entrance. Appliances include a 12-kVA (12 kilovolt amperes) electric range and a 4.5 kVA water heater. There is also a washer/dryer (rated at 5.5 kVA) in the utility room along with a gas furnace using a 1 kW blower motor.

General lighting loads are calculated on the basis of 3 watts per square foot of living space in single-family dwellings. This includes non-appliance duplex receptacles into which table lights, television, etc. may be plugged. Therefore, the area of the building must be calculated first. If the building is under construction, the dimensions can be determined by scaling the working drawings used by the builder. If the residence is an existing building, with no drawings, actual measurements will have to be made on the site.

Using the floor plan of the residence in Fig. 2-3 as a guide, an architect's scale is used to measure the longest width of the building which is 33 feet. The longest length of the building is 48 feet. These two measurements multiplied together gives (33′ x 48′ =) 1584 square feet of living area. However, there is an open carport on the lower left of the drawing. This carport

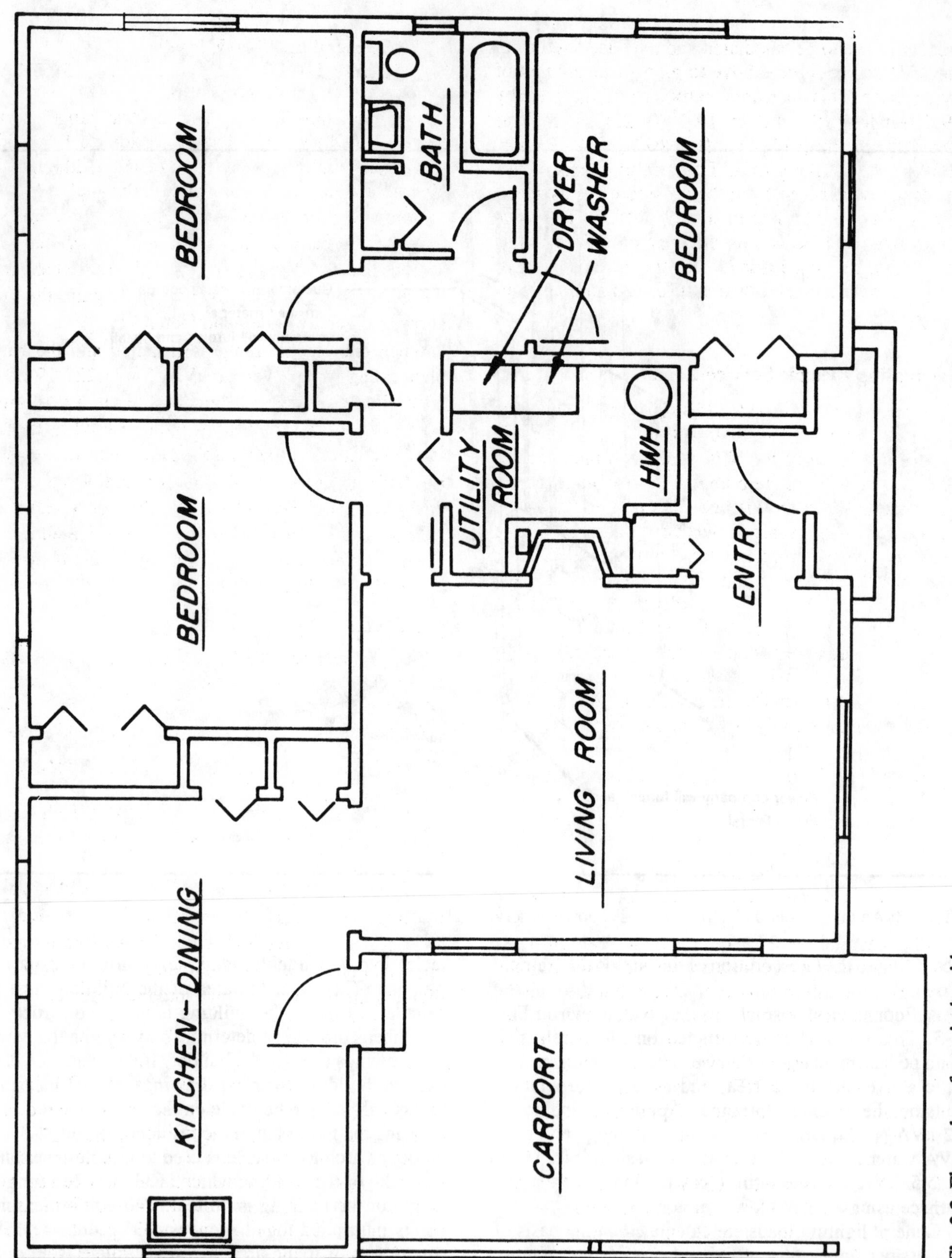

Fig. 2-3: Floor plan of a typical residence.

area will have to be calculated and then deducted from the 1584 sq. ft. figure above to give a true amount of living space. This open area (carport) is 12 feet wide by 19.5 feet long. So, 12′ by 19.5′ = 234 sq. ft. This area deducted from 1584 sq. ft. leaves (1584 - 234) = 1350 sq. ft. of living area. The measurements of this residence were taken from the outside walls as specified in Section 220-3(b) of the 1993 NEC. The same policy must be followed for any type of building when calculating lighting loads by the square-foot method. This rule also applies to warehouses and similar occupancies.

Calculating Electric Service Load

General Lighting Load: 1350 sq. ft. of living space at 3 watts per sq. ft. = 4050 watts.

Section 220-4b of the NEC requires a minimum of two, 120-volt, 20-ampere appliance circuits for small appliance loads in the kitchen, dining area, breakfast nook, and similar areas where small appliances (toasters, coffee makers, etc.) will be used. One 120-volt, 20-ampere circuit is also required for the laundry area which, in the residence in question, is located in the utility room. We will also allow one, 120-volt, 20-ampere circuit for the gas furnace blower.

The above information allows us to calculate the minimum size feeder for the electric service, and the size of the main panelboard or load center.

General Lighting	4050 VA
Small Appliance Load	3000 VA
Laundry Load	1500 VA
Furnace Circuit	1500 VA
Total General Lighting & Appliance Loads	10,050 VA

All residential electrical outlets are never used at one time. There may a rare instance where all the lighting may be on for a short time every night, but if so, all the small appliances, electric range, water heater, furnace, dryer, washer, and the numerous receptacles throughout the house will never be used simultaneously. Knowing this, the NEC allows a diversity or demand factor in sizing electric services. Our calculation continues:

First 3000 VA is rated at 100%	3000 VA
The remaining 7050 VA may be rated 35% (the allowable demand factor). Therefore 7050 x .35 =	2467.5 VA
Net General Lighting & Small Appliance Load	5467.5 VA

We must now look at the electric range, water heater, and dryer loads for this residence before the service calculations can be completed. Although the electric range has a nameplate rating of 12 kVA, seldom will every burner be on high at once. Nor will the oven remain on all the time. When the oven reaches the dialed temperature, the thermostat shuts off the power until it cools down again. Again, the NEC allows a diversity or demand factor.

When one electric range is installed and the nameplate rating is not over 12 kVA, Table 220-19 of the NEC allows a demand factor resulting in a total rating of 8 kVA. Therefore, 8 kVA may be used in the service calculation instead of the nameplate rating of 12 KVA. The electric clothes dryer and water heater, however, must be calculated at 100% when using this method to calculate residential service-entrances. Therefore, if fractions are rounded off, here is the final calculation for the service feeder.

Net General Lighting & Small Appliance Load	5468 VA
Electric Range (using demand factor)	8000 VA
Clothes Dryer	5500 VA
Water Heater	4500 VA
Total Load	23468 VA

Note: Since volts x amperes = watts, the NEC now refers to "watt" as volt-ampere; kilowatt as kilovolt-ampere, etc.

The conventional electric service for residential use is 120/240-volt, 3-wire, single-phase service. Services are sized in amperes and when the wattage is known on single-phase services, amperes may be found by dividing the (highest) voltage into the wattage.

$$23{,}468 \div 240 = 97.7 \text{ amperes}$$

When the net computed load exceeds 10 kVA, the minimum size service conductors and panelboard must be 100 amperes. See Section 230-42 of the NEC.

The neutral conductor in a 3-wire, single-phase service carries only the unbalanced load between the two "hot" legs. Since there are several 240-volt loads in the above calculation, these loads will be balanced and therefore reduce the load on the neutral service feeder conductor. In most cases, the neutral conductor does not have to be as large as two "hot" conductors.

In the above example, the water heater does not have to be included in the neutral conductor calculation, since it is strictly 240 volts with no 120-volt loads. The clothes dryer and electric range, however, have 120-volt lights that will unbalance the current between the phase or "hot" legs. The NEC allows a demand factor of 70% for these two appliances. Using the information, the neutral conductor may be sized accordingly:

General Lighting and Appliance Load ..	5100 VA
Electric Range (8,000 watts x 70% =)..	5600 VA
Clothes Dryer (5,500 watts x 70% =)...	3850 VA

Total 14,550 VA

To find the total phase-to-phase amperes, divide the total wattage by the voltage between phases.

$$14{,}550 \div 240\ V = 60.6\ amperes$$

The service-entrance feeders have now been calculated and must be rated at 100 amperes with a neutral conductor rated for at least 60.6 amperes. If type SE (service-entrance) cable or type USE (Underground Service-Entrance) cable are being used, all that is necessary is to order the required amount of 100-ampere SE cable. The weatherhead, cable straps, and SE connectors are also designated for use on 100-ampere SE cable. But maybe you want to use individual conductors in a protective steel or plastic conduit. What sizes do you need? Tables 320-16 through 310-19 give the current-carrying capacity of most conductors in use today. In this case we would refer to Table 310-16 because we are using three insulated conductors rated at 240 volts within an overall covering; that is, the conductors are either enclosed in a cable or raceway (conduit).

Cable insulation is rated at various ambient temperatures. For most applications, Type TW (rated at 140°F.) is adequate for residential service use. If we wanted to use copper service conductors, first look at the heading under "Type TW" copper in the table, and then follow this column down until we find an AWG size that will carry at least 100 amperes of current. Size #1 copper shows a rating of 102 amperes. Therefore, we would select Size #1 TW copper conductors for the service-entrance feeder. To continue, we also need one conductor for the neutral conductor to carry the unbalanced load. Remember the neutral was sized at 60.6 amperes. Therefore, following the same column downward, size #4 TW copper is rated at 66 amperes and would be the size selected for our service-entrance neutral feeder.

The NEC also allows aluminum conductors for electric services. Aluminum conductors often give connection problems if not installed exactly right, but many contractors consider the savings in costs over copper is worth the extra effort.

If Type TW aluminum conductors are to be used in the residence in question, move over to the appropriate column heading in Table 310-16 and follow down the column. You will note that size 2/0 TW aluminum is rated at 108 amperes; this is adequate for our service-entrance conductors. Using the same technique, our neutral conductor could be size #2 aluminum which is rated at 69 amperes.

Sizing the Load Center

Each circuit in the building must be provided with overload protection either in the form of fuses or circuit breakers. If more than six such devices are needed, you must also provide a means of disconnecting the entire service—either with a main disconnect switch or a main circuit breaker. To calculate the number of fuse holders or circuit breakers required in the residence above, let's look at the general lighting load first. Since we have a total general lighting load of 4050 watts, this can be divided by 120 volts (Ohm's Law states Amperes = W/V) which give us 33.75 amperes. You now have a choice of using either 15-ampere circuits or 20-ampere circuits for the lighting load. Two 20-ampere circuits (2 x 20) gives us a total amperage of 40, so two 20-ampere circuits would be adequate for the lighting. However, two 15-ampere circuits total only 30 amperes and we need 33.75. Therefore, you would need three 15-ampere circuits for the total lighting load.

Many electricians performing residential wiring prefer to use 15- ampere (No. 14 AWG NM Cable) circuits for their lighting because the smaller size wire makes it easier to make up the connections at the lighting fixtures and switches. However, to cut down on voltage drop throughout the house, there are other electricians who prefer the higher-quality 20-ampere circuits (No. 12 AWG NM Cable) for all branch circuits throughout the building.

You will need a minimum of two 20-ampere appliance circuits for the small appliance load, one 20-ampere circuit for the laundry, and another 20-ampere circuit for the gas-fired furnace blower. Thus far, we can count the following branch circuits:

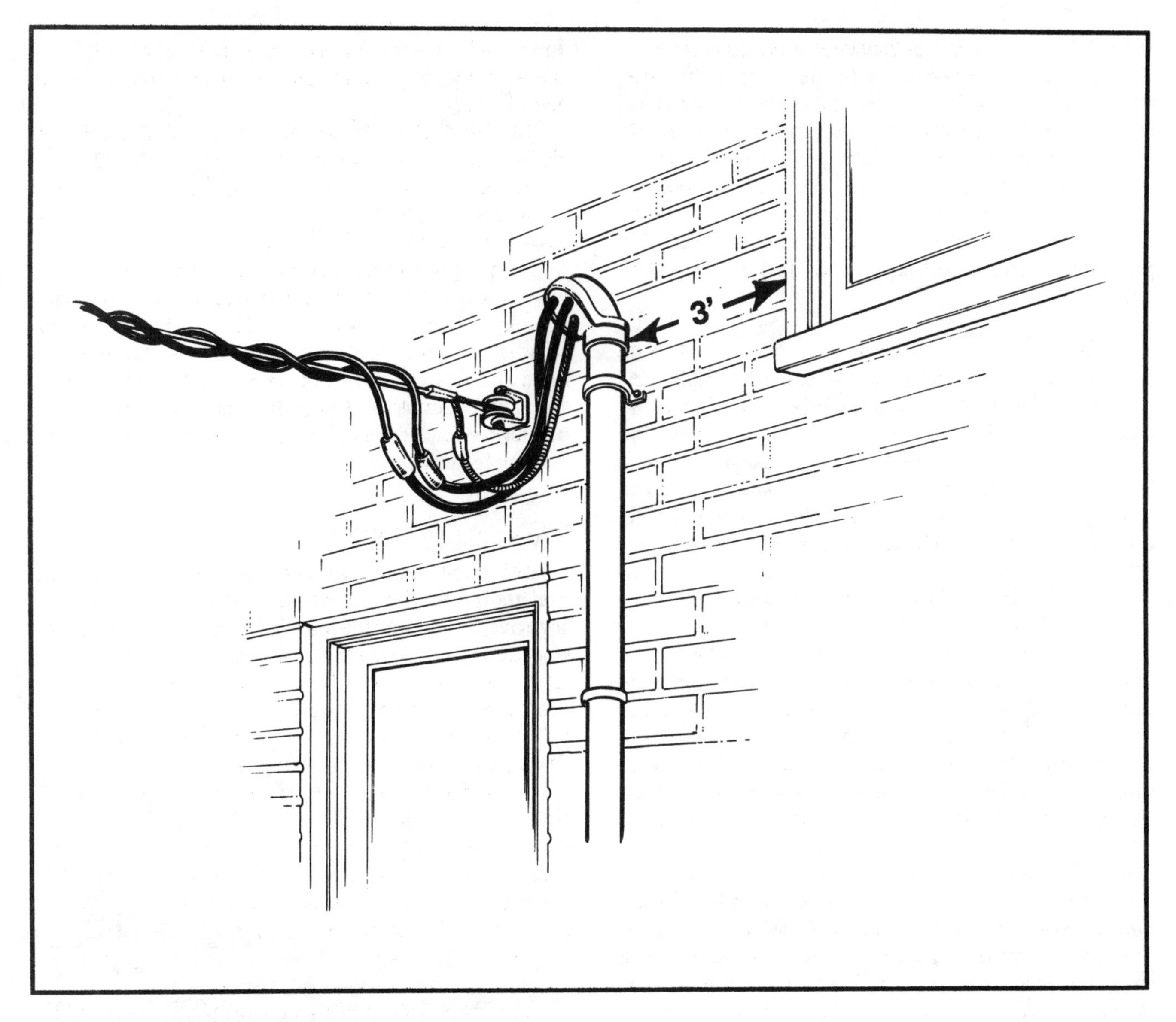

Fig. 2-4: Service drops must have a clearance of not less than 3 feet from building openings such as windows, doors, porches, fire escapes, and similar locations.

General Lighting Load	3
Small Appliance Loads	2
Laundry Load	1
Furnace Load	1
Total	7

Most load centers and panelboards come with an even number of circuit breaker or fuse-holder connectors; that is, 4, 6, 8, 10, etc. The load center for this residence should therefore have either 8 single-pole circuit breakers or 8 single-pole fuse holders for the above loads.

Don't forget about the 240-volt appliances. The electric range will require a 2-pole circuit breaker, as will both the clothes dryer and water heater. These three appliances will therefore require an additional (2 poles x 3 appliances =) 6 spaces in the load center. We could get by with a 14-space circuit-breaker load center with a 100-ampere main circuit breaker. However, it is always better to provide a few extra spaces for future use. Many designers prefer to add 20% extra space in the load center for additional circuits that may be added later. The circuit breakers themselves do not have to be installed, but space should be provided for them. This would put the load center for the house in question up to

16 or 18 spaces. You would need to order a 100-ampere load center with a 100-ampere main circuit breaker. Furthermore, you will need no less than 3 15-ampere, single-pole circuit breakers, 4 20-ampere, single-pole circuit breakers, 1 double-pole circuit breaker for the electric range, 1 double-pole circuit breaker for the clothes dryer, and 1 double-pole circuit breaker for the water heater. Methods of sizing circuit breakers and their respective feeders are covered in Chapter 3.

Installing the Service Entrance

Let's assume that you have selected Type SE (service-entrance) cable, rated at 100 amperes, for your installation. The locations of the service drop, electric meter and the load center should be considered first. It is always wise to consult the local power company to obtain their recommendations; where you want the service drop and where they want it may not coincide. A brief meeting with the power company about the location of the service drop can prevent much grief, confusion, and perhaps expense.

While considering the placement of the service drop, keep in mind that service-drop conductors must not be readily accessible. They must have a clearance of not less than 3 feet from windows, door, porches, fire escapes, or similar locations. See Fig. 2-4. Furthermore, when service-drop conductors pass over rooftops, they must have a clearance of not less than 8 feet from the highest points of roofs. See Fig. 2-5. There are, however, some exceptions. For example, in residential or small commercial wiring installations where the voltage does not exceed 300 volts, and the roof has a slope of not less than 4 inches in 12 inches, a reduction in clearance from 8 feet to no less than 3 feet is permitted. See Fig. 2-6.

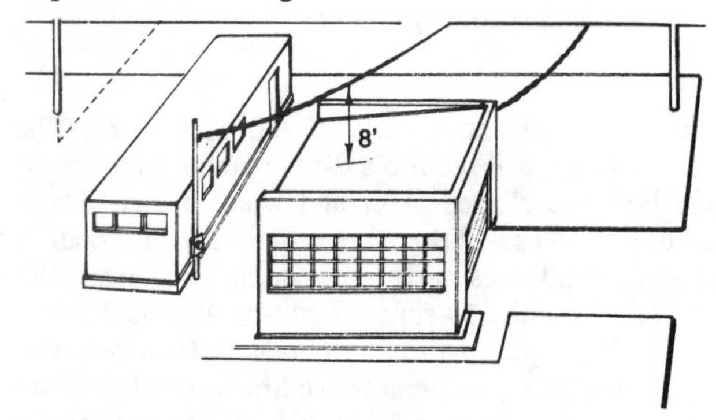

Fig. 2-5: Service conductors must have a clearance of not less than 8 feet over flat rooftops or rooftops with less than 4/12 pitch.

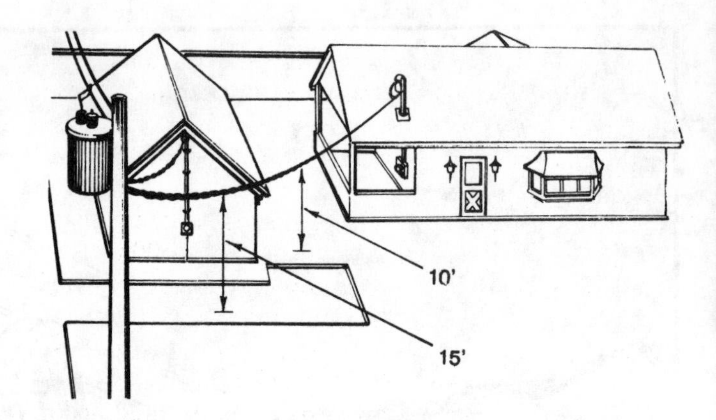

Fig. 2-6: When the roof has not less than 4/12 pitch and service conductors pass over more than 4 feet of roof space, the conductors must be at at least 3 feet above the roof.

When service-entrance conductors do not pass over more than 4 feet of the overhead portion of the roof, and they terminate at a through-the-roof raceway or approved support, the clearance may be further reduced to only 18 inches as shown in Fig. 2-7.

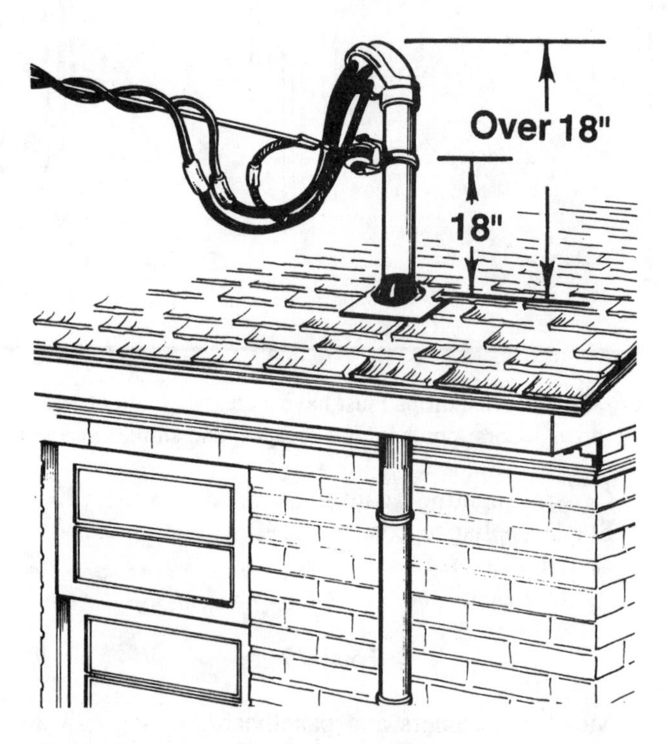

Fig. 2-7: Since these service conductors do not pass over more than 18″ of roof space, the clearance of service conductors over roof may be reduced to a minimum of 18″.

The NE Code also specifies the distance that service-drop conductors must clear the ground. These distances will vary with the surrounding conditions. Fig. 2-8 summarizes these conditions and the required heights.

In general, the 1993 NE Code states that the vertical clearances of all service-drop conductors—600 volts or under—are based on conductor temperature of 60°F (15°C), no wind, with final unloaded sag in the wire, conductor, or cable. Service-drop conductors must be at least 10 feet above the ground or other accessible surfaces at all times. More distance is required under most conditions. For example, if the service conductors pass over residential property and driveways or commercial property not subject to truck traffic, the conductors must be at least 15 feet above the ground. However, this distance may be reduced to 12 feet when the voltage is limited to 300 volts to ground.

In other areas such as over public streets, alleys, roads, parking areas subject to truck traffic, driveways on other than residential property, and other land traversed by vehicles such as cultivated, grazing, forest, and orchard, the minimum vertical distance is 18 feet.

The first example showing how to size the service-entrance for a single-family dwelling is very basic. Most modern homes have many more electrical devices for added comfort. Some would include air conditioning, a garbage disposal, a dishwasher and perhaps a central vacuum system. So let's update the service that we used in our first example with the addition of the following loads:

1 18 amp, 240-volt air conditioner
1 10 amp, 120-volt dishwasher
1 8 amp, 120-volt garbage disposal

	Line A	Neutral	Line B
Loads from first example	97.7	60.6	97.7
Air conditioning system	18.0	-	18.0
Dishwasher	10.0	10.0	-
Garbage disposal	-	8.0	8.0
25% of largest motor (Section 430-24)	4.5	-	4.5
Amperes per line	130.2	78.6	128.2

If these loads are added, the electric service size should be increased to 150 amperes rather than the previously calculated 100 amperes. This means that 150-amp SE cable will have to be used along with all the related components; that is service head, cable straps, weatherproof cable connectors, etc. A 150-ampere load center and main circuit breaker will also be required. The circuit capacity in the load center should be increased to 20 spaces to provide room for another two-pole, 20-ampere circuit breaker and two 15-ampere single-pole circuit breakers, leaving some spare spaces for future use.

Selecting the type of service and the proper materials to use also depends upon the type of occupancy. Figs. 2-9 through 2-14 show a variety of service-entrance applications along with the installation methods and materials required.

Temporary service: The pole-mounted service in Fig. 2-9 could be used for a temporary (90-day) electric service for use during the construction of a building, until the permanent wiring is installed. This same arrangement may also be used to serve an individual mobile home or recreation vehicle (RV). When used as a temporary electric service, follow the rules as specified in NE Code Sections 230 and 305. For mobile home installations, use NE Code Section 230 and Article 550.

If the electric service is used for a mobile home the following guidelines should be used in sizing the service.

Electrical services for mobile homes may be installed either overhead or underground provided a pole or power pedestal is provided at the point of attachment. The mobile home may be connected to the pole or power pedestal with a permanently installed circuit or by means of not more than one 50-ampere mobile home power cord with an integral molded cap. However, if the mobile home is factory equipped with gas or oil-fired central heating and cooling appliances—such as a gas range—the NEC (NE Code Section 550-5) permits the use of a 40-ampere mobile home power cord.

The NEC gives specific instructions for determining the size of the distribution-panel load for each unit. The calculations are based on the size of the mobile home, the small-appliance circuits, and other electrical equipment that will be connected to the service.

Lighting loads are computed on the basis of the mobile home's area: width times length (outside di-

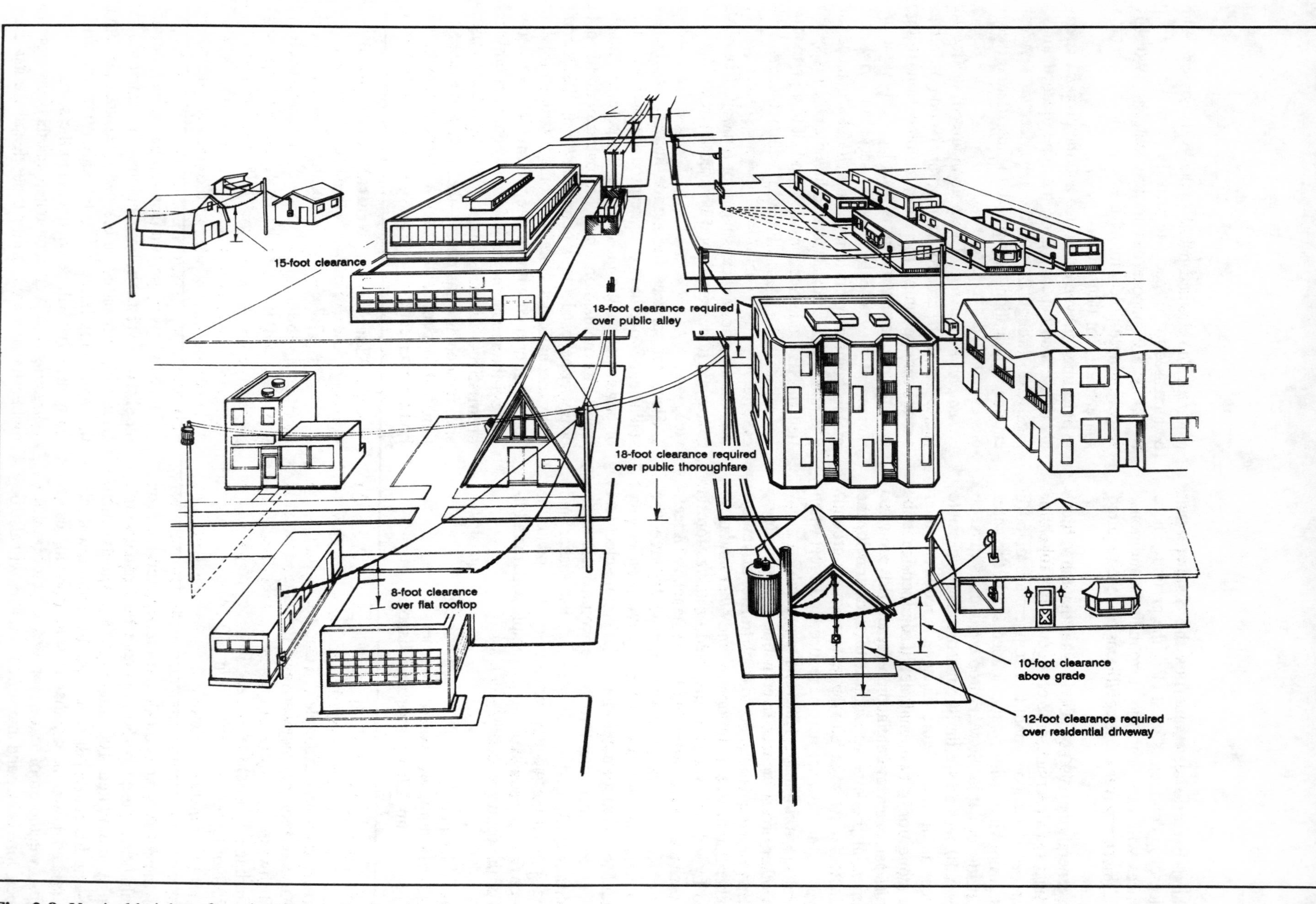

Fig. 2-8: Vertical heights of service-drops at various locations.

mensions exclusive of coupler) times 3 VA per square foot.

Length x width x 3 = lighting load in VA

The number of circuits required for the above lighting load may be further calculated by the following equation:

Length x width x 3 (VA)/120 x 15 (or 20) = Number of 15 - (or 20-) ampere circuits

A minimum of two small-appliance loads must be installed in every mobile home. These loads are computed on the basis of the number of circuits multiplied by 1500 VA for each 20-ampere appliance receptacle circuit.

One method to calculate the service requirements for a typical 70′ x 10′ mobile home is as follows: Using the first equation above, 70′ x 10′ = 700 sq. ft. of living space. This area multiplied by 3 VA per sq. ft. [NE Code Section 550-13(a)] = 2100 VA. Let's assume that this mobile home also has three small appliance circuits; a 2000-VA air conditioner; a 250-VA 120-volt kitchen exhaust fan; a 2500-VA water heater, and an 8000-VA electric range.

Lighting load:
70 x 10 x 3 VA = 2100 VA
Small-appliance load:
1500 VA x 3 circuits = 4500 VA

Total lighting and small appliance loads
2100 VA + 4500 VA = 6600 VA

NE Code Section 550-13(a) states that the first 3000 VA of the lighting and small-appliance loads must be calculated at 100%. However, a demand factor may be applied to the remaining wattage; this demand factor is 35% (0.35). The result of this calculation divided by the voltage (240 volts) gives the current or amperage per phase.

The first 3000 VA, therefore, is deducted from the total lighting and small-appliance loads, leaving a balance of 3600 VA. Multiplying this figure by the demand factor (0.35) gives a demand load of 1260 VA. This demand load combined with the first 3000 VA gives a total demand load for lighting and small-appliance circuits of 4260 VA. Now find the amperage per phase. Ohm's law (I = W/E) may be used to find the amperes per phase:

4260 VA/240 volts = 17.75 amperes per phase

The service calculation is not complete until the large-appliance loads are added:

Water heater: 2500/240 = 10.41 amps/line
Air conditioner: 2000/240 = 8.33 amps/line
Kitchen exhaust fan: 250/120 = 2.08 amps/phase
Electric range: 8000/240 x 0.8 = 26.66 amps/line

Note: NE Code Section 550-13(5) allows a demand factor for electric ranges used in mobile homes. Those ranges with nameplate ratings of between 0 and 10,000 watts rate a demand factor of 0.8.

To summarize the above results, phase A of the 120/240 V single-phase service carries a load of 17.75 + 10.41 + 8.33 + 2.08 + 26.66 = 65.23 amperes. Phase B will be the same except for subtracting 2.08 amperes for the kitchen fan. This fan operates on 120 volts and is connected to phase A only. Therefore phase B will carry a load of 63.15 amperes.

Based on the higher current calculated for either phase, a feeder with a minimum rating of 66 amperes must be used as a power supply for this mobile home.

When a single service drop is to provide electric service for several mobile homes (mobile home park) such as shown in the upper right corner of Fig. 2-8, the service arrangement shown in Fig. 2-10 may be used. In general, a service mast is used to route the service conductors into a weatherproof wire trough. Taps are made from the service conductors to individual meter sockets and mobile home power outlets—each containing a main circuit breaker. Up to six mobile homes may be fed from this type of service arrangement. If more than six are to be serviced, a main disconnect must be provided to shut off power to the wire trough which will then shut down all power outlets simultaneously.

Underground service: Sometimes it is desirable not to attach an overhead service to the building itself, yet the owner may not want the expense of a complete underground service with a pad-mounted transformer. In this case, an overhead service is normally brought to a pole mounted at the edge of the property line several feet away from the building or structure. See Fig. 2-11. A conduit is then installed from the meter base or socket down the pole to at least 18 inches underground. This is provided to protect the conductors against damage as specified in NE Code Section 230-32. The installation must comply with provisions stipulated in NE Code Section 300-5. A direct-burial cable is used from the meter socket to the panelboard inside the building.

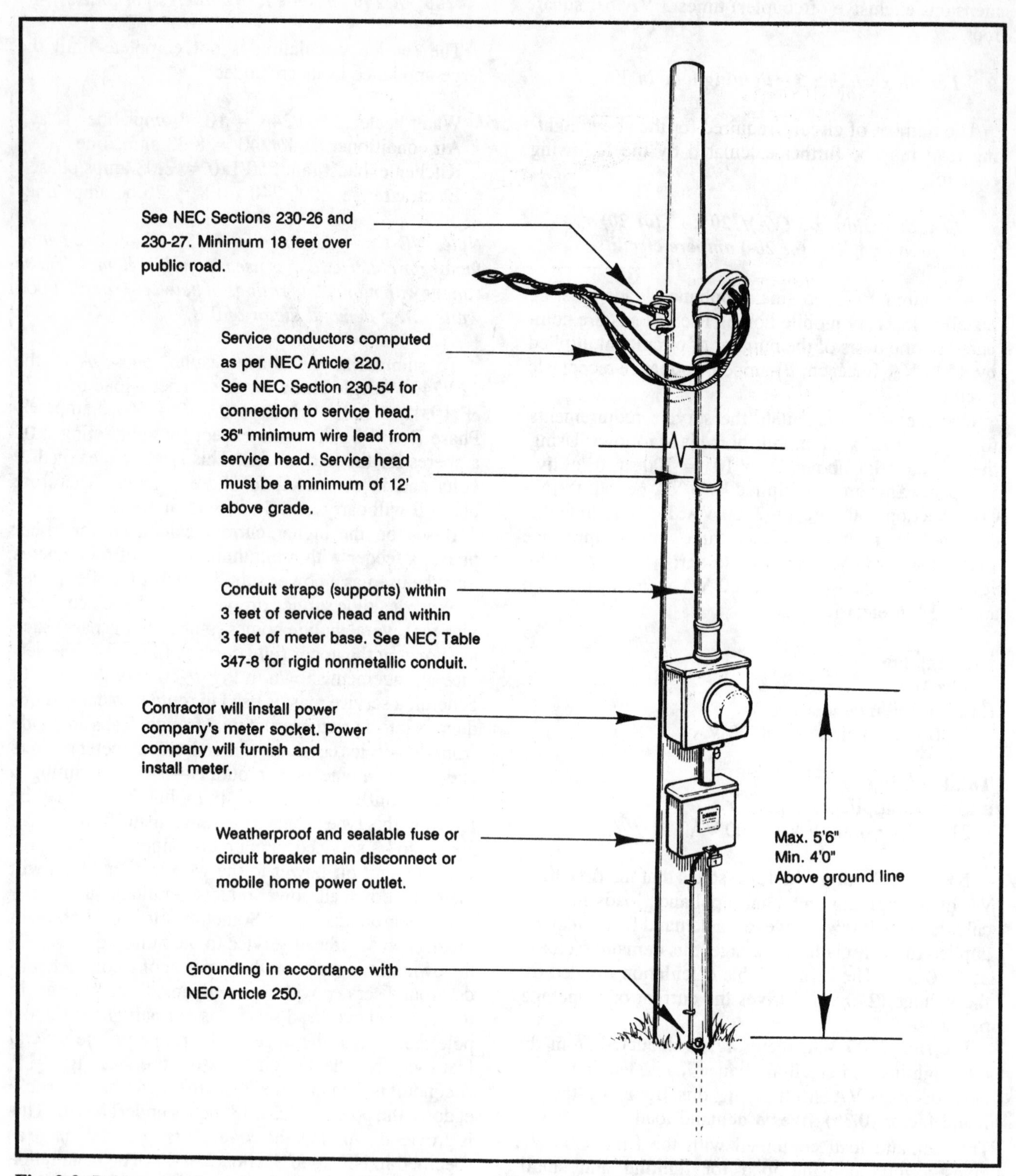

See NEC Sections 230-26 and 230-27. Minimum 18 feet over public road.
Service conductors computed as per NEC Article 220. See NEC Section 230-54 for connection to service head. 36" minimum wire lead from service head. Service head must be a minimum of 12' above grade.
Conduit straps (supports) within 3 feet of service head and within 3 feet of meter base. See NEC Table 347-8 for rigid nonmetallic conduit.
Contractor will install power company's meter socket. Power company will furnish and install meter.
Weatherproof and sealable fuse or circuit breaker main disconnect or mobile home power outlet.
Max. 5'6"
Min. 4'0"
Above ground line
Grounding in accordance with NEC Article 250.

Fig. 2-9: Pole-mounted service drop suited for temporary electric service or for an individual mobile home.

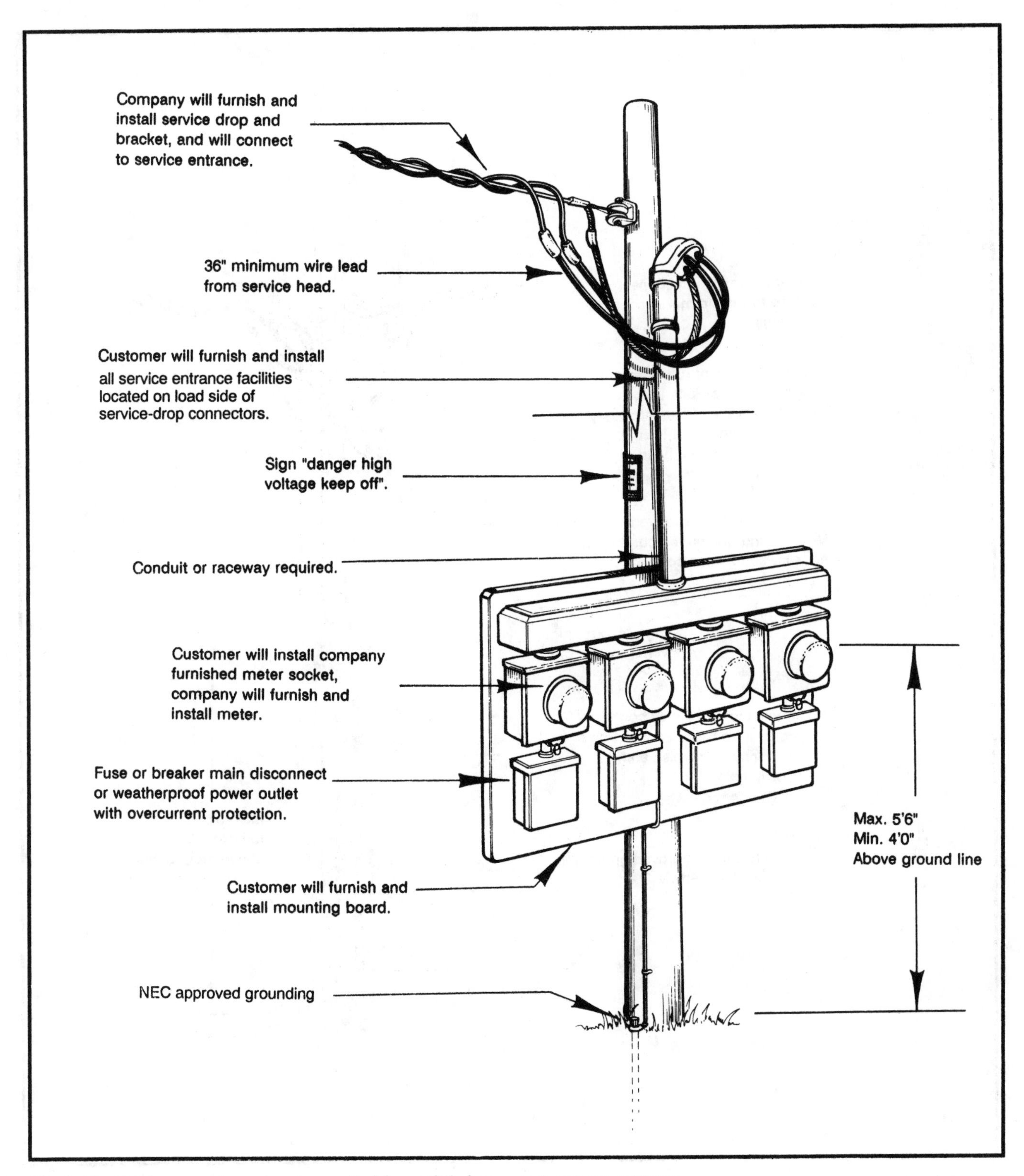
Company will furnish and
install service drop and
bracket, and will connect
to service entrance.
36" minimum wire lead
from service head.
Customer will furnish and install
all service entrance facilities
located on load side of
service-drop connectors.
Sign "danger high
voltage keep off".
Conduit or raceway required.
Customer will install company
furnished meter socket,
company will furnish and
install meter.
Fuse or breaker main disconnect
or weatherproof power outlet
with overcurrent protection.
Max. 5'6"
Min. 4'0"
Above ground line
Customer will furnish and
install mounting board.
NEC approved grounding

Fig. 2-10: Pole-mounted service drop suited for mobile home park service entrance.

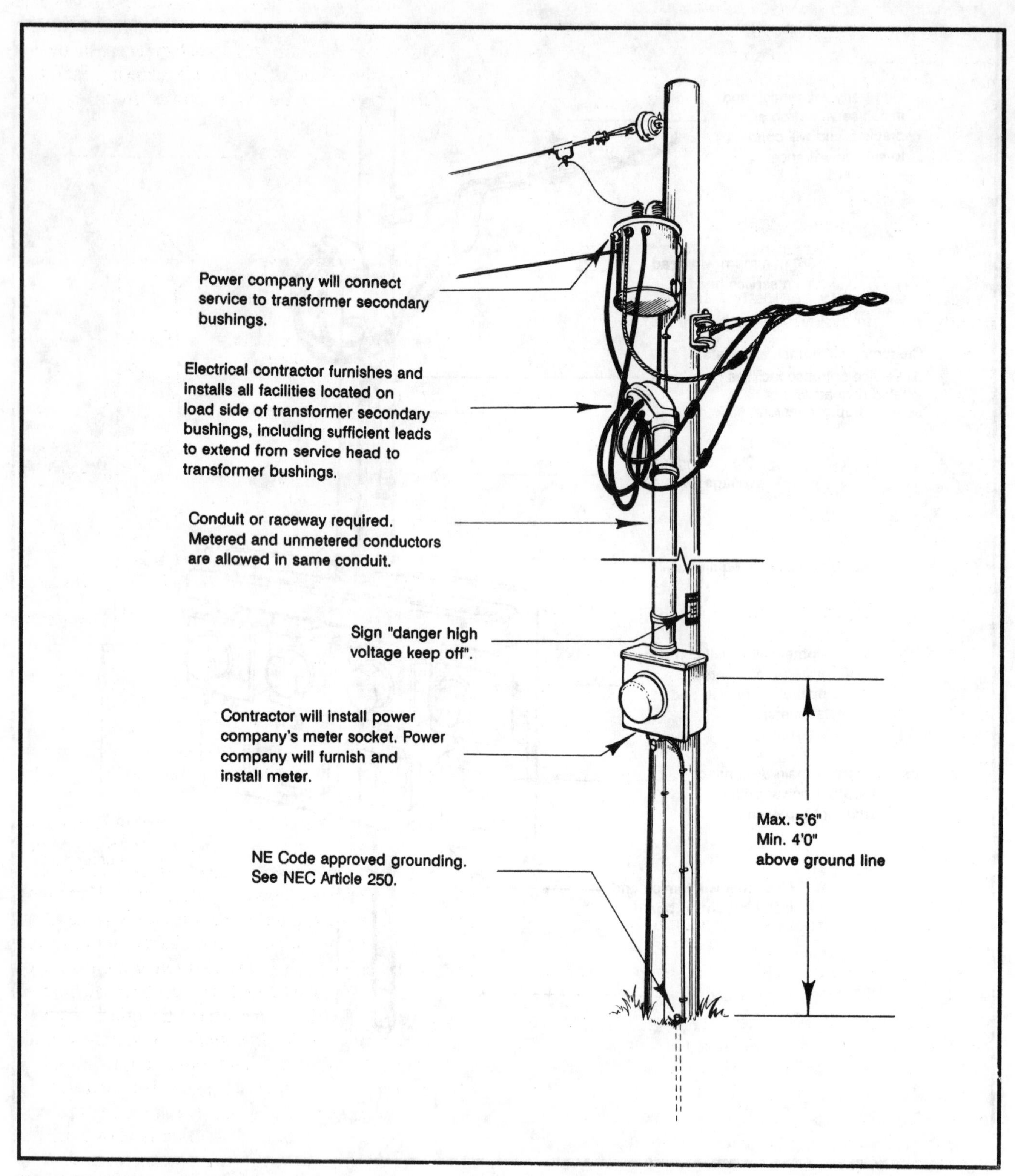

Power company will connect service to transformer secondary bushings.
Electrical contractor furnishes and installs all facilities located on load side of transformer secondary bushings, including sufficient leads to extend from service head to transformer bushings.
Conduit or raceway required. Metered and unmetered conductors are allowed in same conduit.
Sign "danger high voltage keep off".
Contractor will install power company's meter socket. Power company will furnish and install meter.
NE Code approved grounding. See NEC Article 250.
Max. 5'6"
Min. 4'0"
above ground line

Fig. 2-11: Pole-mounted service drop suited for one type of underground service entrance.

These conductors must be of sufficient size to carry the computed load, along with allowances for voltage drop. Regardless of the computed load, the conductors must never be smaller than No. 8 AWG copper or No. 6 AWG aluminum wire. The grounded conductor (neutral) must be sized in accordance with NE Code Section 250-23(b). Where these underground conductors enter the building, they must be installed as specified in NE Code Section 230-6 or protected by a raceway wiring method identified in Section 230-43.

Farm Services: The service for the farm begins at the secondary terminals of the transformer that is connected to the utility company's lines. Two types are usually readily available: single-phase, 3-wire, and three-phase, 4-wire. The single-phase, 3-wire is by far the most common type of service used on farms throughout North American. It will provide satisfactory power for lighting circuits and convenience outlets and for single-phase motors rated up to about 7½ horsepower. When motors larger than 7½ horsepower are to be used for farm equipment such as irrigation pumps, feed grinders, crop-drying fans, etc., a three-phase, 4-wire system is the preferred type to use.

Two types of three-phase service are available in most areas: three-phase, 4-wire wye and three-phase, 4-wire delta. Of the two, the delta system seems to be used the most—mainly because 240-volt single-phase motors and appliances may also be operated satisfactorily on this service. On the other hand, the wye system supplies only 208 volts between phases. However, caution must be employed with the four-wire delta system to make sure the "high leg" which supplies about 180 volts to ground, is not mistakenly connected to a 120-volt outlet.

Household appliances and motors can be purchased for use on the wye system, and where a number of 120-volt motors and appliances are in use, the wye is easier to balance than the delta system. There is also less chance of an improper connection that might be made by an inexperienced person.

Farm services are calculated the same as any other building service; that is, the total loads are listed, a demand or diversity factor is applied (if applicable), and the resulting figures indicate the service size. In sizing such a service, provisions should be made for future additions and loads. It is much less expensive to provide adequate service for these loads at the outset rather than having to enlarge the service entrance each time a new load is added.

When possible, the farm service pole should be located near the center of the electrical load to reduce the size and length of feeder wires and thus reduce the cost of the installation. A typical farm electric service is shown in Fig. 2-12. If the service loads exceed 200 amperes, metering (current) transformers will be located at the top of the pole with all metering facilities usually furnished by the local power company. See Chapter 10 of this book.

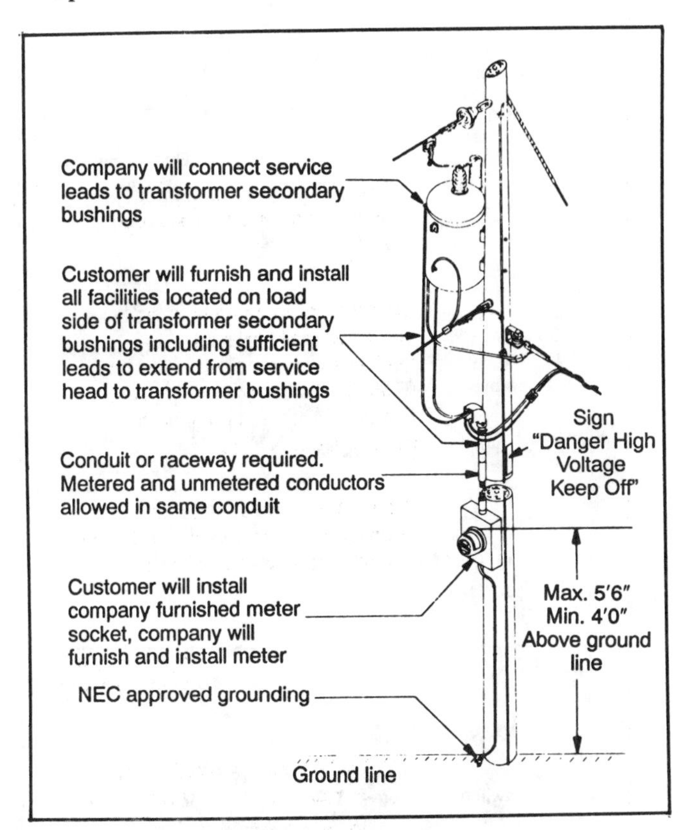

Fig. 2-12: Central farm pole electric service.

Since electrical machines are used in abundance around the modern farm, provisions are usually provided for stand-by electrical power in case of an interruption of the normal power. For example, if the electric service on a large dairy farm shut down, auxiliary power must be provided for the electric milking machines; the cows must be milked twice a day. Power must also be provided to feed grinders because the livestock must eat. Article 700 of the NE Code gives regulations pertaining to the installation of stand-by emergency power for farms and other applications. A 200-ampere farm service-entrance designed for use with an emergency stand-by generator facility is shown in Fig. 2-13. A 400-ampere service is shown in Fig. 2-14. The double-throw (transfer) switch may be either manually operated or of the automatic type; it transfers the circuits from the power company's lines to the generator feeders.

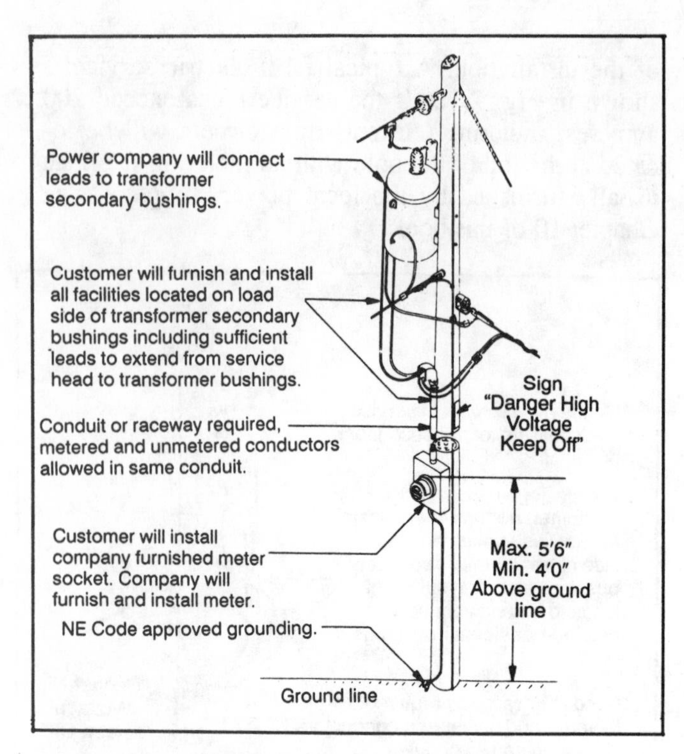

Fig. 2-13: A 200-ampere central farm pole-mounted service-entrance with emergency stand-by generator facilities.

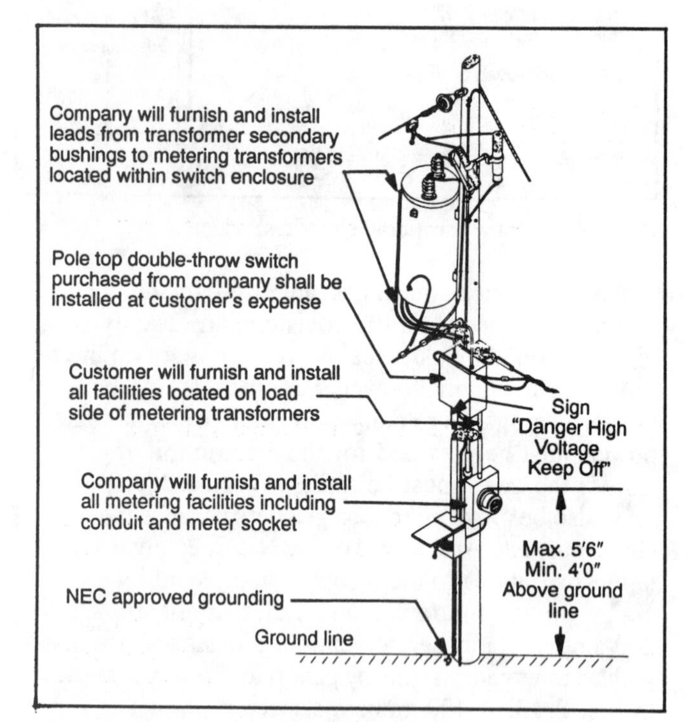

Fig. 2-14: A 400-ampere central farm pole-mounted service-entrance with emergency stand-by generator facilities with pole-top switch.

Since most farms utilize several barns, silos, and outbuildings, many feeders and subpanels are normally required for adequate electrical distribution. Proper design and installation are necessary to keep voltage drop to the minimum and to provide a service within the budget of the farmer's means.

In general, two methods of installing feeder conductors are currently in use: overhead and underground. Of the two, the latter is the type preferred although the initial installation is usually more expensive. Underground wiring, however, offers several advantages over overhead wiring, which usually offsets the additional initial cost. First, ice storms take their toll of overhead lines each year. The lines themselves are designed to take the load of ice on them, but an early snow or ice storm, before all leaves have left adjacent trees, can cause the tree limbs to break and fall across the lines, bringing them down also. Underground wiring does not have this problem. Conductors buried in the ground are also safe from heavy farm equipment that might come into contact with overhead lines, and aesthetic appearance is greatly improved with underground lines.

Service masts: Sometimes it is desirable (or necessary) to utilize a conduit or service mast for the connection of the service conductors. Where a service mast is used for the support of the service conductors (see Fig. 2-15), the mast must be of adequate strength or be supported by braces or guys to withstand safely the strain imposed by the service drop. The fittings must also be approved for the purpose.

Panelboard Location

The NE Code requires that a service be provided with a disconnecting means for all conductors in the building or structure from the service-entrance conductors. This disconnecting means should be located at or near the point where the service conductors enter the building. The disconnecting means must be located at a readily accessible point, either inside or outside the building, and adequate access and working space must be provided all around the disconnecting means. See Fig. 2-16.

Overcurrent protection is required both at the main source and for all individual feeders and branch circuits in order to protect the electrical installation against ground faults and overloads. See Chapter 4 of this book.

Referring again to our small residence that was used for service calculations earlier in this chapter, the only practical location for the load center is in the utility

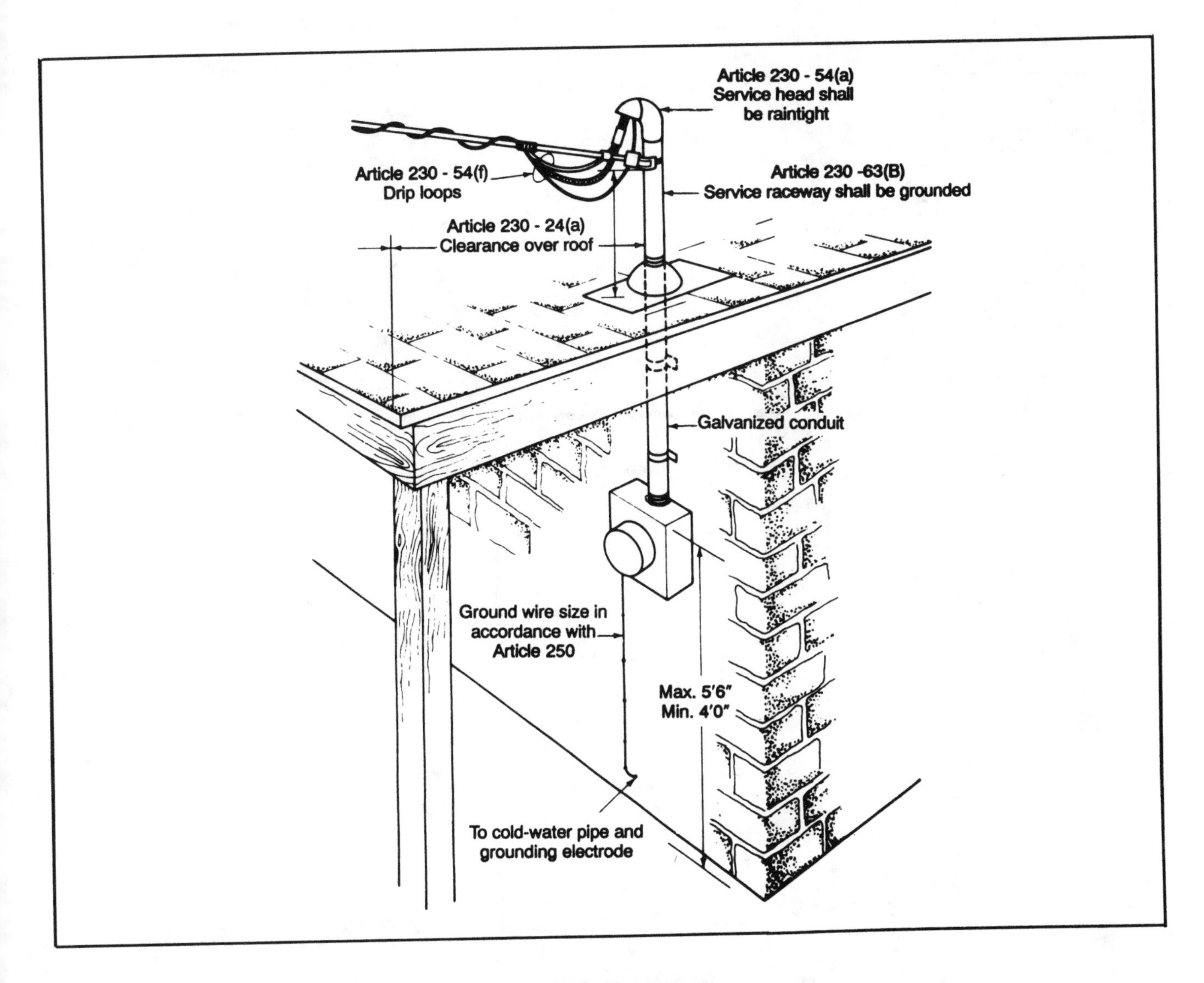

Fig. 2-15: NE Code requirements for service mast installations.

room which is located approximately in the center of the building. Since the NE Code requires that an unfused service cable with a bare neutral terminates immediately into a disconnecting means when it enters the building, some special provisions will have to be made to make this application meet NE Code requirements. In this case, two solutions come to mind: The service conductors from the meter base on the outside of the building may be installed in conduit and routed under the concrete slab to the vicinity of the utility room. An elbow is then installed on the conduit to run the conductors up and into the load center. In other words, service conductors installed in conduit and run under a minimum of 4 inches of concrete are considered to be outside the building.

Another alternative is to install a weatherproof disconnect—either of the fused or circuit breaker type—next to the meter base on the outside of the building and then use three-wire w/ground wire S.E. cable. In this type of cable, both grounded and ungrounded conductors are insulated and since the ungrounded conductors are provided with overcurrent protection before they enter the building, the cable may be routed any place in the building as provided in the NE Code. That is, they may be run in wood partitions, up in the attic and then down to the load center, and in several other ways.

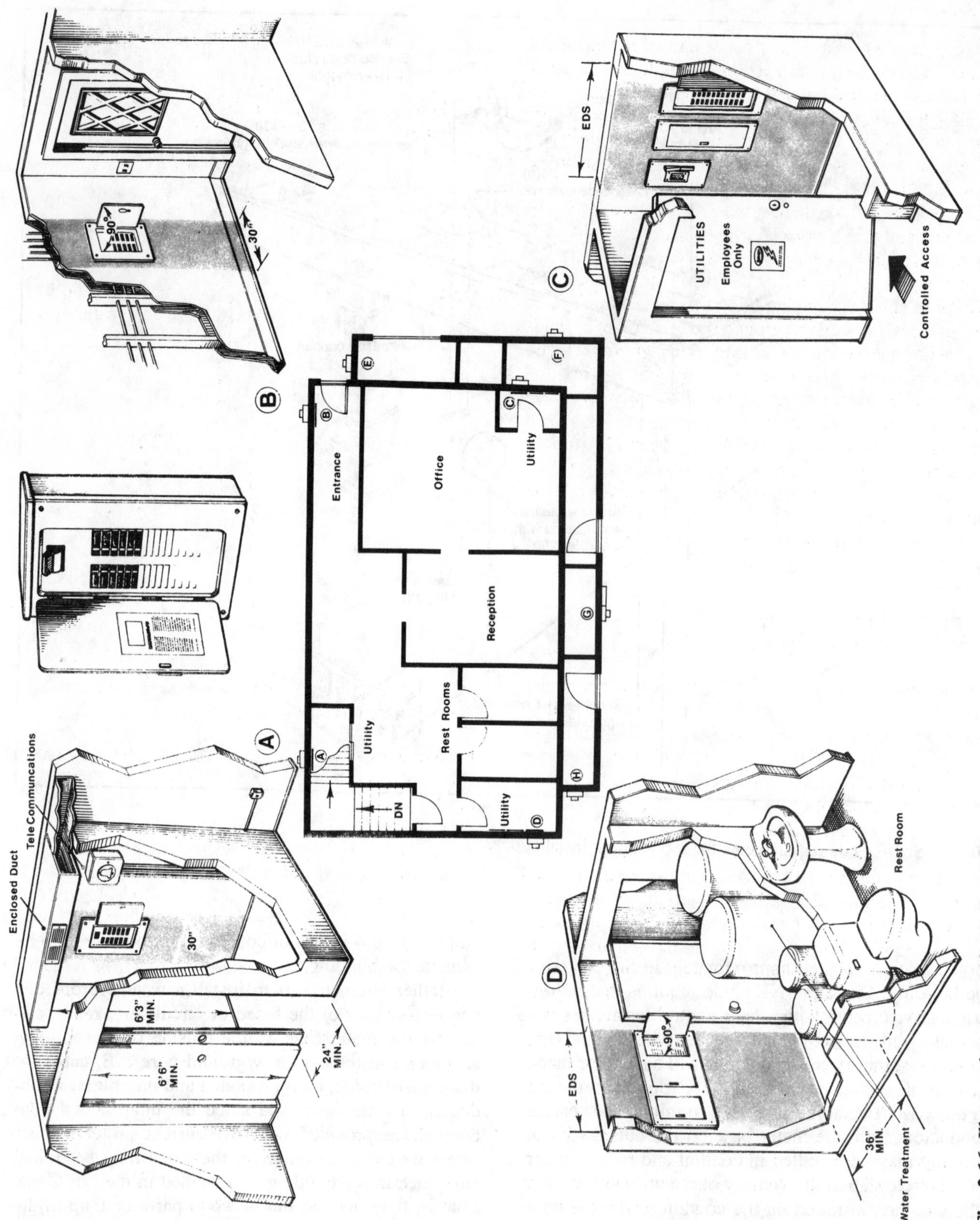

Fig. 2-16: Illustrated summary of panelboard placement.

Grounding

The grounding system is a major part of the electrical system. Its purpose is to protect life and equipment against the various electrical faults that can occur. It is sometimes possible for higher-than-normal voltages to appear at certain points in an electrical system or in the electrical equipment connected to the system. Proper grounding ensures that the high electrical charges that cause these high voltages are channeled to earth or ground before damaging equipment or causing danger to human life.

When we refer to *ground*, we are talking about ground potential or earth ground. If a conductor is connected to the earth or to some conducting body that serves in place of the earth, such as a driven ground rod (electrode) or cold-water pipe, the conductor is said to be *grounded*. The neutral conductor in a three- or four-wire service, for example, is intentionally grounded and therefore becomes a *grounded conductor*. However, a wire used to connect this neutral conductor to a grounding electrode or electrodes is referred to a *grounding conductor*. Note the difference in the two meanings; one is ground*ed*, while the other is ground*ing*.

There are two general classifications of protective grounding:

- System grounding
- Equipment grounding

The system ground relates to the service-entrance equipment and its interrelated and bonded components. That is, system and circuit conductors are grounded to limit voltages due to lighting, line surges, or unintentional contact with higher voltage lines, and to stabilize the voltage to ground during normal operation.

Equipment grounding conductors are used to connect the noncurrent-carrying metal parts of equipment, conduit, outlet boxes, and other enclosures to the system grounded conductor, the grounding electrode conductor, or both, at the service equipment or at the source of a separately derived system. Equipment grounding conductors are bonded to the system grounded conductor to provide a low impedance path for fault current. That facilitates the operation of overcurrent devices under ground-fault conditions.

Article 250 of the NE Code covers general requirements for grounding and bonding. Nearly 75 changes or additions have been made to this Article since the 1990 NE Code was printed. This should be reason enough to carefully read all parts of this Article over several times until you have a thorough understanding of its contents.

To better understand a complete grounding system, let's take a look at a conventional residential system beginning at the power company's high-voltage lines and transformer as shown in Fig. 2-17. The pole-mounted transformer is

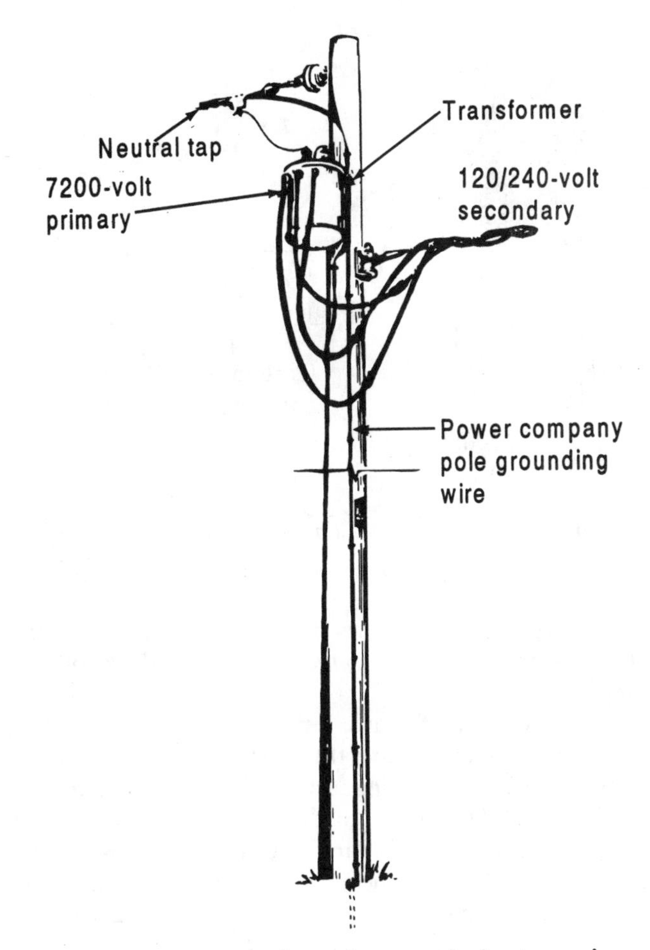

Fig. 2-17: Pole-mounted transformer reducing transmission voltage to usual house current.

fed with a two-wire 7200-volt system which is transformed and stepped down to a 3-wire, 120/240-volt, single-phase electric service suitable for residential use. A wiring diagram of the transformer connections is shown in Fig. 2-18. Note that the voltage between phase A and phase B is 240 volts. However, by connecting a third wire (neutral) on the secondary winding of the transformer—between the other two—the 240 volts are split in half, giving 120 volts between either phase A or phase B and the neutral conductor. Consequently, 240 volts are available for household appliances such as ranges, hot-water heaters, and clothes dryers, while 120 volts are available for lights, small appliances, tvs, and the like.

Referring again to the diagram in Fig. 2-18, conductors A and B are ungrounded conductors, while the neutral is a grounded conductor. If only 240-volt loads were connected, the neutral (grounded conductor) would carry no current. However, since 120-volt loads are present, the neutral will carry the unbalanced load and becomes a current-carrying conductor. For example, if phase A carries 60 amperes and phase B carries 50 amperes, the neutral conductor would carry only (60 - 50 =) 10 amperes. This

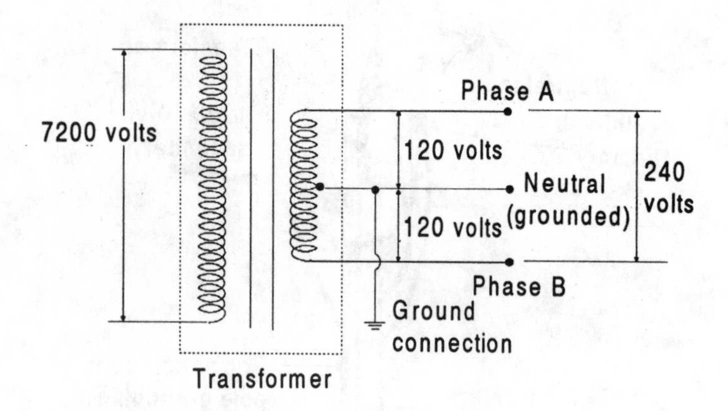

Fig. 2-18: Wiring diagram of 7200-volt to 120/240-volt, single-phase transformer connection.

is why the NE Code allows the neutral conductor in an electric service to be smaller than the ungrounded conductors.

The typical pole-mounted service-entrance is normally routed by messenger cable from a point on the pole to a point on the building being served, terminating in a meter housing. Another service conductor is installed between the meter housing and the main service switch or panelboard. This is the point where most systems are grounded—the neutral bus in the main panelboard. See Fig. 2-19.

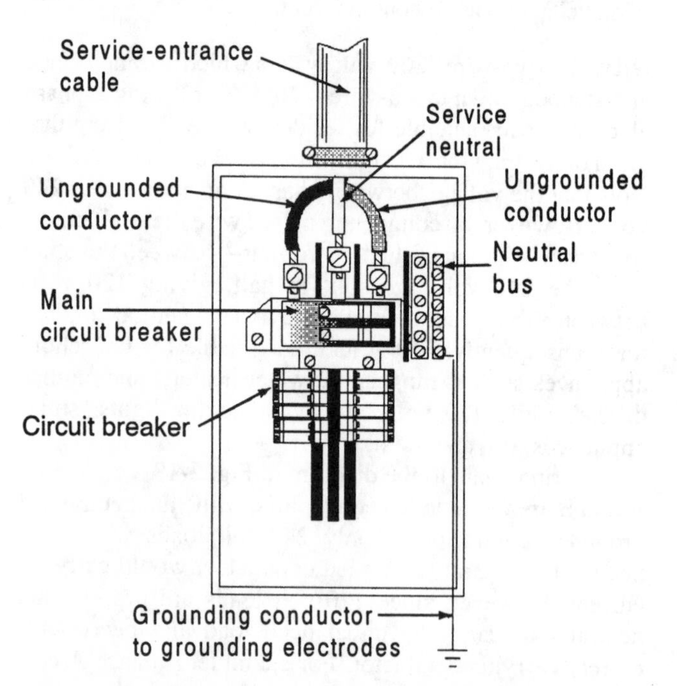

Fig. 2-19: Interior view and connections of typical panelboard.

Grounding Methods

Methods of grounding an electric service are covered in NEC Section 250-81. In general, all of the following (if available) and any made electrodes must be bonded together to form the grounding electrode system:

- An underground water pipe in direct contact with the earth for no less than 10 feet.
- The metal frame of a building where effectively grounded.
- An electrode encased by at least 2 inches of concrete, located within and near the bottom of a concrete foundation or footing that is in direct contact with the earth. Furthermore, this electrode must be at least 20 feet long and must be made of electrically conductive coated steel reinforcing bars or rods of not less than ½-inch diameter, or consisting of a least 20 feet of bare copper conductor not smaller than No. 2 AWG wire size.
- A ground ring encircling the building or structure, in direct contact with the earth at a depth below grade not less than 2½ feet. This ring must consist of at least 20 feet of bare copper conductor not smaller than No. 2 AWG wire size.

In most residential structures, only the water pipe will be available, and this water pipe must be supplemented by an additional electrode as specified in NEC Sections 250-81(a) and 250-83. With these facts in mind, let's take a look at a typical residential electric service, and the available grounding electrodes. See Fig. 2-20.

The residence in Fig. 2-20 has a metal underground water pipe that is in direct contact with the earth for more than 10 feet, so this is one valid grounding source. The house also has a metal underground gas-piping system, but this may not be used as a grounding electrode (NEC Section 250-83(a)). NEC Section 250-81(a) further states that the underground water pipe must be supplemented by an additional electrode of a type specified in Section 250-81 or in Section 250-83. Since a grounded metal building frame, concrete-encased electrode, or a ground ring is not normally available for most residential applications, NEC Section 250-83—*Made and Other Electrodes*—must be used in determining the supplemental electrode. In most cases, this supplemental electrode will consist of either a driven rod or pipe electrode, specifications for which are shown in Fig. 2-21.

An alternate method to the pipe or rod method is a plate electrode. Each plate electrode must expose not less than 2 square feet of surface to the surrounding earth. Plates made of iron or steel must be at least ¼ inch thick, while

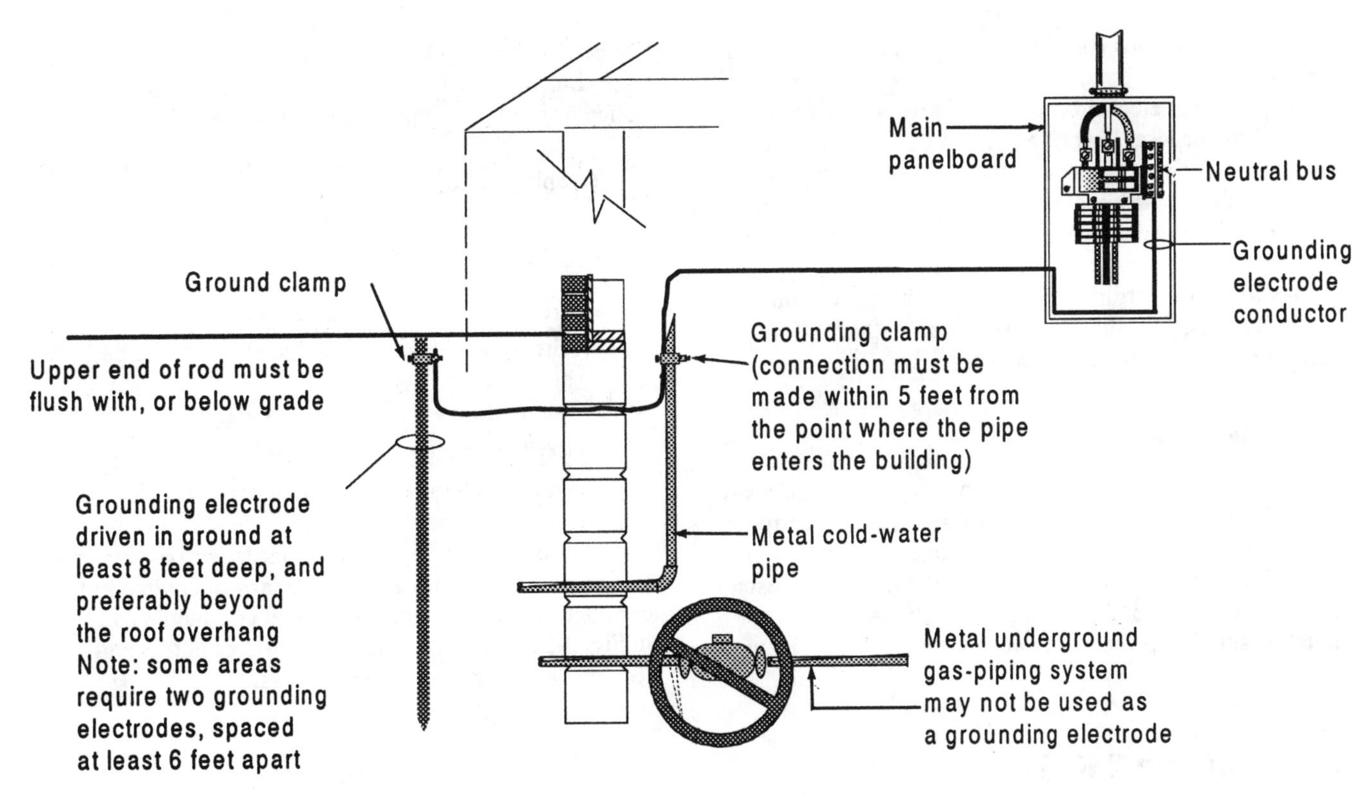

Fig. 2-20: Components of a residential service grounding system.

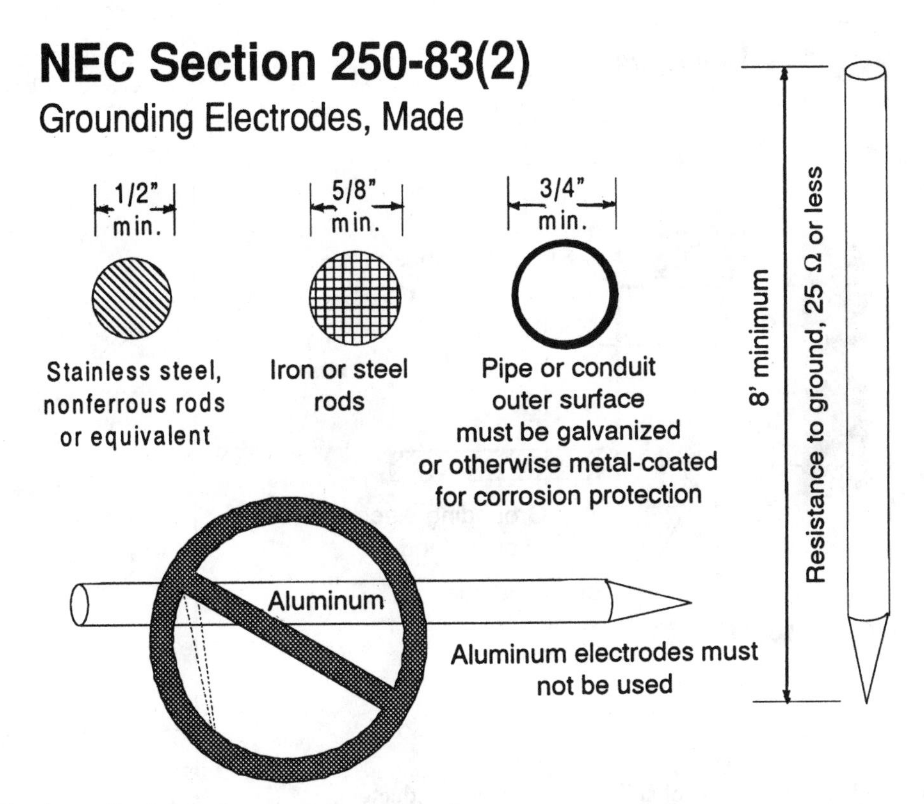

Fig. 2-21: Specifications of rod and pipe grounding electrodes.

plates of nonferrous metal like copper need only be .06 inch thick.

Either type of electrode must have a resistance to ground of 25 ohms or less. If not, they must be augmented by an additional electrode spaced not less than 6 feet from each other. In fact, many locations require two electrodes regardless of the resistance to ground. This, of course, is not an NEC requirement, but is required by some power companies and local ordinances in some cities and counties. Always check with the local inspection authority for such rules that surpass the requirements of the NEC.

Grounding Conductors

The grounding conductor, connecting the panelboard neutral bus to the water pipe and grounding electrodes, must be of either copper, aluminum, or copper-clad aluminum. Furthermore, the material selected must be resistant to any corrosive condition existing at the installation or it must be suitably protected against corrosion. The grounding conductor may be either solid or stranded, covered or bare, but it must be in one continuous length without a splice or joint—except for the following conditions:

- Splices in busbars are permitted.
- Where a service consists of more than one single enclosure, it is permissible to connect taps to the grounding electrode conductor provided the taps are made within the enclosures.
- Grounding electrode conductors may also be spliced at any location by means of irreversible compression-type connectors listed for the purpose or the exothermic welding process. See Fig. 2-22.

The size of grounding conductors depend on service-entrance size; that is, the size of the largest service-entrance conductor or equivalent for parallel conductors. The table in Fig. 2-23 gives the proper sizes of grounding conductors for various sizes of electric services.

NEC Section 250-81
Exception No. 1

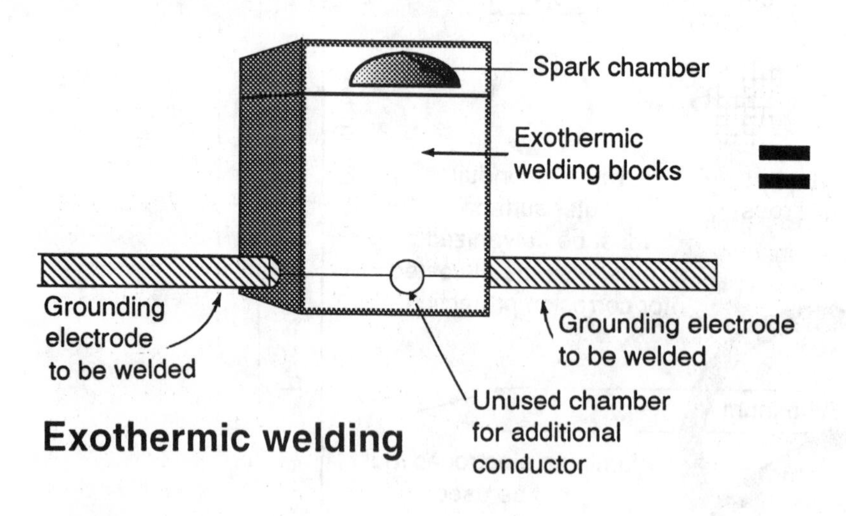

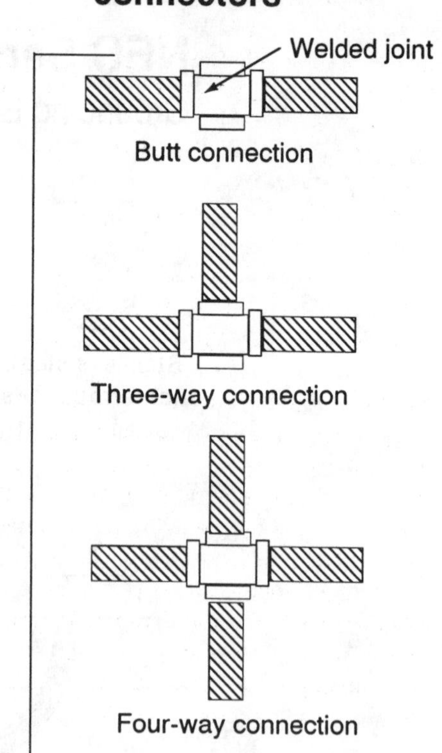

Fig. 2-22: Approved NE Code methods of splicing grounding conductors.

Size of Largest Service-Entrance Conductor or Equivalent for Parallel Conductors		Size of Grounding Electrode Conductor	
Copper	Aluminum or Copper-Clad Aluminum	Copper	Aluminum or Copper-Clad Aluminum
2 or smaller	0 or smaller	8	6
1 or 2	2/0 or 3/0	6	4
2/0 or 3/0	4/0 or 250 kcmil	4	2
Over 3/0 through 350 kcmil	Over 250 kcmil through 500 kcmil	2	0
Over 350 kcmil through 600 kcmil	Over 500 kcmil through 900 kcmil	0	3/0
Over 600 kcmil through 1100 kcmil	Over 900 kcmil through 1750 kcmil	2/0	4/0
Over 1100 kcmil	Over 1750 kcmil	3/0	250 kcmil

Fig. 2-23: Grounding electrode conductor sizes for service-entrance grounding systems.

Equipment Grounding

Fig. 2-24 summarizes the equipment grounding rules for most types of equipment that is required to be grounded. These general NE Code regulations apply to all installations except for specific equipment (special applications) as indicated in the 1993 NE Code. The NE Code also lists specific equipment that is to be grounded regardless of voltage.

In all occupancies, major appliances and many handheld appliances and tools are required to be grounded. The appliances include refrigerators, freezers, air conditioners, clothes dryers, washing machines, dishwashing machines, sump pumps, and electrical aquarium equipment. Other tools likely to used outdoors and in wet or damp locations must be grounded or have a system of double insulation.

Although most appliance circuits require an equipment grounding conductor, the frames of electric ranges, clothes dryers and similar appliances that utilize both 120 and 240 volts may be grounded via the grounded circuit conductor (neutral) under most conditions. In addition, however, the grounding contacts of any receptacles on the equipment must be bonded to the equipment. If these specified conditions are met, it is not necessary to provide a separate equipment grounding conductor, either for the frames or any outlet or junction boxes which are part of the circuit for these applications.

A bonding jumper is sometimes used to assure electrical conductivity between metal parts. When the jumper is installed to connect two or more portions of the equipment grounding conductor, the jumper is referred to as an *equipment bonding jumper*. The rules for the equipment bondng jumper are summarized in Fig. 2-24. Some specific cases in which a bonding jumper is required are also listed below:

- Metal raceways, cable armor, and other metal noncurrent-carrying parts that serve as grounding conductors must be bonded whenever necessary in order to assure electrical continuity.
- When flexible metal conduit is used for equipment grounding, an equipment bonding jumper is required if the length of the ground return path exceeds 6 feet, or the circuit conductors contained within the flexible conduit is rated for over 20 amperes. Otherwise, the conduit itself may be used as the equipment ground with bonding between outlet boxes accomplished by approved fittings.
- A short length of flexible metal conduit that contains a circuit rated over 20 amperes may not serve as a grounding conductor itself, but a separate bonding jumper can be provided in place of a separate equipment grounding conductor for the circuit. This bonding jumper may be installed inside or outside the conduit, but an outside jumper cannot exceed 6 feet in length.

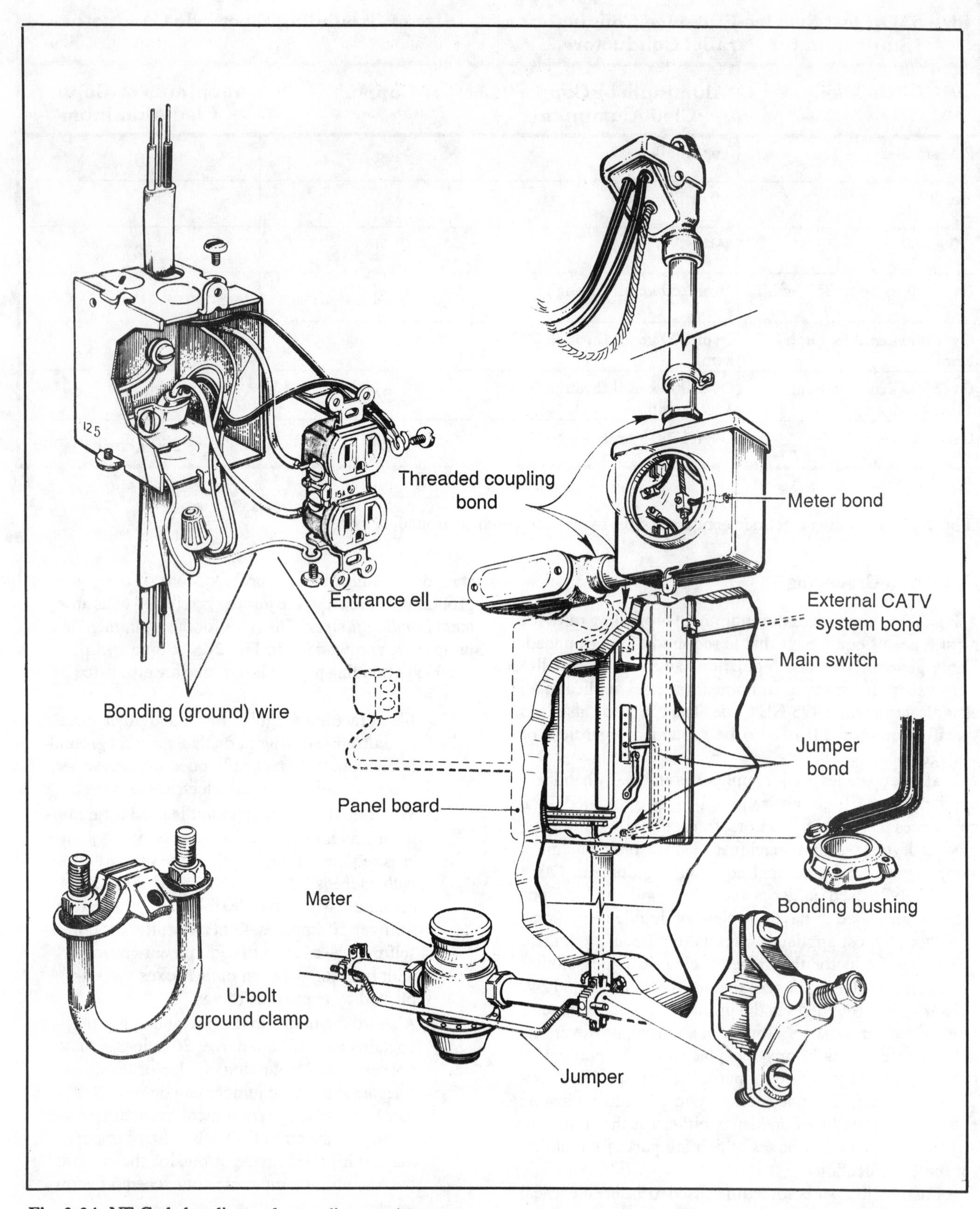

Fig. 2-24: NE Code bonding and grounding requirements.

Additional Service Calculations

Residential Service—Optional Method: A large one-family dwelling has a total area of 3800 square feet and contains the following loads:

- 12 kVA electric range
- 1.0-kilovolt-ampere, 120-volt garbage disposal
- 10-kVA, 240-volt strip heaters
- 6-kVA, 240-volt, 130-gallon hot-water heater
- Two 28-ampere, 240-volt air-conditioner motor-compressors with circuits protected by inverse-time circuits breakers
- Four ½-horsepower, 120-volt blower motors for heating and cooling systems

The branch circuits and service use THW copper conductors. The design problem is summarized in Fig. 2-25.

The lighting, small appliance, and laundry load requires eight 20-ampere circuits. The 12-kVA range is considered as a branch-circuit load of 8-kVA and requires a 40-ampere circuit. The disposal and the four ½-horsepower motors must be supplied by conductors rated at 125% of the full-load current of the motors.

The conductors for the central electric space heating must be rated at least

1.25 x 10,000 VA/240 V = 52.1 amperes

The 6-kVA hot water heater represents a load of 25 amperes at 240 volts. The hot water heater is not considered a continuous load since its capacity is more than 120 gallons.

The branch-circuit ampacity of the motor-compressors must be 125% of the nameplate rated-load current, or 1.25 x 28 amperes = 35 amperes. The rating of the inverse-time circuit breaker serving as the branch-circuit overcurrent protective device must be no larger than 1.75 x 28 amperes = 49 amperes, or a 45-ampere standard size.

The computation of the service load is made like previous examples after the heating and air-conditioning loads have been compared. The load of the two motor-compressors is 2 x 240 volts x 28 amperes = 13,440 volt-amperes. The central electric space heating is considered at the 65% diversified demand, or .65 x 10,000 VA = 6500 VA (volt-amperes). Since the heating load is the smaller load, it is neglected. The four blower motors, however, are part of the heating and cooling system and their load of 4704 volt-amperes must be included at the 100% demand factor.

The "other load" includes the lighting, small appliance, and laundry circuits as well as the nameplate rated-load of all fixed appliances for a total of 34,900 volt-amperes. The applicable demand factors reduce the demand load to 19,960 volt-amperes. The air-conditioning equipment represents an additional 13,440 + 4704 = 18,144 volt-amperes for which no reduction is permitted.

The total service load of 38,104 volt-amperes requires a minimum service rating of

38,104 VA/240 V = 158.8 A

A neutral load of 19,113 volt-amperes, or 79.6 amperes, is contributed by the unbalanced load from the 120-volt and 120/240-volt loads.

No. 2/0 type THW copper conductors are required for the ungrounded conductors and a No. 4 is required for the neutral conductor.

The rating of the feeder overcurrent protective device may not exceed the following:

- Air conditioner protective device 45.0
- Other motor loads [28 x (4704/240)] 47.6
- "Other load" (19,960/240) 83.2
- Maximum rating 175.8 amperes

A 175-ampere rating is the next smaller standard size for the feeder overcurrent protective device.

A 1½-inch conduit is adequate to enclose the conductors and not exceed the 40% fill limitation. A No. 4 copper grounding electrode conductor would be used at the service.

Multi-family Dwelling Calculations

The service load for the multi-family dwelling is not simply the sum of the individual dwelling unit loads because of demand factors that may be applied to multi-family dwellings.

When the standard calculation is used to compute the service load, the total lighting, small appliance, and laundry loads as well as the total load from all electric ranges and electric clothes dryers are subject to the application of demand factors as described in Article 220 of the NE Code. In addition, further demand factors may be applied to the portion of the neutral load contributed by electric ranges and the portion of the total neutral load greater than 200 amperes.

SERVICE LOAD CALCULATION

Residential Service—Optional Method

General lighting = 3800 sq. ft. x 3 VA	11,400
Small appliance load = 2 x 1500 VA	3,000
Laundry = 1 (circuit) x 1500 VA	1,500
Total lighting, small appl. and laundry	15,900 VA
Electric range	12,000
Disposal	1,000
Hot water heater	6,000
Total other load	34,900 VA

Applying Demand Factors

First 10 kVA @100%	10,000
Remainder (34,900 - 10,000) @40%	9,960
Total demand load	19,960 VA
Compressor motors @100% (2 x 240V x 28A)	13,440
Fan-coil motors @100% (4 x 120V x 9.8A)	4,704

Since there are several 240-volt loads, the neutral conductor will carry only the unbalanced load of 19,113 VA

Service Equipment

38,104/240 = 158.8 amperes
Neutral carries 19,113/240 = 79.6 amperes

Service conductors may be two No. 2/0 and one No./4 THW copper conductors. Grounding conductors should be No. 4 AWG. See Table 310-16, Table 250-94, Section 430-63 and also Chapter 9 of the 1993 NE Code.

Fig. 2-25: Summary of residential service calculations using the optional method.

When the optional calculation is used, the total connected load is subject to the application of a demand factor that varies according to the number of individual units in the dwelling.

A summary of the calculating methods for designing wiring systems in multi-family dwellings and the applicable NE Code references are shown in Fig. 2-20. The selection of a calculation method for computing the service load is not affected by the method used to design the feeders to the individual dwelling units.

The rules for computing the service load are also used for computing a main feeder load when the wiring system consists of a service that supplies main feeders which, in turn, supply a number of subfeeders to individual dwelling units.

The standard calculation for computing feeder or service loads may be used for any dwelling, whether it is a one-family, two-family, or multi-family dwelling. When the standard calculation is used for a two-family dwelling or for a multi-family dwelling, the total connected load for each type of load is first computed for the entire dwelling. Then any applicable demand factors are applied. The demand factors depend on the size of the load, or the number of appliances in some cases, but not on the number of dwelling units.

Example of standard calculation: Figure 2-21 summarizes the required calculations for designing the service for a 20-unit apartment building. Each apartment unit has the following loads:

- 850 square feet of living area
- 1.2-kilovolt-ampere, 120-volt dishwasher
- 600-volt-ampere, 120-volt garbage disposal
- 12-kVA electric range

The service is a 120/240 volt, three-wire circuit and uses type THW copper service-entrance conductors.

The general lighting load of 3 volt-amperes per square foot, two small appliance circuits, and a laundry circuit are included for each unit. The total general lighting, small appliance, and laundry load for 20 units is 141,000 volt-amperes. The demand factors for lighting loads reduce this load to a demand load of 49,200 volt-amperes.

The dishwashers and garbage disposals are taken at nameplate rating. The total loads are 24,000 volt-amperes and 12,000 volt-amperes, respectively. A Code provision for applying a demand factor to these appliances is ignored as explained previously.

The total demand load for 20 ranges of 12-kVA rating is given in the NE Code as 35 kVA. The neutral load can be reduced to 70% of the load on the ungrounded conductors, or .7 x 35 kVA = 24,500 VA (volt-amperes).

If it is assumed that the disposals represent the largest rated motor, an additional 25% of the load of one disposal, or .25 x 600 volt-amperes = 150 volt-amperes, is added to the service load.

The total service load is 120,350 volt-amperes for the ungrounded conductors and 109,850 volt-amperes for the neutral load. The line load in amperes at 240 volts is 501 amperes. The calculated neutral load of 458 amperes may be reduced to a load of 200 amperes + 70% of the amount in excess of 200 amperes, or 200 + .7 (458 - 200) = 381 amperes. The minimum sizes for the THW copper service-entrance conductors are 900 kcmil for the ungrounded conductors and 600 kcmil for the neutral. A standard-size 600-ampere overcurrent protective device could be selected since it is the nearest standard size above the ampacity of the conductors. An equipment grounding conductor would be required to be at least a No. 1 copper conductor based on the size of the overcurrent protective device. The disconnecting means must be rated not less than the load current of 501 amperes.

A grounding electrode conductor at the service must be at least a size 2/0 copper conductor. The conduit encloses the two 900 kcmil conductors: one 600 kcmil conductor and one No. 1 equipment grounding conductor. A 4-inch conduit is needed in order to meet the allowable fill limitation of 40% for four conductors.

The optional calculation for multi-family dwellings may be used to compute the service load when each individual dwelling unit is supplied by a single feeder, is equipped with electric cooking equipment, and is equipped with either electric space heating or air conditioning, or both. The primary feature of the optional calculation is that a total connected load is computed for the entire dwelling and then a demand factor based on the number of units is applied.

The size of the neutral may be computed by using the standard calculation. In many cases, this results in a smaller neutral than would be calculated by the optional method.

The optional calculation for multi-family dwellings should not be confused with the optional calculation for computing the feeder load for an individual dwelling unit. If the stated conditions for its use are satisfied, the optional calculation may be used to compute the service load for a multi-family dwelling regardless of the method used for designing the feeders to the individual dwelling units.

Any loads that are supplied from the service or a main feeder but are not from individual dwelling units

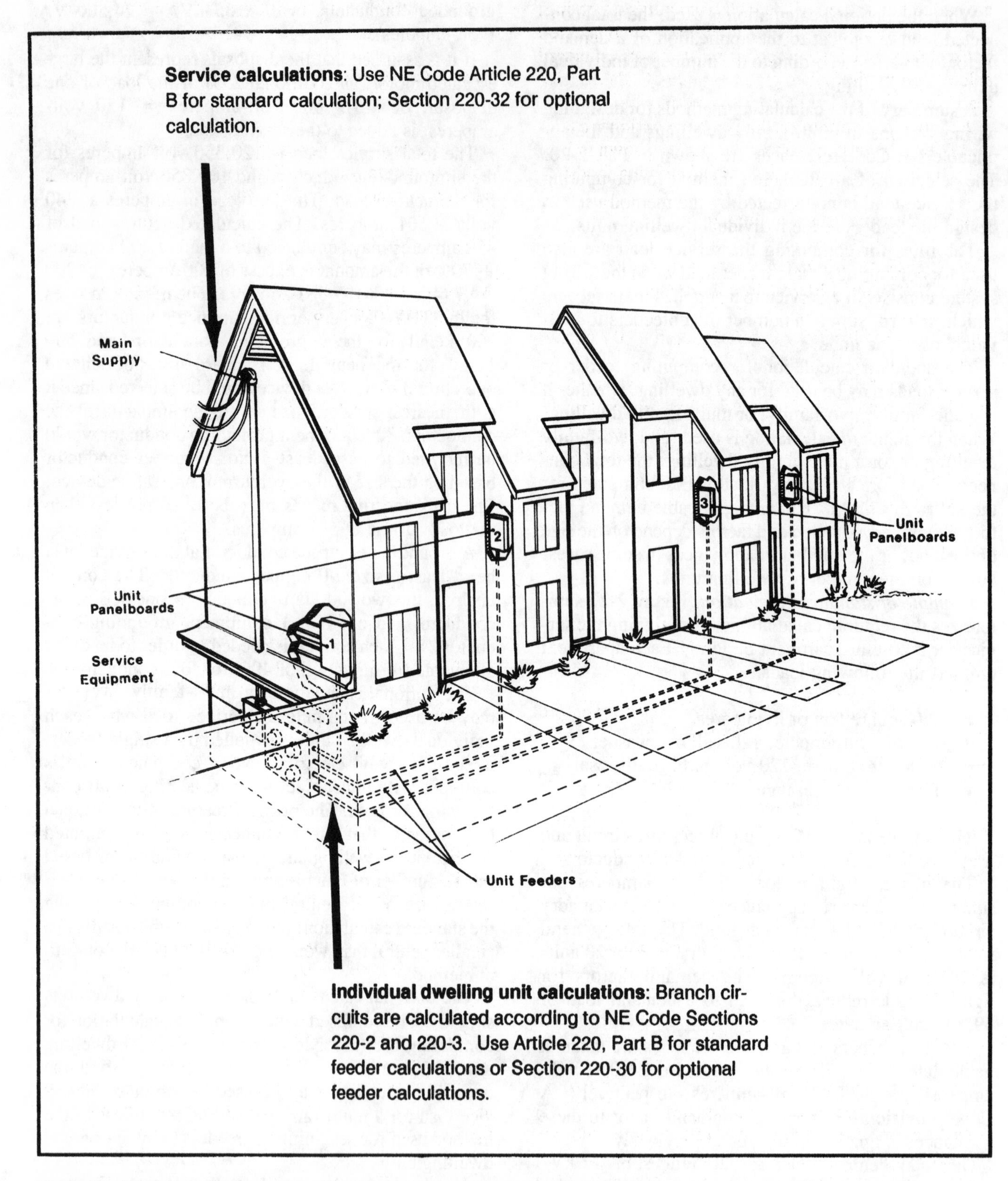
Service calculations: Use NE Code Article 220, Part B for standard calculation; Section 220-32 for optional calculation.
Main Supply
Unit Panelboards
Service Equipment
Unit Panelboards
Unit Feeders
Individual dwelling unit calculations: Branch circuits are calculated according to NE Code Sections 220-2 and 220-3. Use Article 220, Part B for standard feeder calculations or Section 220-30 for optional feeder calculations.

Fig. 2-26. Method used to calculate electric service for multi-family dwellings.

are called house loads. The load from outside lighting, central laundry, and similar facilities are considered house loads and are computed by using the standard calculation method. The result is added to the service load computed by the optional method.

Example of optional calculation: An optional calculation for a 20-unit apartment building is shown in Fig. 2-21. Each apartment contains the following loads:

- 850 square feet of living area
- 12-kVA electric range
- 1.8-kVA, 240-volt space heating
- 15-ampere, 120-volt air conditioner

The 120/240-volt service uses 250 kcmil THW copper conductors in parallel in multiple conduits.

The connected load on the ungrounded conductors is simply the load of each apartment multiplied by 20. The total connected load is 417,000 volt-amperes. A demand factor of 38% is applied to reduce this load to 158,460 volt-amperes.

The line load in amperes at 240 volts is 660 amperes and the minimum number of 250 kcmil conductors needed to carry this load is

660 A/255 A conductor = 2.6 conductors

or, three per line. When the circuit is run in three conduits, each conduit must contain a complete circuit including the neutral conductor.

The neutral ampacity is computed by the standard calculation. The demand load for twenty 12-kVA ranges is 35 kVA. The neutral load for the ranges is .7 x 35 kVA = 24,500 VA. The total neutral load of 73,700 volt-amperes represents a load in amperes of 307 amperes. A further demand factor may be applied to the neutral load over 200 amperes which reduces the total neutral load to 275 amperes. The neutral conductor in each of the three conduits must be capable of carrying 275/3 = 92 amperes; therefore, No. 3 THW copper conductors would ordinarily be selected. However, the smallest conductor size permitted for parallel use is size 1/0. A 2-inch conduit would be required to enclose the two ungrounded conductors and the neutral of each circuit. The grounding electrode conductor must be at least a size 2/0 copper conductor based on the equivalent size of the parallel service-entrance conductors.

Design example with house loads: Assume that the 20-unit apartment complex of the previous example contains the following facilities to allow a basis for a service calculation:

- Central laundry facility with five 5 kVA electric clothes dryers and five 3 kVA, 240-volt washing machines
- 10-horsepower, 240-volt elevator motor with a 60-minute rating.

The additional load computed by the standard calculation would increase the required ampacity of the service-entrance conductors as follows:

Dryer load = 5 x 5 kVA = 25,000 VA
Washer load = 5 x 3 kVA = 15,000 VA
Motor load = 0.90 x 240 V x 50 A = 10,800 VA
Total house load 50,800 VA

The additional load is 50,800 VA/240 volts = 211.7 amperes. Using the results from the previous example, we see that the service load in amperes is 660 + 211.7 = 871.7 amperes. At least four 250 kcmil conductors in parallel would be required to carry the total service load. If laundry receptacles are not installed in the apartment units, the laundry circuit load can be deleted from the service calculation.

Commercial

The design of the electrical circuits for commercial and industrial occupancies is based on specific NE Code rules that relate to the loads present in such occupancies. The design approach is to separate the loads into those for lighting, receptacles, motors, appliances, and other special loads and apply the applicable NE Code rules. In general, the loads are considered continuous unless specific information is given to the contrary.

If the electric load is large, the main service may be a three-phase supply which supplies transformers on the premises. It is not uncommon to have secondary feeders supplying panelboards which, in turn, supply branch circuits operating at different voltages. In this case, the design of the feeder and branch circuits for each voltage is considered separately. The rating of the main service is based on the total load with the load values transformed according to the various circuit voltages if necessary. A feeder circuit diagram is essential when loads of different voltages are present.

In most commercial and industrial occupancies, no demand factors are applied to the loads served. The lighting load in hospitals, hotels and motels, and warehouses, however, is subject to the application of demand factors. In restaurants and similar establishments the load of electric cooking equipment is subject

SERVICE LOAD CALCULATION

Multi-Family Apartment Building

General lighting = 850 sq. ft. x 20 units x 3 VA	51,000
Small appliance load = 2 x 1500 VA x 20	60,000
Laundry = 1 (circuit) x 1500 VA x 20	30,000
Total lighting, small appl. and laundry	141,000 watts
12 kVA electric range x 20	240,000
Disposal x 20 x 600 VA	12,000
1.2 kVA dishwasher x 20	36,000
25% of largest motor = .25 x 600 watts	150
Total large appliance load	288,150 watts
Total connected load	429,150

Applying Demand Factors

First 3 kVA of lighting load @ 100%	3,000
Next (120,000 - 3,000) 117,000 @ 35%	40,950
Remainder (141,000 - 120,000) 21,000 @ 25%	5,250
Total lighting, sm. appl, and laundry demand	49,200 watts
Electric ranges (Section 220-19, 220-22) 20 units	35,000
Remainder at 100%	36,150
Total large appliance demand	72,150
Total demand load	120,350 watts

120,350va/240V = 501 amperes

The neutral for the electric ranges may be sized at 70% of the total demand or 24,500 VA making the total demand for the neutral conductor 109,850 VA, or 458 amperes.

Further demand for neutral conductor: first 200 amperes at 100%; balance at 70%. Therefore, 458 - 200 = 258 amperes x 70% = 181 amperes + 200 = 381 amperes—the demand size for neutral conductor.

Fig. 2-27: Summary of NE Code calculations for 20-unit apartment building.

to a demand factor if there are more than three units. Optional calculation methods to determine feeder or service loads for schools and farms are also provided in the NE Code.

Practical electrical design: In many cases, the equipment ratings that result from strict application of NE Code rules are not adequate for a specific practical situation, and this is recognized by the NE Code. The conductor sizes that result from calculations performed in accordance with NE Code rules do not take into account voltage drop caused by the length of conductors. The NE Code recommends that the total voltage drop from the service to the farthest branch-circuit outlet should not be greater than 5%. This problem and considerations for future expansion are not covered by the NE Code.

The circuit voltages are taken to be nominal voltages based on the main service voltage or a stepped-down voltage at a transformer. These voltages may be higher than the actual voltage at an outlet (which may be lower as a result of voltage drop on the conductors). For example, a 480-volt service may actually supply 460 volts or even 440 volts to the farthest outlet. A conservative design approach would be to use the lower voltage in the calculations and thereby obtain a larger required ampacity for the conductor.

Sizes and ratings: Transformers used in lighting and power circuits are normally considered capable of carrying their full-rated load continuously. Thus, a 10-kilowatt continuous load requires service conductors rated for 1.25 x 10 kVA = 12.5 kVA, but a transformer rated at only 10 kilovolt-amperes. In practice, transformers are manufactured with a limited number of standard ratings. See Chapter 10 of this book.

Fuses used to protect circuits are supplied in the standard sizes listed in the NE Code. If the current exceeds the fuse size for a nontime-delay fuse, the fuse element will open. Many circuit breakers, especially those for low current values, have a fixed rating or setting. Other breakers for larger loads (100 to 6000 amperes) are supplied in standard size frames such as 100, 225, 400, 600, 800, 1200 amperes, etc. The tripping mechanism that determines the rating or setting can be adjusted at the factory and the standard ratings given in the NE Code can be selected (although all ratings are not readily available commercially). The rules that apply to the rating or setting of overcurrent devices refer to the capacity of the tripping mechanism in the breaker if adjustable breakers are used.

Service-entrance panelboards protected by main circuit breakers are also produced with standard capacities that correspond to the standard circuit breaker frame sizes. The circuit breaker setting is adjusted to protect the circuit as determined by NE Code rules.

Typical commercial occupancy calculations: Typical commercial establishments range from small stores with single-phase services to large office buildings with three-phase services and significant motor loads. The branch circuits are designed first, then the feeders if used, and finally the main service.

When transformers are not involved, a relatively simple design problem with a single voltage results. If step-down transformers are used, the transformer itself must be protected by an overcurrent device which may also protect the circuit conductors in most cases. In any design, the rules for the protection of the transformer must be considered to assure a design in complete conformance with NE Code rules.

Switchboards and panelboards used for the distribution of electricity within a building are also subject to NE Code rules. In particular, a lighting and appliance panelboard cannot have more than 42 overcurrent devices to protect the branch circuits originating at the panelboard. This rule could affect the number of feeders required when a large number of lighting or appliance circuits are needed.

Service Installations

The power-riser diagram in Fig. 2-28 was used on a small commercial building. In analyzing this diagram, the following things are apparent:

- A 3-inch PVC (plastic) conduit extends from 6 feet beyond the building to the C/T (current transformer) cabinet.
- A 1-inch empty conduit runs from the C/T cabinet to the meter base mounted on the outside wall of the building. This conduit is for the power company's six meter wires from the current transformer.
- Panel A is mounted above the C/T cabinet.
- A two-pole time clock is mounted next to the panel and is fed with three No. 12 AWG conductors. This is to control the outside sign and building lights.

This power-riser diagram, in itself, is not sufficient to tell the electrical contractor exactly what is to be done. However, the written specificaitons along with the panelboard schedule should give all necessary information for a complete installation without further questions.

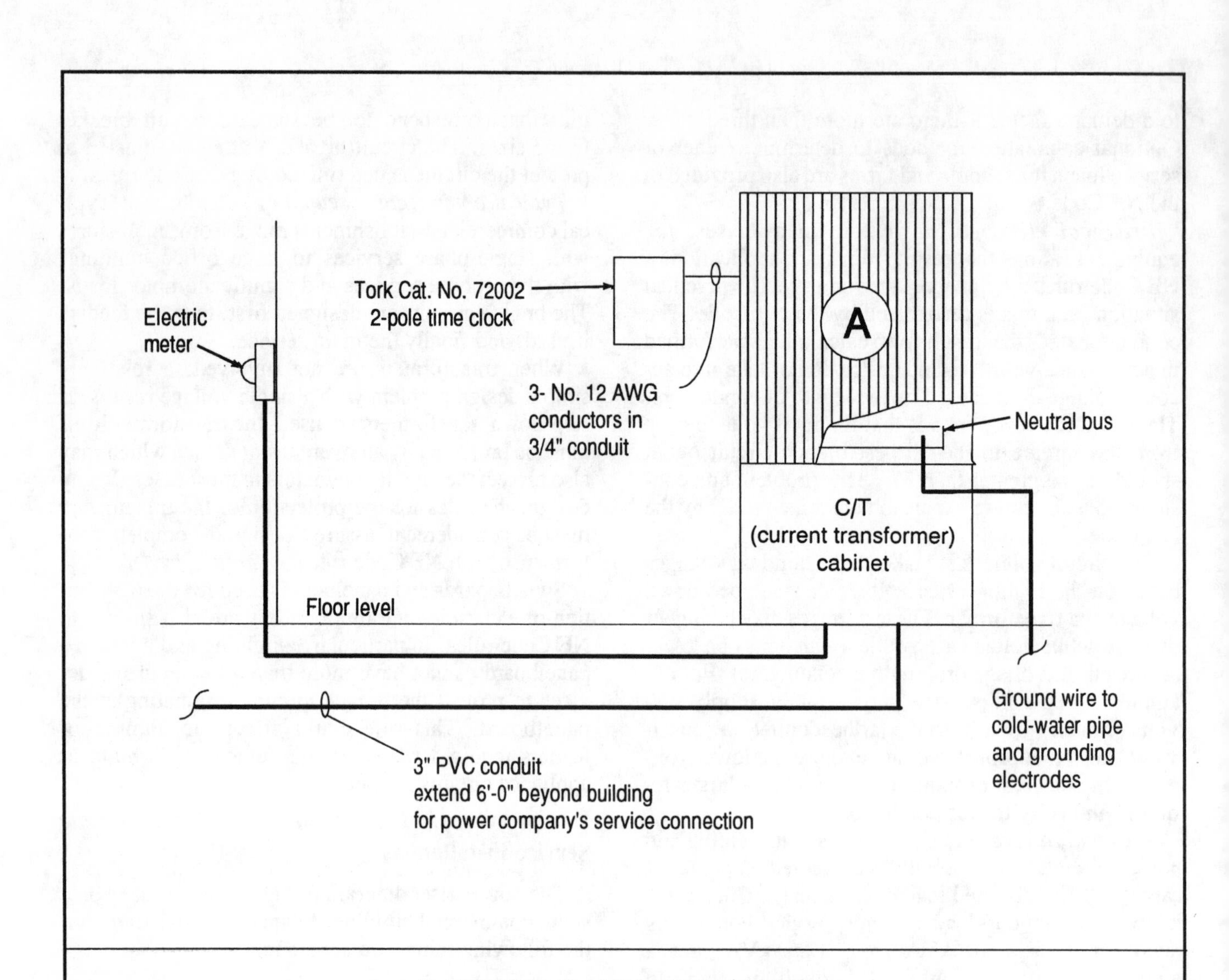

PANELBOARD SCHEDULE

Panel No.	Type Cabinet	Panel mains: Amperes	Panel mains: Volts	Branches: Phase	Branches: 1P	Branches: 2P	Branches: 3P	Branches: Prot	Branches: Frame	Items fed
"A"	Flush	200A	120/240V	3ϕ 4WΔ	-	1	-	20A	70A	Time clock
Square "D" NQOB w/main circuit breaker					-	-	1	20A	70A	A.H.U.
					-	1	-	30A	70A	Water htr.
					-	-	1	30A	70A	Cond. unit
					5	-	-	20A	70A	Lights
					10	-	-	20A	70A	Recepts.
					5	-	-	20A	70A	Spares
					12	-	-	-	-	Provisions only

Fig. 2-28: Power-riser diagram with related panelboard schedule.

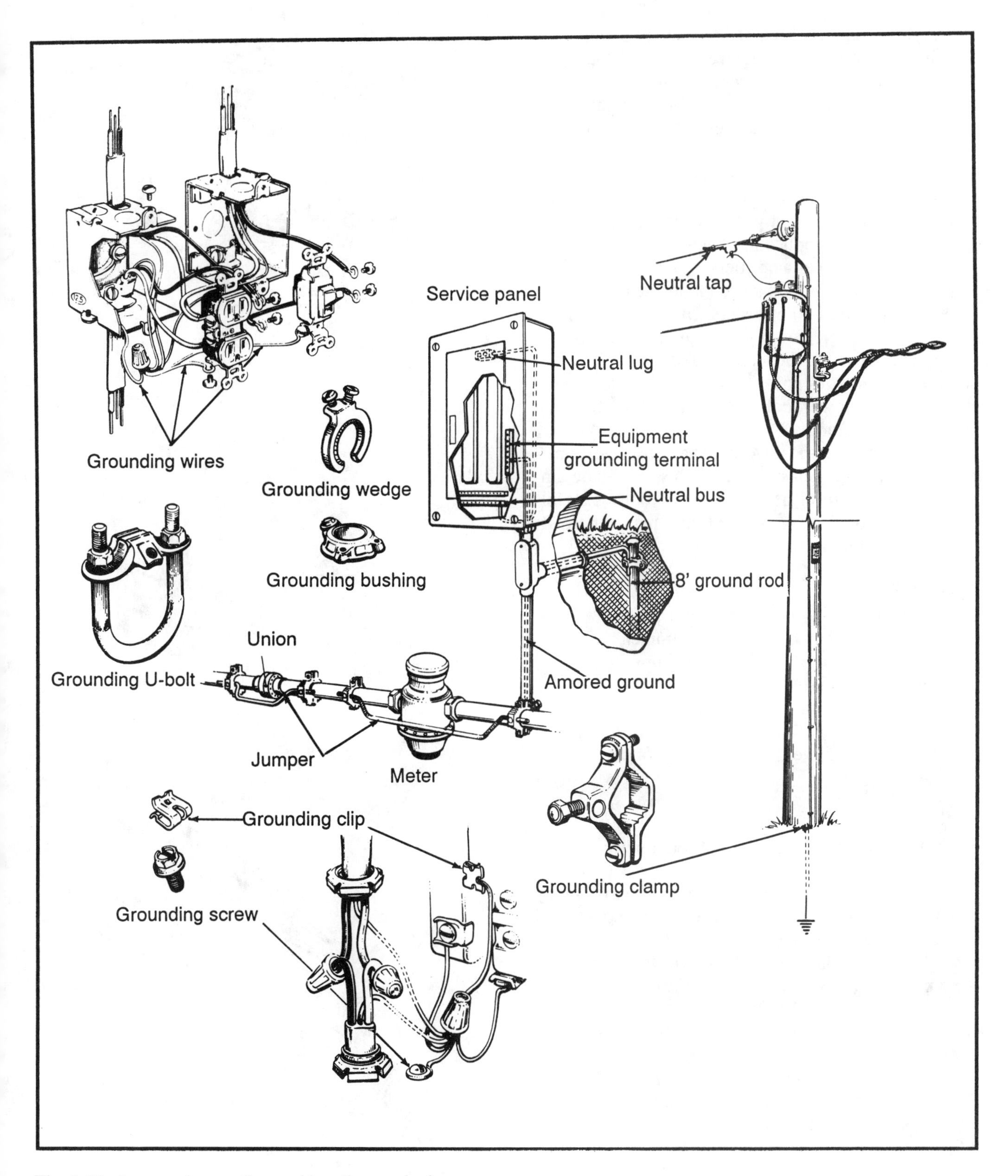

Fig. 2-29: Approved grounding and bonding methods.

Tools of the Trade

Ever since humans became "tool-making animals," we have developed many devices to extend the range and usefulness of our five senses. The discovery of electricity and its practical use have brought about many comforts and conveniences that were almost unthought of a century ago. Electrical systems, however, must be maintained and certain tools are necessary to accomplish this task. Some of these tools are commonly used in many different ways, while others are used only for a specific application.

Some tools serve as amplifiers of our senses and are diagnostic in nature—being used to locate the trouble. Such tools are often called testers. Others, used to turn, hold, cut, hammer, and pry—as an extension of our fingers and muscles—are called hand tools. Tools using a source of extra power in addition to man's muscles are called power tools, but they still do the same things that the hand tools do.

Historically, electrical workers have been responsible for furnishing their own hand tools, while the employer normally furnishes most power tools. The following list should be considered minimum for troubleshooting and maintaining electrical systems.

1 tool box
1 8″ side cutter
2 pr. channel locks
1 10″ screw driver
1 6″ screw driver
1 6″ crescent wrench
1 6′ rule
1 Phillips screw driver
1 diagonal pliers
1 10″ mill file
1 tap wrench
1 small screw driver
1 key hole saw
1 tap wrench
1 claw hammer
1 10″ tin snips
1 pocket knife
1 8″ level
1 combination square
1 hack saw frame
1 long nose pliers
1 fuse puller
1 small socket set
1 volt-ohmmeter

An experienced worker usually will have many items of small hand tools over and above those listed above. They will be tools that he personally likes to use and which have proven to allow him to perform his work more efficiently, in a more workmanlike manner, and with less fatigue. For example, a "Yankee" or ratchet screwdriver or perhaps a cable stripper for stripping NM or BX cable. More often than not, an electrical worker can be judged by the type and condition of the small hand tools which he carries in his personal tool box or tool pouch.

The following illustrations will help acquaint you with some of the more popular hand tools.

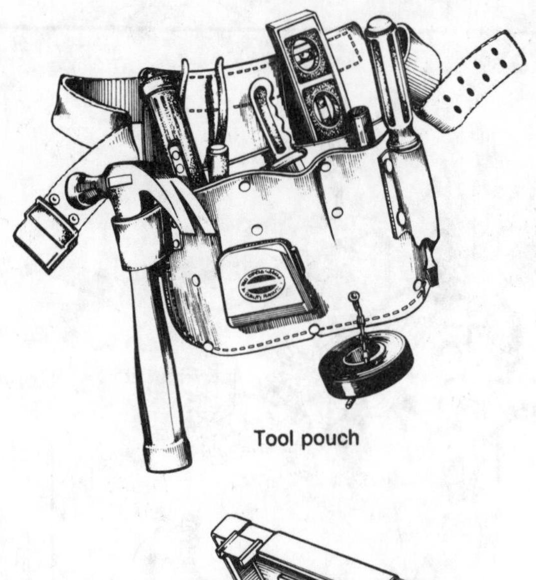
Tool pouch

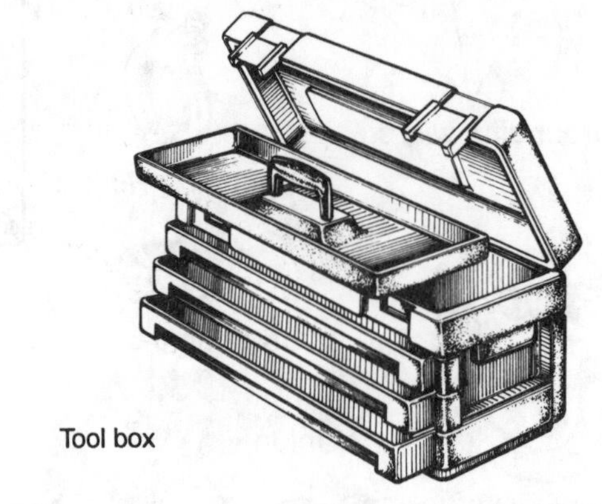
Tool box

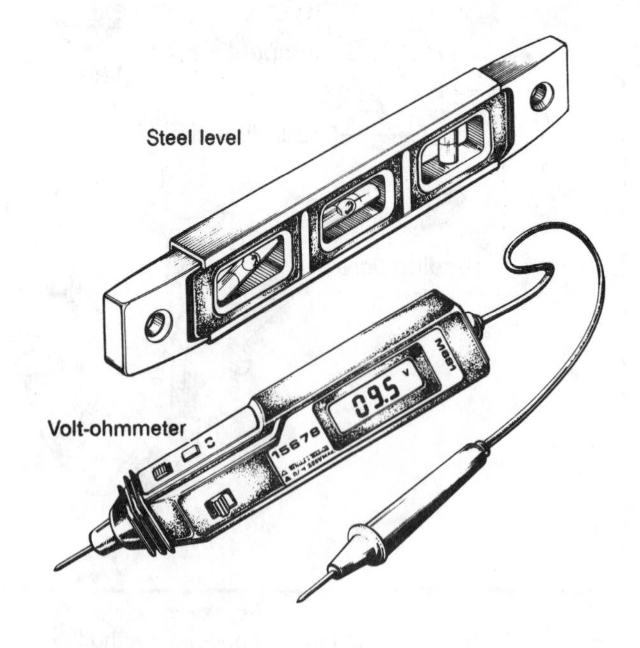

Steel level

Volt-ohmmeter

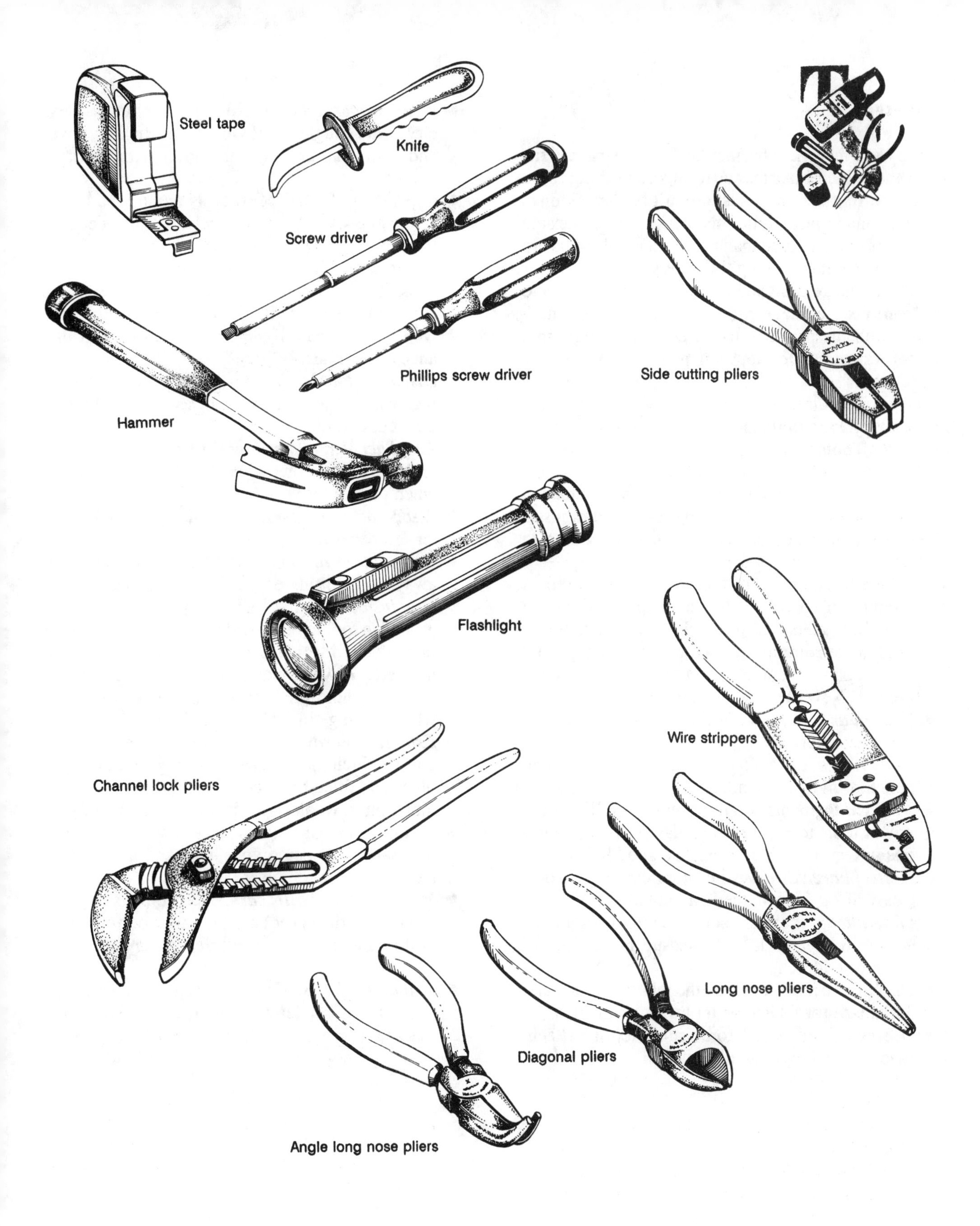
Steel tape
Knife
Screw driver
Phillips screw driver
Side cutting pliers
Hammer
Flashlight
Wire strippers
Channel lock pliers
Long nose pliers
Diagonal pliers
Angle long nose pliers

Safety Reminders

Safety is everyone's business. Much has been written on the subject, but the most important thing to be aware of is that the safety just doesn't happen. Things *have* to be made safe —equipment *has* to be made safe —work areas *have* to be made safe...and, all of us are responsible for maintaining this safety.

Power tools are designed to help cut, drill, sand and grind various materials. Most of these tools are high speed, using sharp cutting tools. Don't be caught in a careless moment. Develop and practice good safety habits.

Basic Safety Instructions for Power Tools

- *Know your power tool:* Read owner's manual carefuly. Learn its applications and limitations as well as the specific potential hazards peculiar to this tool.
- *Ground all tools:* If tool is equipped with a three-prong plug, it should be plugged into a three-hole electrical receptable. If an adapter is used to accommodate two-pronged receptacle, the adapter wire must be attached to a known ground. Never remove the third prong.
- *Keep guards in place:* Make sure they are also in working order.
- *Keep work area clean:* Cluttered areas and benches invite accidents.
- *Avoid dangerous environment:* Don't expose power tools to rain or use in damp, wet, or gaseous or explosive locations. Use good lighting.
- *Keep children away:* All visitors should be kept at a safe distance from the work area.
- *Store idle tools:* When not in use, tools should be stored in dry, high or locked-up place—out of reach of children.
- *Don't force tool:* It will do the job better and safer at the rate for which the tool was designed.
- *Use right tool:* Don't force small tool or attachment to do the job of a heavy tool.
- *Wear proper apparel:* No loose clothing or jewelry to get caught in moving parts. Rubber gloves and footwear are recommended with working outdoors.
- *Use safety glasses:* Most tools require the use of safety glasses. Also face mask if cutting operation is dusty.
- *Don't abuse cord:* Never carry tool by cord or yank it to disconnect it from receptable. Keep cord from heat, oil and sharp edges.
- *Don't overreach:* Keep proper footing and balance at all times.
- *Disconnect tools:* When not in use, before servicing; when changing accessories such as blades, bits, cutters, etc.
- *Avoid accidental starting:* Don't carry plugged-in tool with finger on switch. Be sure switch is off when plugged in.
- *Wear ear protectors,* when using tools for extended periods.
- *Use caution around moving parts:* Keep hands away from cutting edges and all moving parts.
- *Use insulated surfaces:* A double insulated or grounded tool may be made "live" if the bit comes in contact with live wiring in a wall, floor, ceiling, etc. Always check the work area for live wires and hold the tool by the insulated surfaces when making "blind" or plunge cuts.
- *Extension cords:* Use only three-wire extension cords which have three-prong grounding-type plugs and three-pole receptacles which accept the tool's plug. Replace or repair damaged cords.
- *Secure work:* Use clamps or a vise to hold work. It is safer than using your hand and it frees both hands to operate tool.
- *Maintain tools with care:* Keep tools sharp and clean at all times for best and safest performance. Follow instructions for lubricating and changing accessories.
- *Grinding wheels:* Use only grinding wheels with "safe speed" at least as high as "no load rpm" marked on the name plate.

chapter 3

SWITCHES, PANELBOARDS, AND LOAD CENTERS

This chapter is intended to familiarize the reader with terms and concepts which are fundamental to an understanding of distribution equipment (switches and load centers) and its application. Study of the definitions, symbols, diagrams and illustrations will give the reader a background in the language and basic principles associated with electric distribution equipment as it applies to building construction.

Service switches or main distribution panelboards are normally installed at a point immediately where the service-entrance conductors enter the building. Branch circuits and feeder panelboards (when required in addition to the main service panelboard) are usually grouped together at one or more centralized locations to keep the length of the branch-circuit conductors at a practical minimum of operating efficiency and to lower the initial installation costs.

Distribution equipment is generally intended to carry and control electrical current, but is not intended to dissipate or utilize energy. Eight basic factors influence the selection of distribution equipment:

1. *Codes and Standards:* Suitability for installation and use, in conformity with the provisions of the NE Code and all local codes, must be considered. Suitability of equipment may be evidenced by listing or labeling.
2. *Mechanical Protection:* Mechanical strength and durability, including the adequacy of the protection provided must be considered.
3. *Wiring Space:* Wire bending and connection space is provided according to UL standards in all distribution equipment. When unusual wire arrangements or connections are to be made, then extra wire bending space, gutters, and terminal cabinets should be investigated for use.
4. *Electrical Insulation:* All distribution equipment carries labels showing the maximum voltage level that should be applied. The electrical supply voltage should always be equal to, or less than the voltage rating of distribution equipment; never more.
5. *Heat:* Heating effects under normal conditions of use and also under abnormal conditions likely to arise in service must be constantly considered. Ambient heat conditions, as well as wire insulation ratings, along with the heat rise of the equipment must be evaluated during selection.
6. *Arcing Effects:* The normal arcing effects of overcurrent protective devices must be considered when the application is in or near combustible materials or vapors. Enclosures are selected to prevent or contain fires created by normal operation of the equipment. Selected locations of

equipment must be made when another location may cause a hazardous condition.

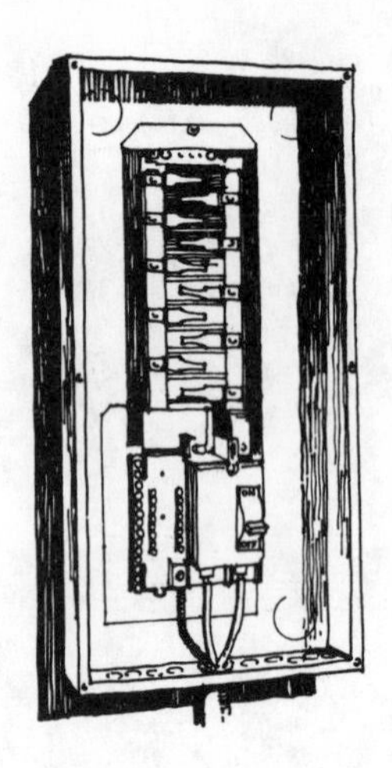

Fig. 3-1: Panelboards consist of a metal enclosure containing bus bars and overcurrent protective devices.

7. *Classification:* Classification according to type, size, voltage, current capacity, interrupting capacity and specific use must be considered when selecting distribution equipment. Loads may be continuous or noncontinuous and the demand factor must be determined before distribution equipment can be selected.
8. *Personnel Protection:* Other factors which contribute to the practical safeguarding of persons using or likely to come in contact with the equipment must be considered. The equipment selected for use by only qualified persons may be different from equipment used or applied where unqualified people may come in contact with it.

In electrical wiring installations, overcurrent protective devices, consisting of fuses or circuit breakers, are factory assembled in a metal cabinet, the entire assembly commonly being called a *panelboard* as shown in Fig. 3-1.

Sometimes the main service-disconnecting means will be made up on the job by the workers by assembling individually enclosed fused switches or circuit breakers on a length of metal auxiliary gutter, as shown in Fig. 3-2. Note that the various components are connected by means of short conduit nipples in which the insulated conductors are fed. Other services will consist of one large panelboard, often called a main distribution panelboard, which gives a neater appearance. See Fig. 3-3.

Panelboards consist of assemblies of overcurrent protective devices, with or without disconnecting devices, placed in a metal cabinet. The cabinet includes a cover or trim with one or two doors to allow access to the overcurrent and disconnecting devices and, in some types, access to the wiring space in the panelboard.

Panelboards fall into two mounting classifications: (1) flush mounting, wherein the trim extends beyond the outside edges of the cabinet to provide a neat finish with the wall surface (Fig. 3-4), and (2) surface mounting, wherein the edge of the trim is flush with the edge of the cabinet as shown in Fig. 3-5.

Panelboards fall into two general classifications with regard to overcurrent devices: (1) circuit breaker and (2) fused. Small circuit breaker and fusible panelboards commonly referred to as *load centers* are manufactured for use in residential and small commercial and industrial occupancies.

Enclosures

The majority of overcurrent devices (fuses and circuit breakers) are used in some type of enclosure; that is, panelboards, switchboards, motor control centers, individual enclosures, etc.

NEMA has established enclosure designations because individually enclosed overcurrent protective devices are used in so many different types of locations, weather and water conditions, dust and other contaminating conditions, etc. A designation such as "NEMA 12" indicates an enclosure type to fulfill requirements for a particular application. The NEMA designations were recently revised to obtain a clearer and more precise definition of enclosures for panelboards, safety switches, and motor starters needed to meet various standard requirements.

Some of the revisions in the NEMA designations are: The NEMA Type 1A (semi-dust tight) has been dropped. The NEMA 12 enclosure now can be substituted in many installations in place of the NEMA 5. The advantage of this substitution is that the NEMA 12 enclosure is much less expensive than the NEMA 5 enclosure. NEMA Type 3R as applied to circuit breaker enclosures is a lighter weight, less expensive rainproof enclosure than the other "Weather Resistant" enclosure types. The table in Fig. 3-6 lists a brief explanation of the NEMA enclosure specifications.

Safety Switches

Most manufacturers of safety switches have at least two complete lines to meet industrial, commercial and residential requirements. Both types usually have visi-

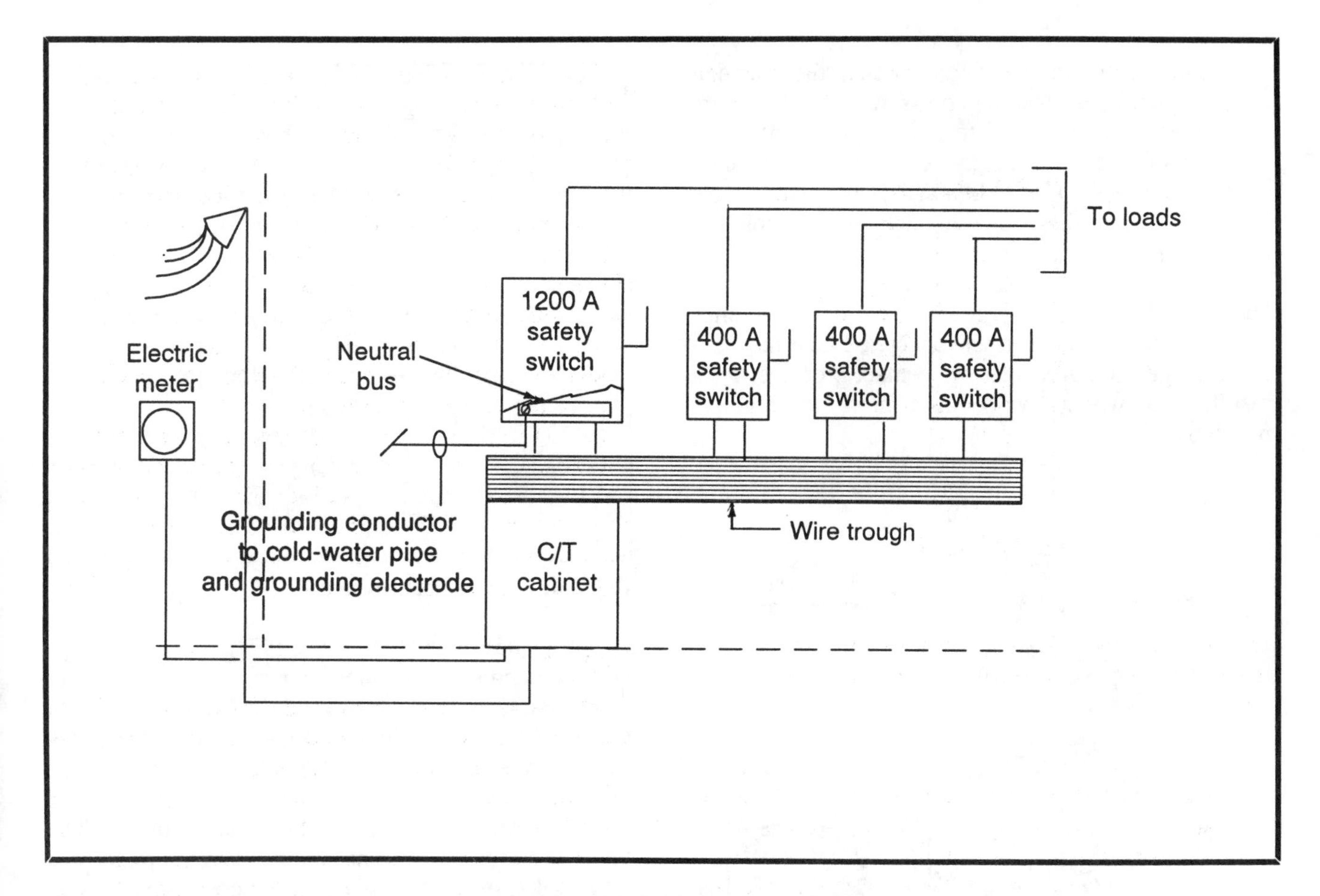

Fig. 3-2: An electric service made up of wire trough and individual safety switches.

ble blades and safety handles. With visible blades, the contact blades are in full view so you can clearly see you're safe. Safety handles are always in complete control of the switch blades, so whether the cover is open or closed, when the handle is in the "OFF" position the switch is always "OFF"; that is, on the load side of the switch. The feeder or line side of the switch is still "hot" (energized) so when working with safety switches keep this in mind. See Fig. 3-6.

Heavy duty switches are intended for applications where price is secondary to safety and continued performance. This type of switch is usually subjected to frequent operation and rough handling. Heavy duty switches are also used in atmospheres where a general duty switch would be unsuitable. Heavy duty switches are widely used by most heavy industrial applications; motors and HVAC equipment will also be controlled by such switches. Most heavy duty switches are rated 30 through 1200 amperes, 240 to 600 volts (ac-dc). The switches with horsepower ratings are able to interrupt approximately six times the full-load, motor-current ratings. When equipped with Class J or Class R fuses (see Chapter 4), many heavy duty safety switches are UL listed for use on systems with up to 200,000 amperes available fault current.

Heavy duty switches are available with NEMA 1, 3R, 4, 4X, 5, 7, 9 and 12 enclosures.

Heavy duty safety switches find use mostly in industrial applications where heavy continuous loads are encountered, and where a shut-down could be costly to the owners. The blades and contacts in heavy duty switches will stand more heat than those in general duty switches, and therefore, can be relied upon to handle critical loads with more reliability than those of the general duty type. This type of switch also finds use in high-rise office buildings.

Switch Contacts: There are two types of switch contacts used in today's safety switches. One is the "butt" contact similar to those used in circuit breaker devices; the other is a knife-blade and jaw type. The knife-blade types are considered to be superior to other types on the market.

All current-carrying parts of safety switches are usually plated with tin, cadmium or nickel to reduce heating by keeping metal oxidation at a minimum. Switch blade and jaws are made of copper for high conductivity. With knife-blade construction, the jaws distribute a uniform clamping pressure over the entire blade-to-jaw contact surface. In the event of high-current fault, the electromagnetic forces which develop tend to squeeze the jaws tightly against the blade. In the butt type contact, these forces tend to force the contacts apart, causing them to burn severely.

Fuse clips are also plated to control corrosion and keep heating to a minimum. All heavy duty fuse clips have steel reinforcing springs to increase their mechanical strength and give a firmer contact pressure. As a result, fuses will not work loose due to vibration or rough handling.

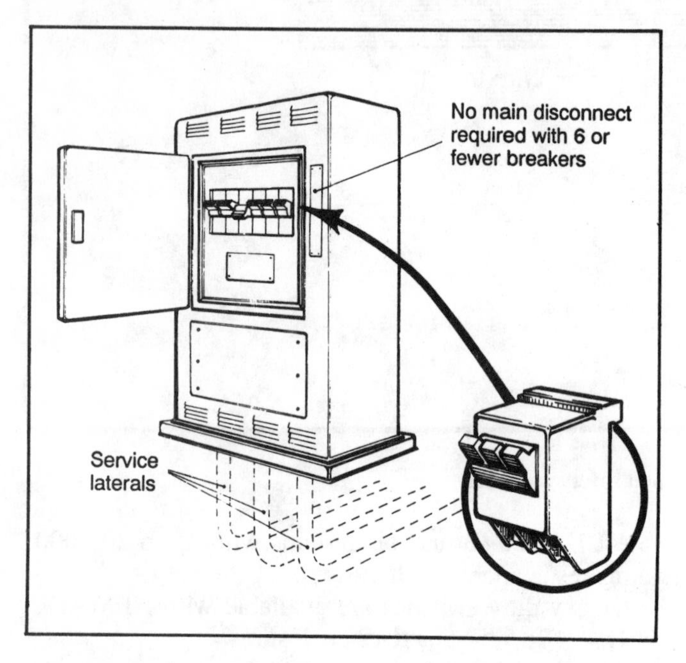

Fig. 3-3: A main distribution panel (MDP) contains all service components in one cabinet—giving a neater appearance, and oftentimes taking up less floor space.

Insulating Materials: As the voltage rating of switches is increased, arc suppression becomes more difficult and the choice of insulation material becomes a more critical problem. Arc suppressors used by many manufacturers consist of a housing made of insulation material and one or more magnetic suppressor plates. All arc suppressors are tested to assure proper control and extinguishing of arcing.

Operating Mechanism: Heavy duty safety switches have spring driven, quick-make, quick-break mechanisms. A quick-breaking action is necessary if a switch is to be safely switched "OFF" under a heavy load. The spring action, in addition to making the operation quick-make, quick-break, firmly holds the switch blades in an "ON" or "OFF" position. The operating handle is an integral part of the switching mechanism, so if the springs should fail the switch can still be operated. When the handle is in the "OFF" position the switch is always "OFF."

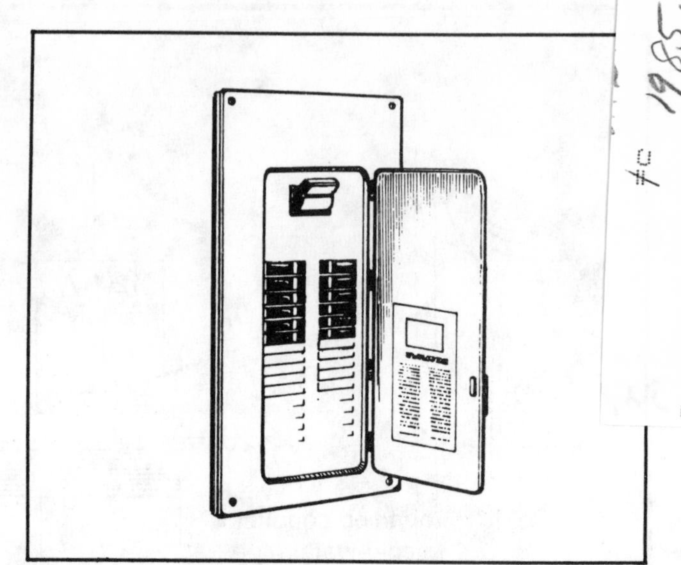

Fig. 3-4: A flush-mounted panelboard.

A one-piece cross bar is usually employed to offer direct control over all blades simultaneously. The one-piece cross bar means stability and strength, plus proper alignment for uniform blade operation.

Dual cover interlocks are also standard on all heavy duty switches. The dual interlock prevents the enclosure door from being opened when the switch is "ON" and also keeps the switch from being turned "ON" while the door is open.

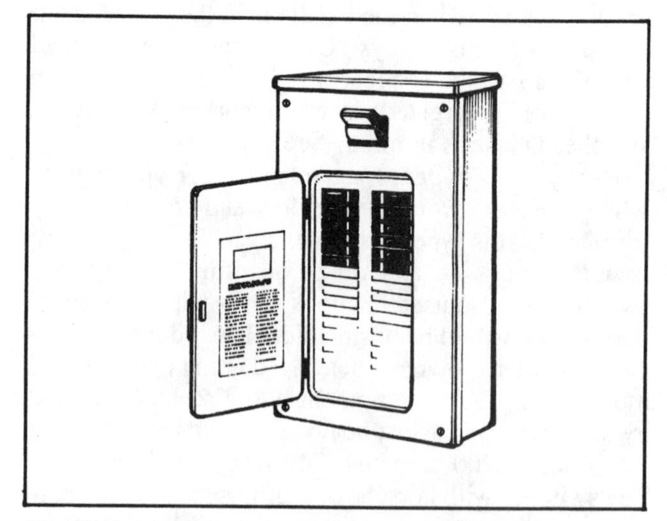

Fig. 3-5: A surface-mounted panelboard has a neat appearance when mounted on the surface of a wall.

The following chart offers a brief explanation of the NEMA enclosure specifications:

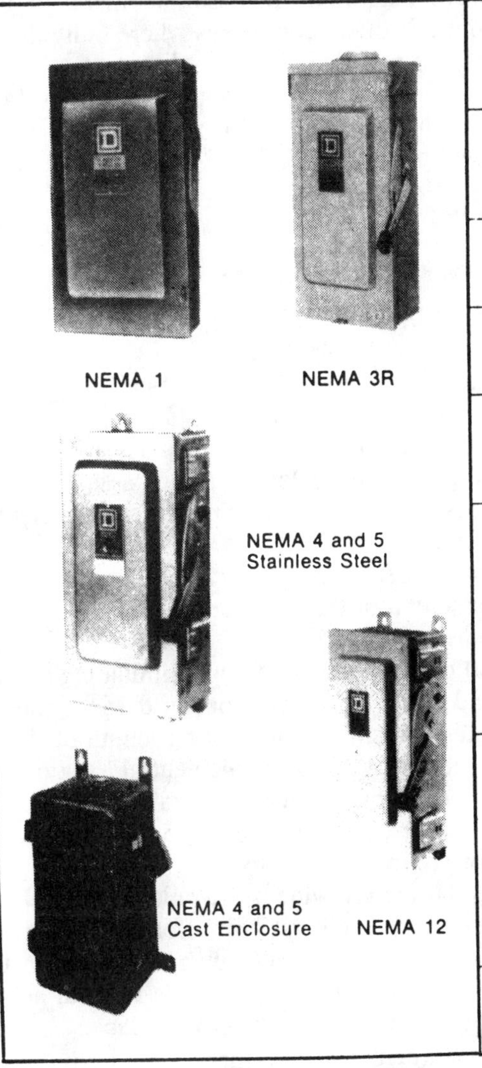

NEMA 1

NEMA 3R

NEMA 4 and 5
Stainless Steel

NEMA 4 and 5
Cast Enclosure

NEMA 12

ENCLOSURE	EXPLANATION
NEMA Type 1 General Purpose	To prevent accidental contact with enclosed apparatus. Suitable for application indoors where not exposed to unusual service conditions.
NEMA Type 3 Weatherproof (Weather Resistant)	Protection against specified weather hazards. Suitable for use outdoors.
NEMA Type 3R Raintight	Protects against entrance of water from a rain. Suitable for general outdoor application not requiring sleetproof.
NEMA Type 4 Watertight	Designed to exclude water applied in form of hose stream. To protect against stream of water during cleaning operations, etc.
NEMA Type 5 Dusttight	Constructed so that dust will not enter the enclosed case. Being replaced in some equipment by Square D NEMA 12 Types.
NEMA Type 7 Hazardous Locations A, B, C or D Class I — Air Break Letter or letters following type number indicates particular groups of hazardous locations per N.E.C.	Designed to meet application requirements of National Electrical Code for Class I, Hazardous locations (Explosive atmospheres). Circuit interruption occurs in air.
NEMA Type 9 Hazardous Locations E, F or G Class II Letter or letters following type number indicates particular groups of hazardous locations per N.E.C.	Designed to meet application requirements of National Electrical Code for Class II Hazardous Locations (Combustible Dusts, etc.).
NEMA Type 12 Industrial Use	For use in those industries where it is desired to exclude dust, lint, fibers and filings, or oil or coolant seepage.

Fig. 3-6: NEMA classifications of safety switches.

General Duty Safety Switches: General duty switches are for residential and light commercial applications where the price of the device is a limiting factor. General duty switches are meant to be used where operation and handling are moderate and where the available fault current is less than 10,000 amperes. Some examples of general duty switch applications would be: residential HVAC equipment, water heaters, electric clothes dryers, light duty fan-coil circuit disconnects for commercial projects, and the like.

General duty switches are rated up to 600 amperes at 240 volts (ac only) in general purpose (NEMA 1) and rainproof (NEMA 3R) enclosures. These switches are horsepower rated and capable of opening a circuit with approximately six times a motor's full-load current rating.

All current-carrying parts of general duty switches are plated with either tin or cadmium to reduce heating. Switch jaws and blades are made of plated copper for high conductivity. A steel reinforcing spring increases the mechanical strength of the jaws and assures a firm contact pressure between blade and jaw.

Double-Throw Safety Switches: Double-throw switches are used as transfer switches and are not intended as motor circuit switches; therefore, they are not horsepower rated. Three lines of double-throw switches are normally available: 82,000 line and 92,000 line, plus a "DTU" rainproof manual transfer switch.

The 82,000 line of switches is available as either fused or unfused devices. These switches have quick-make, quick-break action, plated current-carrying parts, a key controlled interlock mechanism and screw-type lugs. Arc suppressors are supplied on all switches rated above 250 volts.

The 92,000 line switches are manually operable and not quick-make, quick-break. They are available as either fused or unfused devices in NEMA 1 enclosures only.

Main Lugs Load Centers

Main lugs load centers provide distribution of electrical power where a main disconnect with overcurrent protection is provided separately from the load center. All terminals are suitable for aluminum or copper conductors.

All main lugs load centers 125A and up have interiors which are reversible for either top or bottom feed. The cover does not need to be reversed when the interior is reversed. Load centers of 125A and above are available in 14-inch wide boxes for more wire-bending space. Most single-phase, 3-wire load centers are also approved for three-phase grounded "B" systems at 240V ac. A main lugs load center can also be converted to a main breaker load center simply by plugging on any QO or Q1 circuit breaker and back-feeding that breaker with the line conductors. After the main breaker is inserted, the number of spaces available for branch circuits is equal to the number of spaces originally in the load center, less the number of spaces taken by the main breaker. (Space = the area required by a single-pole standard size breaker. Examples: QO120 takes 1 space; QO220 takes 2 spaces.)

The 600A type HQ2 mains rated load center is available for garden, town house and other types of two to six unit apartment complexes where individual metering is not required. It is UL listed as service entrance equipment only, and the neutral is factory bonded to the box. The 600A mains rated load center will accept plug-on FA or Q2 2-pole breakers through 200A.

Main Breaker Load Centers

QO main breaker load centers have many time saving and money saving features:

The main breakers are factory installed, cutting installation costs. There are no lugs to remove, no screws, no nuts, or washers to misplace and no expensive main breakers to lose. Factory assembled main disconnects also assure a proper and safe electrical connection.

The main breaker and neutral terminals are located at the same end of the load center and are adjacent to one another.

This allows straight-in wiring, eliminating awkward bends and space cramping loops in the incoming service cables and saving on the total length of the conductors needed. Also, since the neutral location is not in the branch wiring gutter, the load gutters carry only branch breaker connections and are not cluttered with neutral or ground conductors.

Boxes 14 inches wide are available in 100-225A main breaker load centers. This offers more side gutter space for wiring, and permits flush load center installation between 16-inch centered studs without an extra mounting support to hold the box in place.

Each load center has a separate solid cover with a door. This means that instead of stocking a complete surface and a complete flush device, you can stock only one box with interior and a surface or flush cover.

The neutral bars for branch circuits have alternating lugs rated No. 14-8 and No. 14-4 wire size. The bars are UL listed only for one conductor per hole, and are suitable for copper or aluminum conductors.

The line side terminals of the main breakers are suitable for use with copper or aluminum conductors.

The door completely covers both the branch breakers and the main breaker, leaving no exposed breaker handles to detract from the appearance of the load center or to be accidentally switched.

Split Bus Load Centers

The NEC allows the use of a maximum of six main disconnects in a common enclosure. Split-bus QO load centers have the bus split or divided into sections which are insulated from each other to provide an economical service entrance device in applications not requiring a single main disconnect.

The line or main section of the load center has provisions for up to six main disconnects for the heavier 240V appliances, subfeeders and lighting main disconnects. The lower section contains provisions for lighting and 120V appliance circuits and is fed by the lighting main disconnect which is located in the main bus section. All split-bus devices are provided with factory installed wires connected to the lower section. These wires can then be field connected to the lighting main disconnect in the main section.

The newest of the 200A split-bus load centers has provisions for a field installable lighting main using a plug-on breaker 60 through 125 amperes. Any of these breakers can be used without changing factory installed lighting main wire described above. A 14-inch wide box is also standard with this device for installing between wall studs on 16-inch centers. The interior is reversible for either top or bottom feed. The cover does not need to be reversed when the interior is reversed. The lighting section will hold 12 full size breakers or a maximum of 24 circuits.

Riser Panels

Riser panels, consisting of a main lugs only load center with an extended gutter of over six inches, are ideally suited for high rise office buildings and for apartment complexes. They are available with 6, 8 and 12 circuit load centers. The box, interior and covers are sold separately so they can be installed at the most convenient time during construction.

Another type of riser panel is the feed through load center. In this panel, the main bus bars have lugs at both ends and therefore actually become part of the riser system. When using these in high rise buildings, the savings in riser wire length needed can be considerable. The branch breakers merely plug on to the main bus bars. Feed through riser panels have no main disconnect.

A standard load center may be converted in the field to a riser panel by adding one of the appropriate auxiliary gutters which may be attached to either the right or the left side of the load center.

Panelboards

Switchgear manufacturers offer complete lines of lighting and distribution panelboards, most of which are available either unassembled from distributor stock or factory assembled. All types should be UL listed and meet Federal Specification WP-115a.

NQO panelboards are rated for use on the following ac services: 120/240V, single-phase, 3-wire; 240V, three-phase, 3-wire delta; 240V, three-phase delta with grounded B phase and 120/208V, three-phase, 4-wire wye. They carry no dc rating. NQO panelboards are available either factory assembled or unassembled.

This type panelboard is suitable for use in industrial buildings, schools, office and commercial buildings and institutions when the largest branch breaker does not exceed 150 amperes and the system voltage is not greater than 240 volts ac.

NQO panelboards have maximum mains ratings of 400A main breaker or main lugs. Branch circuit breakers may be catalog prefix QO, QO-H, QH, Q1 or Q1-H, 1, 2 or 3 pole—having a maximum rating of 150A and featuring plug-on bus connections. QO and Q1 circuit breakers are standard with 10,000 AIC rating and QH breakers with 65,000 AIC rating. Other ratings for specific applications are also available.

Branch circuit breakers with ground fault circuit interruption may also be supplied in Type NQO panelboards. Rated 10,000 AIC symmetrical, these GFI devices provide UL Class A (5 milliampere sensitivity) ground fault protection as well as overload and short circuit protection for branch circuit wiring.

NQO unassembled panelboards are available as follows:

- branch breakers
- interior with solid neutral
- box, either 14″ wide x 4″ deep, 14″ wide x 5¾″ deep or 20″ wide x 5¾″ deep
- mono-flat front with door and flush lock
- accessories

NQO factory assembled panelboards are identical in construction to the unassembled type. Main ratings and branch circuits are the same. Unlike unassembled panelboards, however, the branches are factory installed.

NQO construction—assembled and unassembled boxes are constructed of galvanized steel. A variety of knockouts is provided in each end wall. Interiors having maximum 225A main lugs rating are available in 14″ x 4″ deep, 14″ side x 5¾″ deep or 20″ wide x 5¾″ deep boxes. 14″ wide x 5¾″ deep or 20″ wide x 5¾″ deep boxes are required for panelboards having a main circuit breaker. Boxes for interiors having 400A mains (breakers or lugs) are 20″ wide x 5¾″ deep.

Interiors for standard width panelboards having a maximum rating of 225 amperes are of the "single bus" construction. In this construction, 1, 2 and 3 pole catalog prefix Q1 breakers extend the full width of the panelboard and cannot be mounted opposite each other. QO, QO-H and QH circuit breakers twin mount on the bus assembly. In other words, a 3 pole QO requires three pole spaces, but a 3 pole Q1 requires six QO spaces.

400 ampere interiors utilize a "double row bus" construction. This type of construction consists of 2 sets of bus bars mounted on a single pan. The respective phase busses of each set are paralleled with each other by means of insulated, solid connectors. QO and Q1 breakers mount on a one-for-one basis (i.e. a 3-pole QO requires the same spaces as a 3-pole Q1).

All current carrying parts are plated for maximum corrosion resistance and minimum heating at contact surfaces. Main lugs are Underwriters' Laboratories listed for use with either copper or aluminum cable. Main lugs may be replaced by the appropriate Anderson type VCEL crimp lug, when required. Lug catalog numbers and crimp tool type are called out on the panelboard wiring diagram. Box type lugs for circuits on both branch breakers and the solid neutral permit maximum convenience and speed in wiring and are also Underwriters' Laboratories listed for use with either copper or aluminum cable.

Fig. 3-7 gives a summary of 1993 NE Code rules governing the application of switches, panelboards, and switchgear.

Troubleshooting

Strictly speaking, the first requirement in a completely satisfactory maintenance program for electrical panelboards and switches is good equipment, properly installed. No one can do a good maintenance job on equipment that is either not appropriate for the job that it is doing or equipment that has been installed haphazardly with no eye to future maintenance requirements. If such conditions exist, they should be brought to the attention of the proper party and corrected rather than try to establish a maintenance program for them.

The second requirement for a good maintenance program is proper maintenance personnel. Persons who must maintain equipment should have a thorough knowledge of the equipment's operation and have the ability to be able to make thorough inspections and minor repairs of that equipment.

The third requisite of a good maintenance program is the establishment of preventive maintenance. This is an all-inclusive phrase for the continuing inspection of equipment, the report and recording of the condition of the equipment and the repair of the equipment.

The term "preventive maintenance" has come to mean a system of routine inspections of equipment properly recorded for future reference on some type of inspection records. More specifically, the term stands for the heading off of possible future equipment difficulties by making minor repairs in advance of major operating difficulties. In electrical panelboards and switches specifically, a simple tightening of a lug in one period of time can prevent a serious short-circuit or a heated terminal at a later time. Because the aspect of recording all such inspections, some have gained the impression that preventive maintenance is merely a system of records. Actually the records supplement the inspection and are designed to take the place of the maintenance man's memory. They are a simple supplement to a preventive maintenance program and in such extreme cases where only one maintenance man services a particularly small plant or building, records might be entirely disregarded, though this is not recommended. However, where a number of maintenance personnel are operating they are vital to the proper operation of the maintenance inspection routine.

Because of the high diversity of electrical apparatus, many maintenance people have the mistaken attitude that electrical apparatus is different from other production machinery and will operate under almost any conditions, but it is exactly opposite the truth. Electrical equipment can be damaged more easily by operating conditions than almost any other piece of equipment. Water, dust, heat, cold, humidity, lack of humidity, vibration, and countless other conditions can affect the proper operation of electrical equipment. Because of this, there are four cardinal rules to follow in maintaining electrical apparatus. These are:

- Keep it clean
- Keep it dry
- Keep it tight
- Keep it friction-free

One of the greatest problems existing in panelboards, switchgear, load centers, and the like, is

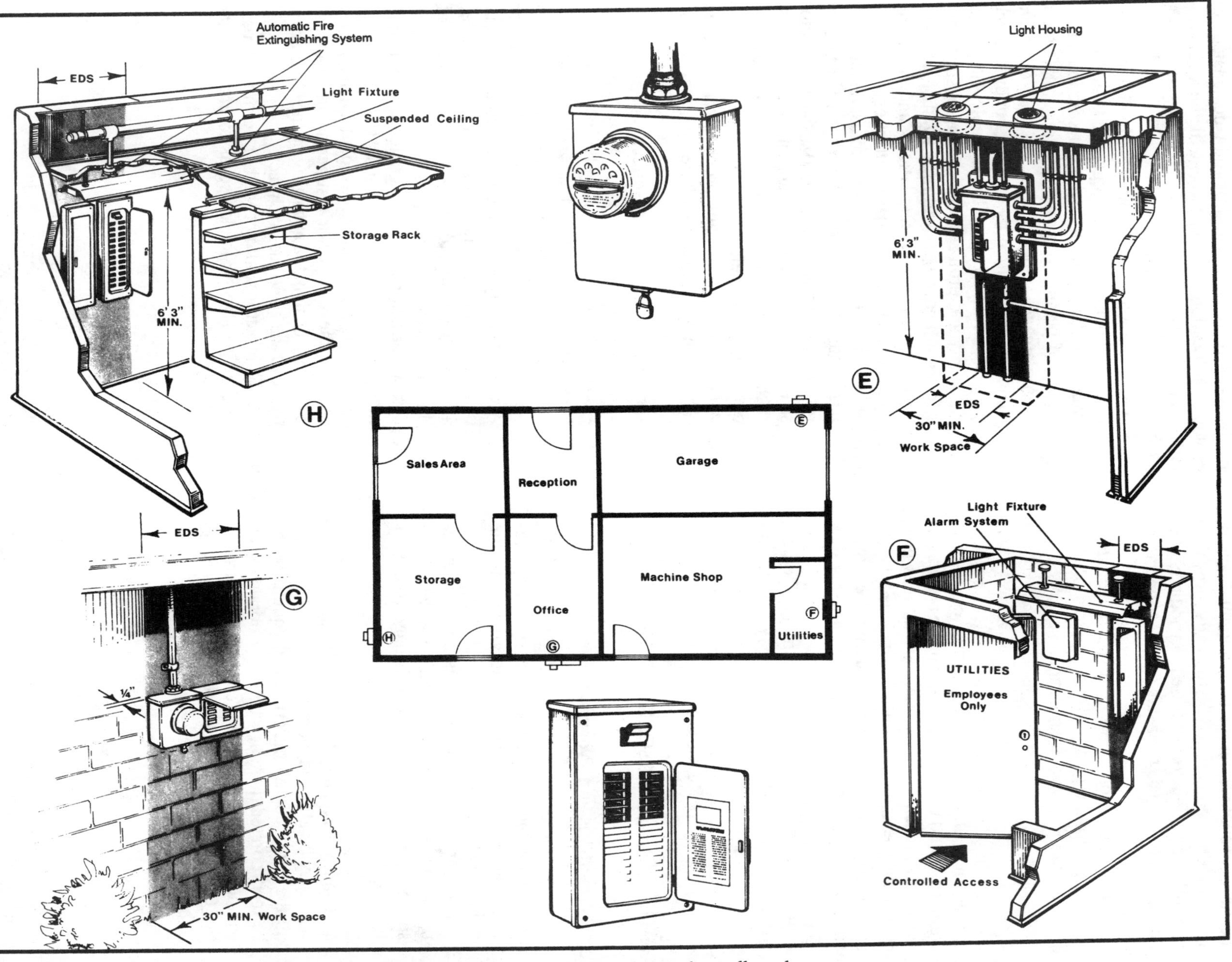

Fig. 3-7: Summary of NE Code rules governing the installation of safety switches and panelboards.

loose connections at both the main lugs and at circuit breakers and/or fuse blocks. Loose connections cause overheating and eventual failure of the terminals and/or conductors. This condition is especially prevalent when aluminum conductors are used. Periodic checks should therefore be made to ensure that all connections are tight.

A loose neutral conductor may be detected by 120-volt lights dimming, or 120-volt motors or other appliances running slower than normal.

Safety

Architectural and engineering firms, contractors and workers should take an active part in the prevention of job-site accidents. For example, certain rules governing safety precautions should be included in the written specifications, and then the architect/engineer or their field inspectors should make certain that they are carried out.

A sample specification included in the General Conditions of a project specification may read as follows:

ARTICLE 10

PROTECTION OF PERSONS AND PROPERTY

Safety Precautions and Programs

10.1 The Contractor shall be responsible for initiating, maintaining and supervising all safety precautions and programs in connection with the work.

10.2 The Contractor shall take all reasonable precautions for the safety of, and shall provide all reasonable protection to prevent damage, injury or loss to:

1. all employees on the Work and all other persons who may be affected thereby;
2. all the Work and all materials and equipment to be incorporated therein, whether in storage on or off the side, under the care, custody or control of the Contractor or any of his Subcontractors or Sub-subcontractors;
3. and other property at the site or adjacent thereto, including trees, shrubs, lawns, walks, pavements, roadways, structures and utilities not designated for removal, relocation or replacement in the course of construction.

10.3 The Contractor shall comply with all applicable laws, ordinances, rules, regulations and orders of any public authority having jurisdiction for the safety of persons or property or to protect them from damage, injury or loss. He shall erect and maintain, as required by existing conditions and progress of the Work, all reasonable safeguards for safety and protection, including posting danger signs and other warnings against hazards, promulgating safety regulations and notifying owners and users of adjacent utilities.

10.4 When the use or storage of explosives or other hazardous materials or equipment is necessary for the execution of the work, the Contractor shall exercise the utmost care and shall carry on such activities under the supervision of properly qualified personnel.

10.5 All damage or loss to any property referred to in other Clauses caused in whole or in part by the Contractor, any Subcontractor, any Sub-subcontractor, or anyone directly or indirectly employed by any of them, or by anyone for whose acts any of them may be liable, shall be remedied by the Contractor, except damage or loss attributable to faulty Drawings or Specifications or to the acts or omissions of the Owner or Architect or anyone employed by either of them or for whose acts either of them may be liable, and not attributable to the fault or negligence of the Contractor.

10.6 The Contractor shall designate a responsible member of his organization at the site whose duty shall be the prevention of accidents. This person shall be the Contractor's superintendent unless otherwise designated in writing by the Contractor to the Owner and the Architect.

10.7 The Contractor shall not load or permit any part of the Work to be loaded so as to endanger its safety.

chapter 4

OVERCURRENT PROTECTION

All electrical circuits, and their related components, are subject to destructive overcurrents. Harsh environments, general deterioration, accidental damage or damage from natural causes, excessive expansion or overloading of the electrical system are factors which contribute to the occurrence of such overcurrents. Reliable protective devices prevent or minimize costly damage to transformers, conductors, motors, equipment, and the other many components and loads that make up the complete electrical system. Therefore, reliable circuit protection is essential to avoid the severe monetary losses which can result from power blackouts and prolonged downtime of facilities. To protect electrical conductors and equipment against abnormal operating conditions and their consequences, protective devices are used in circuits. The fuse and circuit breaker are two such devices.

The table in Fig. 4-1 summarizes the rules for overcurrent devices. When fuses serve as the overcurrent protective device, the NE Code specifically considers plug fuses and cartridge fuses.

Fuses

A fuse is the simplest device for opening an electric circuit when excessive current flows because of an overload or such fault conditions as grounds and short circuits. A "fusible" link or links encapsulated in a tube and connected to contact terminals comprise the fundamental elements of the basic fuse. Electrical resistance of the link is so low that it simply acts as a conductor, and every fuse is intended to be connected in series with each phase conductor so that current flowing through the conductor to any load must also pass through the fuse. The continuous current rating of the fuse in amperes establishes the maximum amount of current the fuse will carry without opening. When circuit current flow exceeds this value, an internal element (link) in the fuse melts due to the heat of the current flow and opens the circuit. Fuses are manufactured in a wide variety of types and sizes with different current ratings, different abilities to interrupt fault currents, various speeds of operation (either quick-opening or time-delay opening), different internal and external constructions, and voltage ratings for both low-voltage (600 volts and below) and medium-voltage (over 600 volts) circuits.

Voltage Rating: Most low voltage power distribution fuses have 250-V or 600-V ratings (other rating are 125 V and 300 V). The voltage rating of a fuse must be at least equal to the circuit voltage. It can be higher but never lower. For instance, a 600-V fuse can be used in a 208-V circuit. The voltage rating of a fuse is a function of or depends upon its capability to open a circuit under an overcurrent condition. Specifically, the voltage rating determines the ability of the fuse to suppress the internal arcing that occurs after a fuse link melts and an arc is produced. If a fuse is used with a voltage rating lower than the circuit voltage, arc suppression will be impaired and, under some fault current conditions, the

NEC Rules for Overcurrent Protection

Application	Rule	NEC Section
Protection Required	Each ungrounded service-entrance conductor must have overcurrent protection. Device must be in series with each ungrounded conductor.	Section 230-90(a)
Number of Devices	Up to six circuit breakers or sets of fuses may be considered as the overcurrent device.	Section 230-90(a)
Location in Building	The overcurrent device must be part of the service disconnecting means or be located immediately adjacent to it.	Section 230-91
Accessibility	In a property comprising more than one building under single management, the ungrounded conductors supplying each building served shall be protected by overcurrent devices, which may be located in the building served or in another building on the same property, provided they are accessible to the occupants of the building served. In a multiple-occupancy building each occupant shall have access to the overcurrent protective devices.	Section 230-91
Location in Circuit	The overcurrent device must protect all circuits and devices, except equipment which may be connected on the supply side including: 1) Service switch, 2) Special equipment, such as lightning arresters, 3) Circuits for emergency supply and load managaement (where separately protected.), 4) Circuits for fire alarms or fire pump equipment (where separately protected.), 5) Meters, with all metal housings grounded, (600 volts or less.), 6) Control circuits for automatic service equipment if suitable overcurrent protection and disconnecting means are provided.	Section 230-94

Fig. 4-1: Summary of NE Code rules for overcurrent protective devices.

fuse may not safely clear the overcurrent. This could cause equipment damage and perhaps even be dangerous to life and property.

Ampere Rating: Every fuse has a specific ampere rating. In selecting the ampacity of a fuse, consideration must be given to the type of load and code requirements. The ampere rating of a fuse should normally not exceed current carrying capacity of the circuit. For instance, if a conductor is rated to carry 20 A, a 20-A fuse is the largest that should be used in the conductor circuit. However, there are some specific circumstances when the ampere rating is permitted to be greater than the current carrying capacity of the circuit. A typical example is the motor circuit; dual-element fuses generally are permitted to be sized up to 175% and non-time-delay fuses up to 300% of the motor full-load amperes. Generally, the ampere rating of a fuse and switch combination should be selected at 125% of the load current. There are exceptions, such as when the fuse-switch combination is approved for continuous operation at 100% of its rating.

A protective device must be able to withstand the destructive energy of short-circuited currents. If a fault current exceeds a level beyond the capability of the protective device, the device may actually rupture and cause severe damage. Thus, it is important in applying a fuse or circuit breaker to use one which can sustain the largest potential short-circuit currents. The rating that defines the capacity of a protective device to maintain its integrity when reacting to fault currents is termed its interrupting rating. The interrupting rating of most branch-circuit, molded case, circuit breakers typically used in residential service entrance boxes is 10,000 A. The rating is usually expressed as "10,000 amperes interrupting capacity (AIC)." Larger, more expensive circuit breakers may have AIC's of 14,000 A or higher. In contrast, most modern, current-limiting fuses have an interrupting capacity of 200,000 A and are commonly used to protect the lower rated circuit breakers. The 1993 NE Code, Section 110-9, requires equipment intended to break current at fault levels to have an interrupting rating sufficient for the current that must be interrupted.

Time Delay: The time-delay rating of a fuse is established by standard UL tests. All fuses have an inverse time current characteristic. That is, the fuse will open quickly on high currents and after a period of time delay, on low overcurrents. Specific types of fuses are made to have specially determined amounts of time delay. The basic UL requirement on time delay for Class RK-1, RK-15, and J fuses which are marked "time delay" is that the fuse must carry a current equal to five times its continuous rating for a period not less than ten seconds. UL has not developed time-delay tests for all fuse classes. Fuses are available for use where time delay is needed along with current limitation on high-level short circuits. In all cases, manufacturers' literature should be consulted to determine the degree of time delay in relation to the operating characteristics of the circuit being protected.

Types of Fuses

Plug Fuses: Plug fuses have a screw-shell base and are commonly used in dwellings for circuits that supply lighting, heating, and appliances. Plug fuses are supplied with standard screw bases (Edison-base) or Type S bases. An Edison-base fuse consists of a strip of fusible (capable of being melted) metal in a small porcelain or glass case, with the fuse strip, or link, visible through a "window" in the top of the fuse. The screw base corresponds to the base of a standard medium-base incandescent lamp. Edison-base fuses are permitted only as replacements in existing installations; all new work must use the S-base fuses (NE Code Section 240-51(b).

- "Plug fuses of the Edison-base type shall be used only for replacements in existing installations where there is no evidence of overfusing or tampering."

The chief disadvantage of the Edison-base plug fuse is that it is made in several ratings from 0 to 30A, all with the same size base—permitting unsafe replacement of one rating by a higher rating.Type S fuses were developed to reduce the possibility of over-fusing a circuit (inserting a fuse with a rating greater than that required by the circuit). There are 15 classifications of Type S fuses based on current rating: 0-30 amperes. Each Type S fuse has a base of a different size and a matching adapter. Once an adapter is screwed into a standard Edison-base fuseholder, it locks into place and is not readily removed without destroying the fuseholder. As a result, only a Type S fuse with a size the same as that of the adapter may be inserted. Two types of plug fuses are shown in Fig. 4-2; a Type S adapter is also shown.

Plug fuses also are made in time-delay types that permit a longer period of overload flow before operation, such as on motor inrush current and other higher-than-normal rated currents. They are available in ratings up to 30 amperes, both in Edison-base and Type S. Their principle use is in motor circuits, where the starting inrush current to the motor is much higher than the running, or continuous, current. The time-

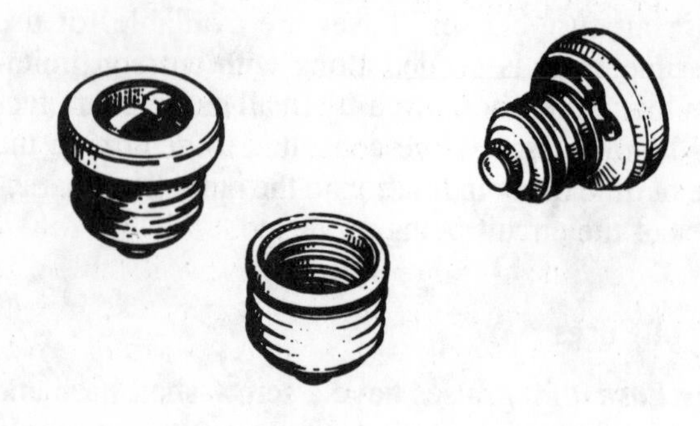

Fig. 4-2: Standard Edison base fuse on left; Type S with adapter shown on right.

delay fuse will not open on the inrush of high-starting current. If, however, the high current persists, the fuse will open the circuit just as if a short circuit or heavy overload current had developed. All Type S fuses are time-delay fuses.

Plug fuses are permitted to be used in circuits of no more than 125 volts between phases, but they may be used where the voltage between any ungrounded conductor and ground is not more than 150 volts. The screwshell of the fuseholder for plug fuses must be connected to the load side circuit conductor; the base contact is connected to the line side or conductor supply. A disconnecting means (switch) is not required on the supply side of a plug fuse.

The plug fuse is a *nonrenewable* fuse; that is, once it has opened the circuit because of a fault or overload, it cannot be used again or renewed. It must be replaced with a new fuse of the same rating and characteristics for safe and effective restoration of circuit operation.

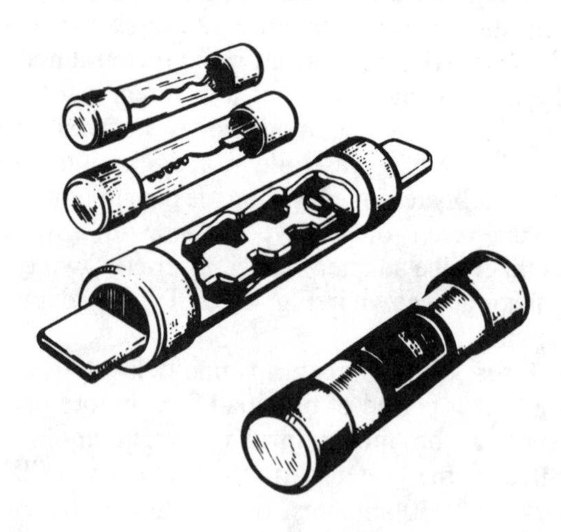

Fig. 4-3: Several types of cartridge fuses.

Cartridge Fuses

In most industrial and commercial applications, cartridge fuses are used because they have a wider range of types, sizes, and ratings than do plug fuses. Many cartridge fuses are also provided with a means to renew the fuse by unscrewing the end caps and replacing the links. In large industrial installations where thousands of cartridge fuses are in use to protect motors and machinery, this replacement feature can be a significant savings over a period of a year. Various types of cartridge fuses are shown in Fig. 4-3.

Single-element cartridge fuses: The basic component of a fuse is the link. Depending upon the ampere rating of the fuse, the single-element fuse may have one or more links. They are electrically connected to the end blades (or ferrules) and enclosed in a tube or cartridge surrounded by an arc-quenching filler material.

Under normal operation, when the fuse is operating at or near its ampere rating, it simply functions as a conductor. However, if an overload current occurs and persists for more than a short interval of time, the temperature of the link eventually reaches a level that causes a restricted segment of the link to melt; as a result, a gap is formed and an electric arc established. however, as the arc causes the link metal to burn back, the gap becomes progressively larger. Electrical resistance of the arc eventually reaches such a high level that the arc cannot be sustained and is extinguished; the fuse will have then completely cut off all current flow in the circuit. Suppression or quenching of the arc is accelerated by the filler material.

Single-element fuses have a very high speed of response to overcurrents. They provide excellent short-circuit component protection. However, temporary harmless overloads or surge currents may cause nuisance openings unless these fuses are oversized. They are best used, therefore, in circuits not subject to heavy transient surge currents and the temporary overload of circuits with inductive loads such as motors, transformers, and solenoids. Because single-element fuses have a high speed-of-response to short-circuit currents, they are particularly suited for the protection of circuit breakers with low interrupting ratings.

Dual-element cartridge fuses: Unlike single-element fuses, the dual-element fuse can be applied in circuits subject to temporary motor overload and surge currents to provide both high performance short-circuit and overload protection. Oversizing in order to prevent nuisance openings is not necessary. The dual-element fuse contains two distinctly separate types of elements. Electrically, the two elements are series connected. The fuse links similar to those used in the single-element fuse perform the sort-circuit protection function; the overload element provides protection against low-level overcurrents or overloads and will hold an overload that is five times greater than the ampere rating of the fuse for a minimum time of ten seconds.

The overload section consists of a copper heat absorber and a spring-operated trigger assembly. The heat-absorber strip is permanently connected to the short-circuit link and to the short-circuit link on the opposite end of the fuse by the S-shaped connector of the trigger assembly. The connector electronically joins the one short-circuit link to the heat absorber in the overload section of the fuse. These elements are joined by a "calibrated" fusing alloy. An overload current causes heating of the short-circuit link connected to the trigger assembly. Transfer of heat from the short-circuit link to the heat absorbing strip in the mid-section of the fuse begins to raise the temperature of the heat absorber. If the overload is sustained, the temperature of the heat absorber eventually reaches a level that permits the trigger spring to "fracture" the calibrated fusing alloy and pull the connector free. The short-circuit link is electrically disconnected from the heat absorber, the conducting path through the fuse is opened, and overload current is interrupted. A critical aspect of the fusing alloy is that it retains its original characteristic after repeated temporary overloads without degradation.

Dual-element fuses may also be used in circuits other than motor branch circuits and feeders, such as lighting circuits and those feeding mixed lighting and power loads. The low-resistance construction of the fuses offers cooler operation of the equipment, which permits higher loading of fuses in switch and panel enclosures without heat damage and without nuisance openings from accumulated ambient heat.

Fuse Marking

It is a requirement of the NE Code that cartridge fuses used for branch-circuit or feeder protection must be plainly marked, either by printing on the fuse barrel or by a label attached to the barrel, showing the following:

- Ampere rating
- Voltage rating
- Interrupting rating (if other than 10,000A)
- "Current limiting," where applicable
- The name or trademark of the manufacturer

Underwriters' Laboratories' Fuse Classes

Fuses are tested and listed by Underwriters' Laboratories Inc. in accordance with established standards of construction and performance. There are many varieties of miscellaneous fuses used for special purposes or for supplementary protection of individual types of electrical equipment. However, here we will chiefly be concerned with those fuses used for protection of branch circuits and feeders on systems operating at 600 volts or below. Cartridge fuses in this category include UL Class H, K, G, J, R, T, CC, and L.

Overcurrents

An overcurrent is either an overload current or a short-circuited current. The overload current is an excessive current relative to normal operating current, but one which is confined to the normal conductive paths provided by the conductors and other components and loads of the electrical system.

A short circuit (Fig. 4-4) is probably the most common cause of electrical problems. It is an undesired current path that allows the electrical current to bypass the load on the circuit. Sometimes the short is between two wires due to faulty insulation, or it can occur between a wire and a grounded object, such as the metal frame of a motor.

Overloads

Overloads are most often between one and six times the normal current level. Usually, they are caused by harmless temporary surge currents that occur when motors are started up or transformers are energized. Since they are of brief duration, any temperature rise is trivial and has no harmful effect on the circuit compo-

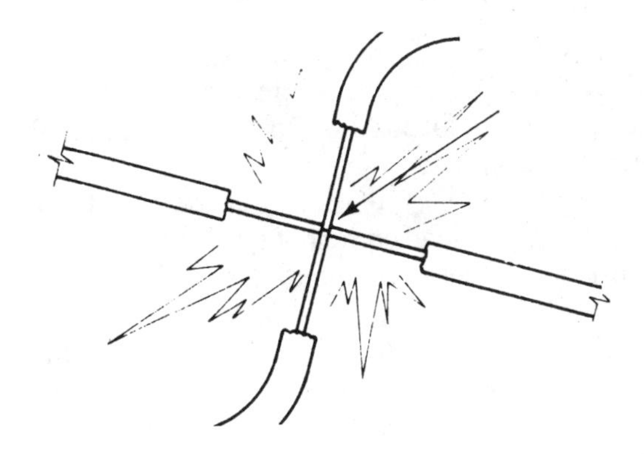

Fig. 4-4: A short circuit is probably the most common cause of electrical problems.

nents. Therefore, it is important that protective devices (fuses and circuit breakers) do not react to them.

Continuous overloads can result from defective motors (such as worn motor bearings), overloaded equipment, or too many loads on one circuit. Such sustained overloads are destructive and must be cut off by protective devices before they damage the electrical distribution system or affect the system loads. However, since they are of relatively low magnitude compared to short-circuit currents, removal of the overload current within a few seconds will generally prevent equipment damage. A sustained overload current results in overheating of conductors and other components and will cause deterioration of insulation, which may eventually result in severe damage and short circuits if not interrupted.

Guide for Sizing Fuses

General quidelines for sizing fuses are given here for most circuits that will be encountered on conventional systems. Some specific applications may warrant other fuse sizing; in these cases, the load characteristics and appropriate NE Code sections should be considered. The selections shown here are not, in all cases, the maximum or minimum ampere ratings permitted by the NE Code. Demand factors as permitted per the NE Code are not included here.

Dual-Element Time-Delay Fuses

Main Service: Each ungrounded service entrance conductor shall have a fuse in series with a rating not higher than the ampacity of the conductor, except as permitted in Art. 230-90 of the NE Code. The service fuses shall be part of the service disconnecting means or be located immediately adjacent thereto (Art. 230-91).

Feeder Circuit With No Motor Load: The fuse size must be at least 125% of the continuous load plus 100% of the non-continuous load. Do not size larger than ampacity of conductor.

Feeder Circuit With All Motor Loads: Size the fuse at 150% of the full load current of the largest motor plus the full-load current of all motors.

Feeder Circuit With Mixed Loads: Size fuse at sum of 150% of the full-load current of the largest motor plus 100% of the full-load current of all other motors plus 125% of the continuous, non-motor load plus 100% of the non-continuous, non-motor load.

Branch Circuit With No Motor Load: The fuse size must be at least 125% of the continuous load plus 100% of the non-continuous load.

Motor Branch Circuit With Overload Relays: Where overload relays are sized for motor running overload protection, the following provide backup, ground fault, and short-circuit protection: Motor 1.15 service factor or 40 degrees C. rise; size fuse at 125% of motor full-load current or next higher standard size.

Motor less than 1.15 service factor or over 40 degrees C. rise; size the fuse at 115% of the motor full-load current or the next higher standard fuse size.

Motor Branch Circuit With Fuse Protection Only: Where the fuse is the only motor protection, the following fuses provide motor running overload protection and short-circuit protection. Motor 1.15 service factor or 40 degrees C. rise; size the fuse at 100% to 125% of the motor full load current.

Motor less than 1.15 service factor or over 40 degrees C. rise: size fuse at 100% to 115% of motor full load current.

Large Motor Branch Circuit: Fuse larger than 600 amps. For large motors, size KRP-C HI-CAP time-delay fuse at 150% to 225% of the motor full load current, depending on the starting method; that is, part-winding starting, reduced voltage starting, etc.

Non-Time-Delay Fuses

Main Service: Service-entrance conductors shall have a short-circuit protective device in each ungrounded conductor, on the load side of, or as an integral part of, the service-entrance switch. The protective device shall be capable of detecting and interrupting all values of current in excess of its trip setting or melting point, which can occur at its location. A fuse rated in continuous amperes not to exceed three times the ampacity of the conductor, or a circuit breaker with a trip setting of not more than six times the ampacity of the conductors shall be considered as providing the required short-circuit protection.

Feeder Circuit With No Motor Loads: The fuse size must be at least 125% of the continuous load plus 100% of the non-continuous load. Do not size larger than the ampacity of the wire.

Feeder Circuit With All Motor Loads: Size the fuse at 300% of the full-load current of the largest motor plus the full-load current of all other motors.

Feeder Circuit With Mixed Loads: Size fuse at sum of 300% of full load current of largest motor plus 100% of full-load current of all other motors plus 125% of the continuous, non-motor load plus 100% of non-continuous, non-motor load.

Branch Circuit With No Motor Load: The fuse size must be at least 125% of the continuous load plus 100% of the non-continuous load.

Motor Branch Circuit With Overload Relays. Size the fuse as close to but not exceeding 300% of the motor running full-load current. Provides ground fault and short-circuit protection only.

Motor Branch Circuit With Fuse Protection Only. Non-time-delay fuses cannot be sized close enough to provide motor running overload protection. If sized for motor overload protection, non-time-delay fuses would open due to motor starting current.

When sizing fuses for a given application, a schematic drawing of the system will help tremendously. Such drawings do not have to be detailed, just a single-line schematic, such as the one shown in Fig. 4-5 will suffice.

Coordination

"Coordination" is the name given to the time-current relationship among a number of overcurrent devices connected in series, such as fuses in a main feeder, subfeeder, and branch circuits. Safety is the prime consideration in the operation of fuses; however, coordination of the characteristics of fuses has become a very important factor in the large and complex electrical system of the present time. Every fuse should be properly rated for continuous current and overloads and for the maximum short-circuit current the electrical system could feed into a fault on the load side of the fuse—but this is not enough. It might still be possible for a fault on a feeder to open the main service fuse before the feeder fuse opens. Or, a branch-circuit fault might open the feeder fuse before the branch-circuit device opens. Such applications are said to be uncoordinated, or nonselective—the fuse closest to the fault is not faster operating than one farther from the fault. When a fault on a feeder opens the main service fuse instead of the feeder fuse, all of the electrical system is taken out of service instead of just the one faulted feeder. Effective coordination minimizes the extent of electrical outage when a fault occurs. It therefore minimizes loss of production, interruption of critical continuous processes, loss of vital facilities, and possible panic.

Selective coordination is the selection of overcurrent devices with time/current characteristics that assure clearing of a fault or short circuit by the device nearest the fault on the line side of the fault. A fault on a branch circuit is cleared by the branch-circuit device. The sub-feeder, feeder, and main service overcurrent devices will not operate. Or, a fault on a feeder is opened by the feeder fuse without opening any other fuse on the supply side of the feeder. With selective coordination, only the faulted part of the system is taken out of service, which represents the condition of minimum outage.

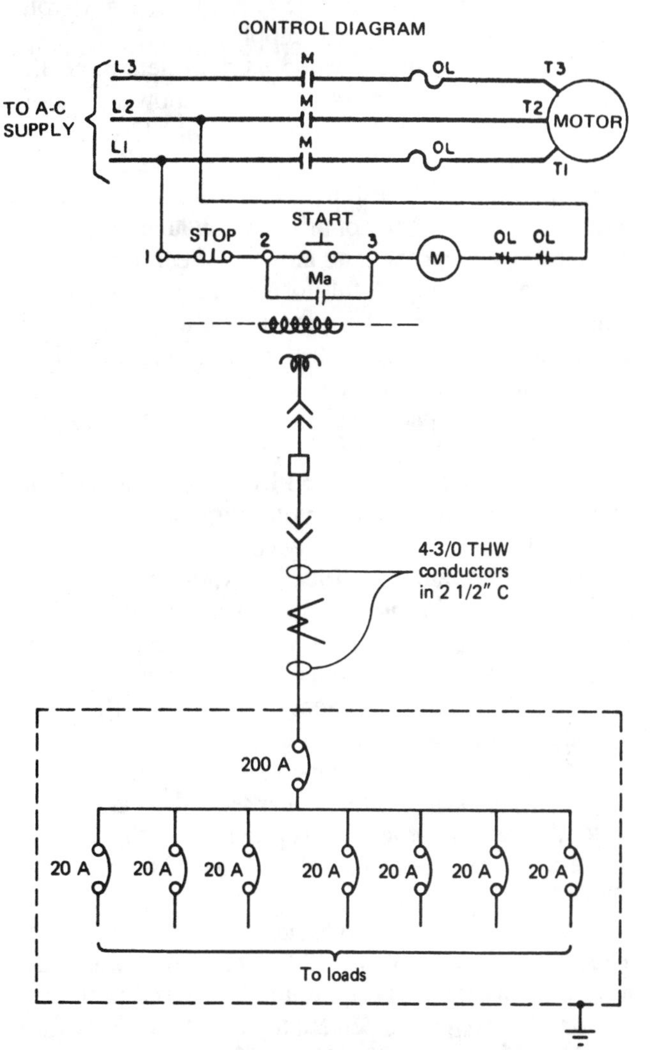

Fig. 4-5: Typical single-line schematic diagram.

With proper selective coordination, every device is rated for the maximum fault current it might be called upon to open. Coordination is achieved by studying the curve of current vs. time required for operation of each device. Selection is then made so that the device nearest any load is faster operating than all devices closer to the supply, and each device going back to the service entrance is faster operating than all devices closer to the supply. The main service fuses must have the longest opening time for any branch or feeder fault.

Let's determine the conductor size, the motor overload protection, the branch-circuit short circuit and ground-fault protection, and the feeder protection for one 25-horsepower squirrel-cage induction motor (full voltage starting, nameplate current 31.6 amperes, service factor 1.15, Code letter F), and two 30-

horsepower wound-rotor induction motors (nameplate primary current 36.4 amperes, nameplate secondary current 65 amperes 40 degree C rise), on a 460-volt, 3-phase, 60-Hertz electrical supply.

The full-load current value used to determine the ampacity of conductors for the 25-horsepower motor is 34 amperes [Section 430-6(a) and Table 430-150]. A full-load current of 34 amperes x 1.25 = 42.5 amperes (Section 430-22). The full-load current value used to determine the ampacity of primary conductors for each 30-horsepower motor is 40 amperes [Section 430-6(a) and Table 430-150]. A full-load primary current of 40 amperes x 1.25 = 50 amperes (Section 430-22). A full-load secondary current of 65 amperes x 1.25 = 81.25 amperes [Section 430-23(a)].

The feeder ampacity will be 125 percent of 40 plus 40 plus 34, or 124 amperes (Section 430.24).

Overload: Where protected by a separate overload device, the 25-horsepower motor with nameplate current of 31.6 amperes, must have overload protection of not over 39.5 amperes [Section 430-6(a) and 430-32(a)(1)]. Where protected by a separate overload device, the 30-horsepower motor, with nameplate current of 36.4 amperes, must have overload protection of not over 45.5 amperes [Section 430-6(a)(1)]. If the overload protection is not sufficient to start the motor or to carry the load, it may be increased according to Section 430-34. For a motor marked "thermally protected," overload protection is provided by the thermal protector [see Sections 430-7(a)(12) and 430-32(a)(2)].

Branch-Circuit Short-Circuit and Ground Fault: The branch circuit of the 25-horsepower motor must have branch-circuit short circuit and ground-fault protection of not over 300 percent for a nontime-delay fuse (Table 430-152) or 3.00 x 34 = 102 amperes. The next smallest standard size fuse is 100 amperes. The fuse size may be increased to 110 amperes (Section 430-52, Exception No. 1 and Section 240-6). Where the maximum value of branch-circuit short circuit and ground-fault protection is not sufficient to start the motor, the value for a nontime-delay fuse may be increased to 400 percent [See Section 430-52, Exception No. 2 (a)]. If time-delay fuse is to be used, see Section 430-52, Exception No.2(b).

Feeder Circuit: The maximum rating of the feeder short-circuit and ground-fault protection device is based on the sum of the largest branch-circuit protective device (110-ampere fuse) plus the sum of the full-load currents of the other motors or 110 plus 40 plus 40 = 190 amperes. The nearest standard fuse which does not exceed this value is 175 amperes [Section 430-62(a)].

NE Code Articles pertaining to fuses and overcurrent protection include:

Article 240	Section 430-C
Section 430-53	Section 430-D
Section 450-3	

Circuit Breaker

A circuit breaker resembles an ordinary toggle switch, and it is probably the most widely used means of overcurrent protection today. On an overload, the circuit breaker opens itself or *trips*. In a tripped position, the handle jumps to the middle position (Fig. 4-6). To reset, turn the handle to the OFF position and then turn it as far as it will go beyond this position; finally, turn it to the ON position.

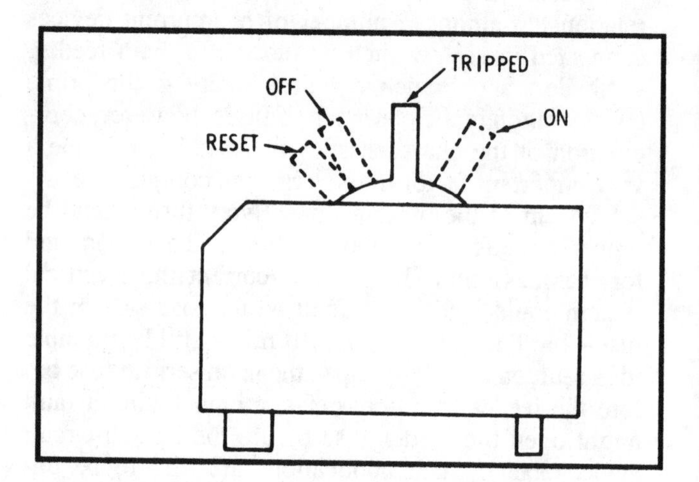

Fig. 4-6: Operating characteristics of a typical circuit breaker.

On a conventional 120/240 volt, single-phase service, one single-pole breaker protects a 120-volt circuit, and one double-pole breaker protects a 240-volt circuit. Three-phase services will require a 3-pole circuit breaker for three-phase circuits. The breakers are rated in amperes, just like fuses, although the particular ratings are not exactly the same as those for fuses. Circuit-breaker enclosures are described in Chapter 3.

NE Code Articles and Sections pertaining to circuit breakers include:

Section 230-70	Section 230-208
Article 240	Section 384-B
Article 380	Section 410-81(b)
Section 430-52	Section 430-58
Section 430-110	

Ground Fault Protection for People

Ground fault protection for people is a subject of interest to all to work with and use electrically-operated equipment—both personally and professionally. A ground fault exists when an unintended path is established between an ungrounded conductor and ground. This situation can occur not only from worn or defective electrical equipment but also from accidental misuse of equipment that is in good working order.

Will a conventional overcurrent device (fuse or circuit breaker) detect a ground fault and open the circuit before irreparable harm is done? Before answering this question, a study of the effects of current on the human body is in order.

*Effects of current on the human body:*Hand-to-hand body resistance of an adult lies between 1,000 and 4,000 ohms, depending on moisture, muscular structure and voltage. The average value is 2,100 ohms at 240V, ac and 2,000 ohms at 120V, ac.

Using Ohm's law, the current resulting from the above average hand-to-hand resistance values is 114 milliamperes (.114 amps) at 240V, ac and 43 milliamperes (.043 amps) at 120V, ac. The effects of 60 Hz alternating current on a normal healthy adult are as follows (note the current is in milliamperes):

- More than 5 ma—generally painful shock
- More than 15 ma—sufficient to cause "freezing" to the circuit for 50% of the population
- More than 30 ma—breathing difficult (possible suffocation)
- 50 to 100 ma—possible ventricular fibrillation or very rapid uncoordinated contractions of the ventricles of the heart resulting in loss of synchronization between heartbeat and pulse beat. Once ventricular fibrillation occurs in man, it usually continues and death will ensue within a few minutes.
- 100 to 200 ma—certain ventricular fibrillation
- Over 200 ma—severe burns; muscle contractions. The heart is more likely to stop than fibrillate.

Now, will a conventional overcurrent device open a circuit before irreparable harm is done? NO! Here's why.

The current that would flow from a defective electric drill, for example, through the metal housing and through the human body to ground would be 43 milliamperes, calculated using 2,800 ohms as average body resistance. Using 1,000 ohms as body resistance, the current flow would be 120 milliamperes. The 43 milliamperes is only 0.29% of the current required to open a 15 ampere circuit breaker or fuse, and yet it approaches the current level which may produce ventricular fibrillation. Obviously, the standard circuit breaker or fuse will not open the circuit under such low levels of current flow.

Ground Fault Circuit Interrupters (GFCI): "People protector" devices are built as Class A devices in accordance with UL Standard No. 943 for Ground Fault Circuit Interrupters. See Section 426-30 and Section 426-32 of the 1993 NEC. UL defines a Class A device as one that "will trip when a fault current to ground is 6 milliamperes or more." The tripping time of such units cannot exceed the value obtained by the equation,

$$T = 20/(I)1.43$$

where T is time in seconds and I is the ground fault current in milliamperes. Also, Class A devices must not trip below 4 milliamperes.

Class A GFCI's provide a self contained means of testing the ground fault circuitry, as required by UL. To test, simply push the test button and the device will respond with a trip indication. UL required that the current generated by the test circuit shall not exceed 9 milliamperes. Also, UL required the device to be functional at 85% of the rated voltage.

Knowing these facts, OSHA and other codes require the use of ground fault circuit interrupters on certain circuits; namely, all construction sites where temporary service is in use for electric power tools. The laundry and bath areas of residential occupancies also require the use of GFCI's for their duplex receptacles, as well as outside receptacles.

Troubleshooting

Good maintenance of circuit breakers and their enclosures is very important and necessary to obtain the best service and performance. In order to keep circuit breakers in proper operating condition, the scheduling of maintenance inspections is important.

Since every circuit breaker failure represents a potential hazard to other equipment on the system, it is difficult to calculate risks involved in prolonging maintenance inspections. One of the chief causes of circuit breaker failure is high heat caused by loose connections at the load side of the breaker. Another cause of circuit breaker failure is defective loads; that is, electrical equipment that cycles too frequently which in turn causes the circuit breaker to overheat. Loading the circuit breaker to the limit with continuous loads will also cause heating. Therefore, in most circuit-breaker failures, heat will be found to be the biggest culprit.

Safety

All users of switching equipment have their own safety rules and regulations. Be sure you know what they are and practice them. There are no short cuts to safety. The greatest hazard or potential accident exists during the switching operation, clearing a circuit for maintenance or repair. Always be particularly careful at this time as well as during the maintenance inspection.

Before starting work on any breaker be sure to check as follows:

- Breaker has been tripped open.
- Any disconnects or safety switches are open.
- Breaker is checked with voltage tester.
- All control circuits are open.
- Make sure all bushings are grounded.
- Breaker and circuits are tagged.
- There is no feed back from other devices.
- When leaving a "dead" panel or other electrical apparatus, for any reason (say, for lunch or to get needed materials), always check the system again when you return to make sure it is still "dead."

chapter 5

BRANCH CIRCUITS AND FEEDERS

Article 210 of the NE Code deals with branch circuits while Article 215 covers the requirements for feeders. In general, a feeder conductor must have an ampacity not lower than required to supply the load as computed in Parts B, C, and D of NE Code Article 220. The minimum sizes are as follows:

- The ampacity of feeder conductors supplying specific circuits other than service-entrance conductors must not be less than 30 amperes if the feeder supplies (1) two or more 2-wire branch circuits supplied with a 2-wire feed; (2) more than two 2-wire branch circuits supplied by a 3-wire feeder; (3) two or more 3-wire branch circuits supplied by a 3-wire feeder; or (4) two or more 4-wire branch circuits supplied by a 3-phase, 4-wire feeder.
- When conductors are used in relation to service-entrance conductors their ampacity must not be lower than that of the service-entrance conductors if the feeder conductors carry the total load supplied by service-entrance conductors with an ampacity of 55 amperes or less. See the service calculations in Chapter 2.

The NE Code recognizes several wiring methods for use in building construction. All of them must be used only in areas as specified in the NE Code. Wiring methods fall into the following three categories:

- Cable systems
- Raceway systems
- Other systems

The following table represents a brief summary of each type of wiring method.

Summary of Cable/Conductor Methods

Open wiring on insulators	Article 320
Messenger supported wiring	Article 321
Concealed knob-and-tube wiring	Article 324
Integrated gas space cable	Article 325
Medium voltage cable	Article 326
Flat conductor cable	Article 328
Mineral-insulated cable	Article 330
Armored cable	Article 333
Metal-clad cable	Article 334
Nonmetallic-sheathed cable	Article 336
Shielded nonmetallic cable	Article 337
Service-entrance cable	Article 338
Underground feeder cable	Article 339
Power/control tray cable	Article 340

In residential wiring systems where NM cable may be subject to damage—like on exposed basement walls—electrical metallic tubing (EMT) or other types of raceway systems might be used. However, raceway systems are more often used in commercial and industrial applications. The following table gives a brief summary of raceway systems.

Summary of Raceway Systems

Cable trays	Article 318
Electrical nonmetallic tubing	Article 331
Intermediate metal conduit	Article 345
Rigid metal conduit	Article 346
Rigid nonmetallic conduit	Article 347
Electrical metallic tubing	Article 348
Flexible metallic tubing	Article 349
Liquidtight flexible metal conduit	Article 351
Surface raceway	Article 352
Underfloor raceway	Article 354
Cellular metal floor raceway	Article 356
Cellular concrete floor raceways	Article 358
Wireway	Article 362
Busways	Article 364

To meet certain special installation requirements, the NE Code allows a number of other wiring methods and techniques. These include the following:

Other Wiring Methods

Cable trays	Article 318
Nonmetallic extensions	Article 342
Multioutlet assembly	Article 353
Flat cable assemblies	Article 363
Cablebus	Article 365
Auxiliary gutters	Article 374

Cable Systems

Nonmetallic sheathed cable, underground feeder cable, service-entrance cable, armored cable, and mineral-insulated metal-sheathed cable are discussed in Chapter 1 of this book. Other cables recognized and approved by the NE Code for use in various electrical systems include the following.

Power and control tray cable: Type TC power and control tray cable is a factory assembly of two or more insulated conductors, with or without associated bare or covered grounding conductors under a nonmetallic sheath, approved for installation in cable trays, in raceways, or where supported by a messenger wire. The use of this cable is limited to industrial establishments—mainly for control circuits—where the conditions of maintenance and supervision assure that only qualified persons will service the installation. See Fig. 5-1.

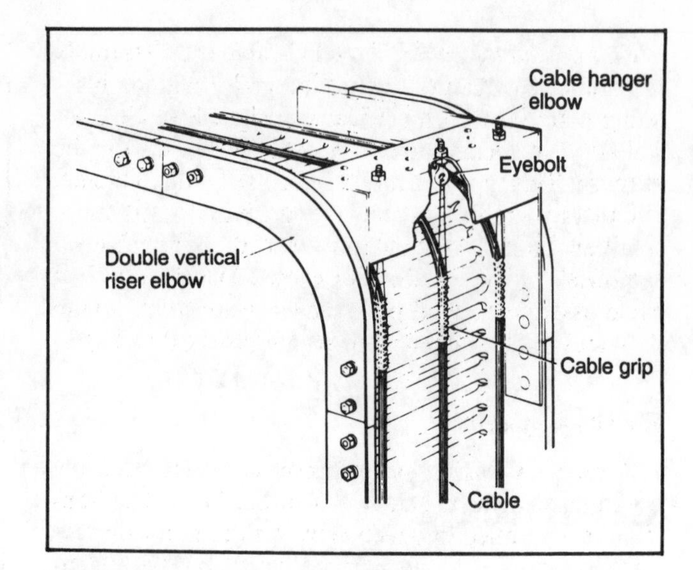

Fig. 5-1: Typical cable tray installation.

Shielded nonmetallic cable: Type SNM cable is a factory assembly of two or more insulated conductors in an extruded core or moisture-resistant, flame-resistant nonmetallic material, covered with an overlapping spiral metal tape and wire shield and jacketed with an extruded moisture-, flame-, oil-, corrosion-, fungus-, and sunlight-resistant nonmetallic material. Type SNM cable may be used where operating temperatures do not exceed the rating marked on the cable, in cable trays or in raceways, and in hazardous locations where permitted in Articles 500 through 516 of the NE Code.

Metal-clad cable: Type MC cable is a factory assembly of one or more conductors, each individually insulated and enclosed in a metallic sheath or interlocking tape or a smooth or corrugated tube. This type of cable may be used for services, feeders, and branch circuits; power, lighting, control, and signal circuits; indoors or outdoors; where exposed or concealed; direct buried; in cable tray; in any approved raceway; as open runs of cable; as aerial cable on a messenger; in hazardous locations as permitted in Articles 501, 502, and 503 of the NE Code; in dry locations; and in wet locations under certain conditions as specified in the NE Code. See Fig. 5-2.

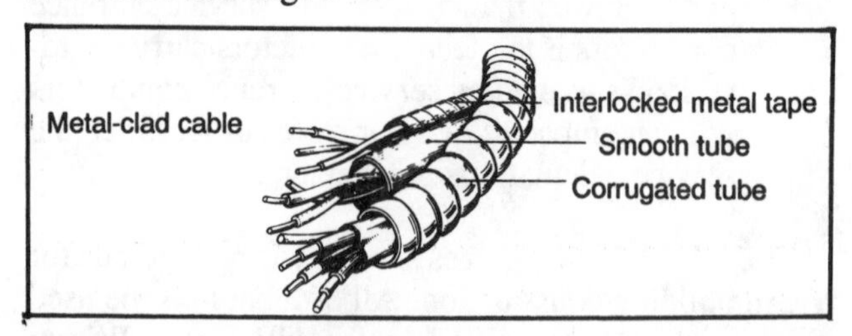

Fig. 5-2: Type MC cable.

Flat conductor cable: Type FC cable is an assembly of parallel conductors formed integrally with an insulating material web specifically designed for field installation in metal surface raceway approved for the purpose. This type of cable may be used only as branch circuits to supply suitable tap devices for lighting, small appliances, or small power loads. Flat cable assemblies shall be installed for exposed work only. Flat cable assemblies shall be installed in locations where they will not be subjected to severe physical damage.

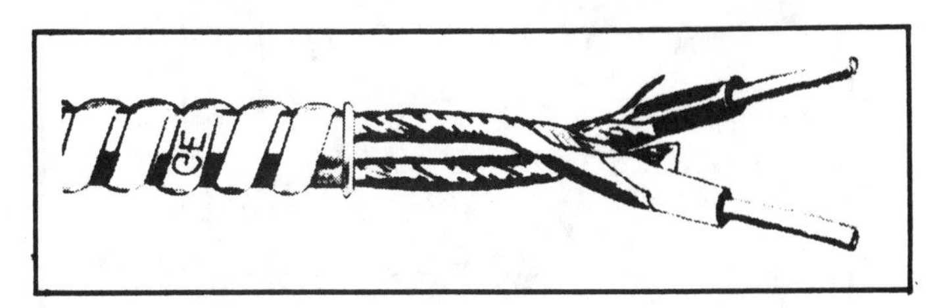

Fig. 5-4: Type NM cable, sometimes called "Romex."

Raceway Systems

A raceway wiring system consists of an electrical wiring system in which two or more individual conductors are pulled into a conduit or similar housing for the conductors after the raceway system has been completely installed. The basic raceways include rigid steel conduit, electrical metallic tubing (EMT), and PVC (polyvinyl chloride) plastic. Other raceways include surface metal moldings and flexible metallic conduit. Examples of various wiring systems are shown in Figs. 5-3 through 5-15.

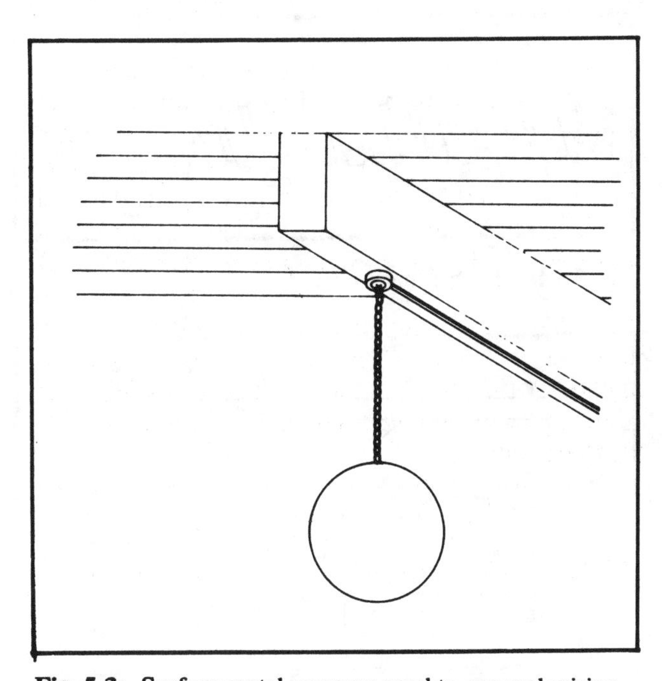

Fig. 5-3: Surface metal raceway used to conceal wiring on solid wood deck.

Fig. 5-5: Type AC (armored cable), frequently called "BX."

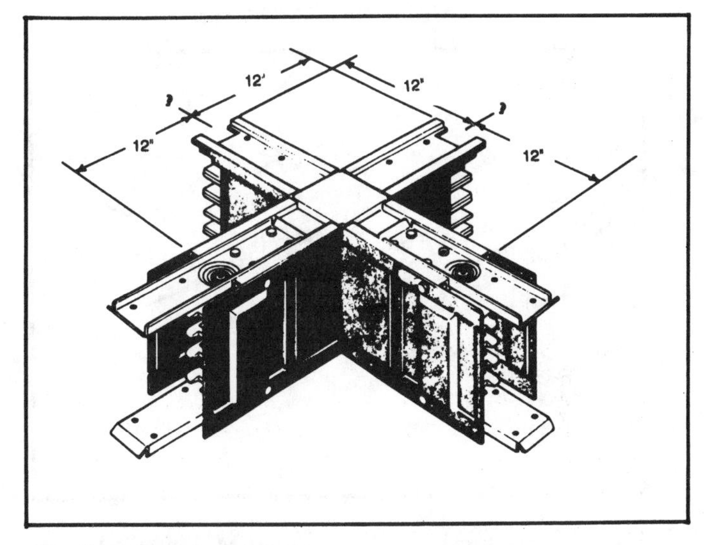

Fig. 5-6: Plug-in busduct.

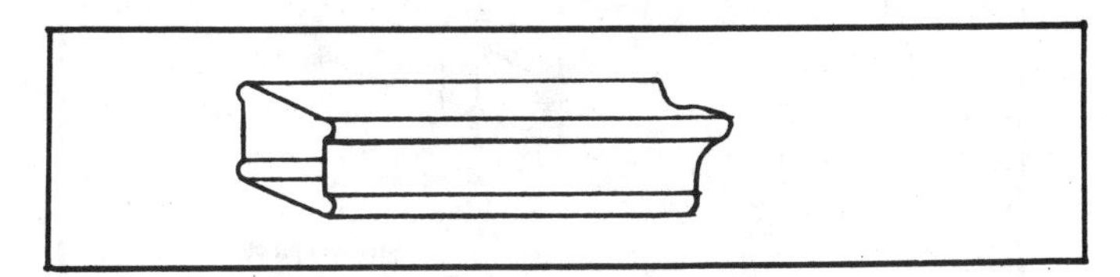

Fig. 5-7: Segment of Wiremold (surface metal raceway).

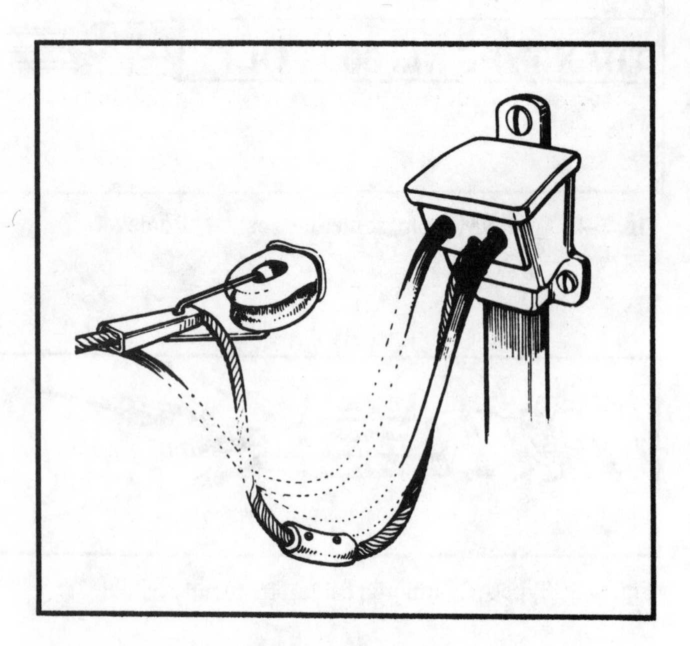

Fig. 5-8: Type SE cable used for feeder from panelboard to aerial messenger cable for feeding outbuildings.

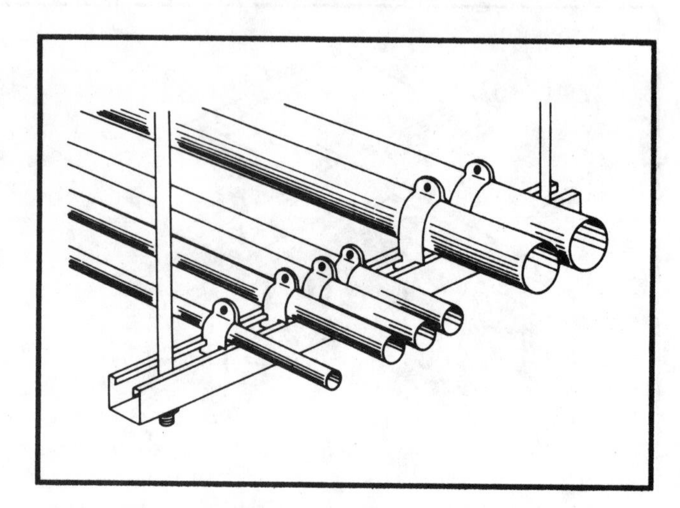

Fig. 5-9: Electrical metallic conduit (EMT) installed on trapeze hangers.

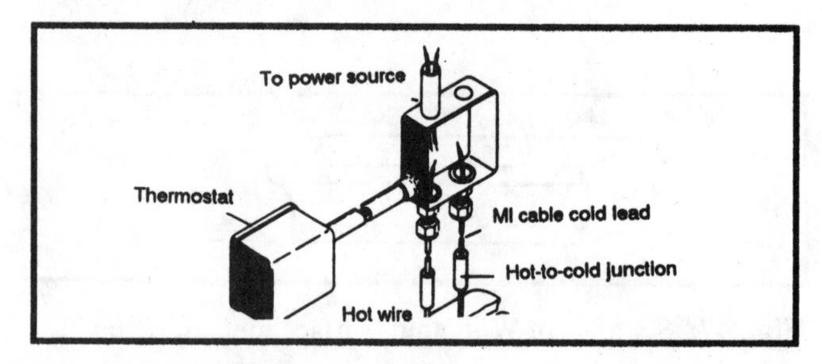

Fig. 5-10: Type MI cable used for cable heating system.

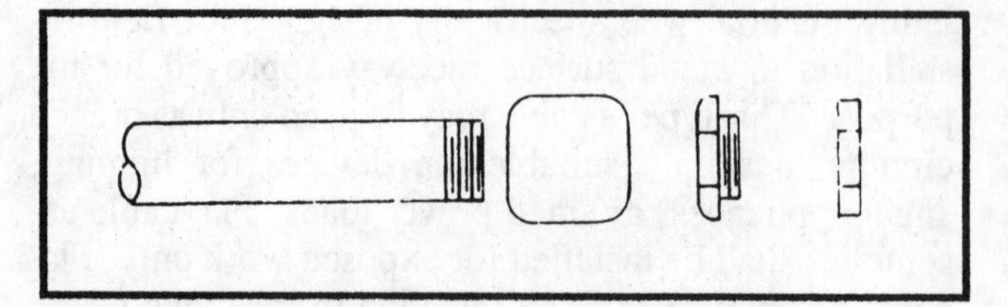

Fig. 5-11: Rigid metal conduit with fittings.

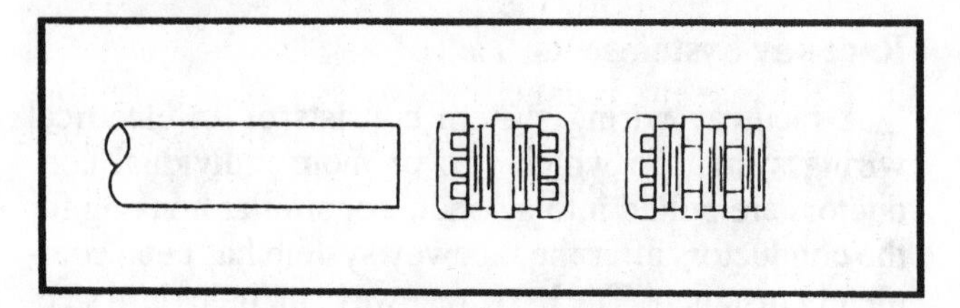

Fig. 5-12: Electrical metallic tubings with fittings.

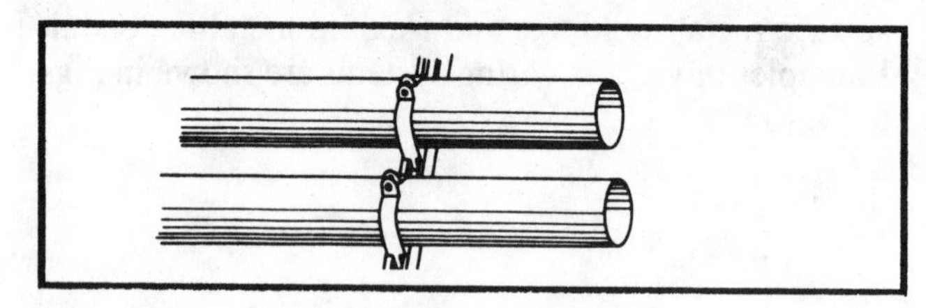

Fig. 5-13: Rigid PVC (Plastic) conduit.

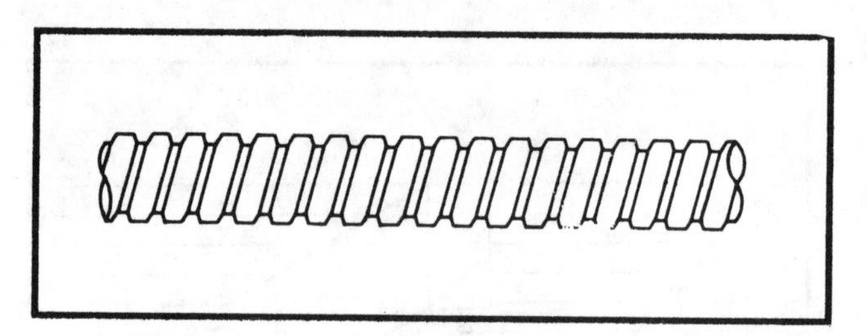

Fig. 5-14: Flexible metal conduit.

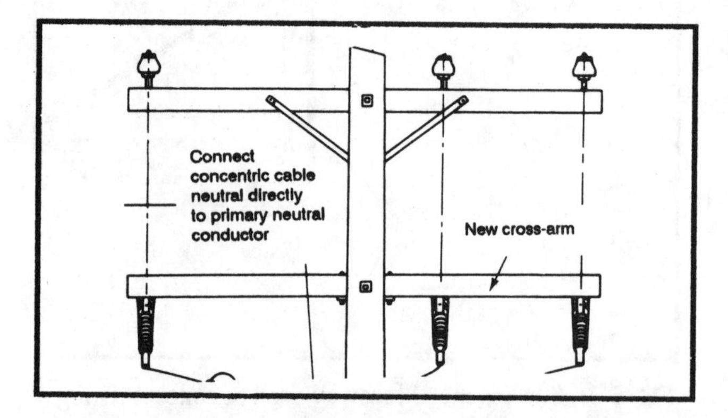

Fig. 5-15: Overhead "knob-and-tube" wiring; that is, separate conductors on insulators.

Wiring in Hazardous Locations

Articles 500 through 504 of the NE Code cover the requirements of electrical equipment and wiring for all voltages in locations where fire or explosion hazards may exist. Locations are classified depending on the properties of the flammable vapors, liquids, gases or combustible dusts or fibers that may be present, as well as the likelihood that a flammable or combustible concentration or quality is present.

Any area in which the atmosphere or a material in the area is such that the arcing of operating electrical contacts, components, and equipment may cause an explosion or fire is considered as a hazardous location. In all such cases, explosion-proof equipment, raceways, and fittings are used to provide an explosion-proof wiring system.

Hazardous locations have been classified in the NE Code into certain class locations. Various atmospheric groups have been established on the basis of the explosive character of the atmosphere for the testing and approval of equipment for use in the various groups.

Class I Locations: Those locations in which flammable gases or vapors may be present in the air in quantities sufficient to produce explosive or ignitable mixtures are classified as Class I locations. Examples of such locations are interiors of paint spray booths where volatile, flammable solvents are used, inadequately ventilated pump rooms where flammable gas is pumped, and drying rooms for the evaporation of flammable solvents.

Class II Locations: Class II locations are those that are hazardous because of the presence of combustible dust. Class II, Division I locations are areas where combustible dust, under normal operating conditions, may be present in the air in quantities sufficient to produce explosive or ignitable mixtures; examples re working areas of grain-handling and storage plants and rooms containing grinders or pulverizers. Class II, Division 2 locations are areas where dangerous concentrations of suspended dust are not likely, but where dust accumulations might form.

Class III Locations: These locations are those areas that are hazardous because of the presence of easily ignitable fibers or flyings, but such fibers and flyings are not likely to be in suspension in the air in these locations in quantities sufficient to produce ignitable mixtures. Such locations usually include some parts of rayon, cotton, and textile mills, clothing manufacturing plants, and woodworking plants.

In Class I and Class II locations the hazardous materials are further divided into groups; that is, Groups A, B, C, D in Class I and Groups E, and G in Class II.

Once the class of an area is determined, the conditions under which the hazardous material may be present determines the division. In Class I and Class II, Division 1 locations, the hazardous gas or dust may be present in the air under normal operating conditions in dangerous concentrations. In Division 2 locations the hazardous material is not normally in the air, but it might be released if there is an accident or if there is faulty operation of equipment.

The information in Figures 5-16 through 5-22 gives a summary of the various classes of hazardous locations, as defined by the NE Code.

Explosion-Proof Equipment

The wide assortment of explosion-proof equipment now available makes it possible to provide adequate electrical installations under any of these hazardous conditions. However, the electrician must be thoroughly familiar with all NE Code requirements and know what fittings are available, how to install them properly, and where and when to use the various fittings.

A floor plan for a hazardous area is shown in Fig. 5-23. In general, either rigid metal conduit, intermediate metal conduit (IMC), or type MI cable is required for all hazardous locations, except for special flexible terminations and as otherwise permitted in the NE Code. The conduit must be threaded with a standard conduit cutting die that provides ¾-inch taper per foot. The conduit should be made up wrench tight in order to minimize sparking in the event fault current flows through the raceway system (NE Code Section 500-2). Where it is impractical to make a threaded joint tight, a bonding jumper should be used. All boxes, fittings, and joints shall be threaded for connection to the conduit system and shall be an approved, explosion-proof type (Fig. 5-24). Threaded joints shall be made up with at least five threads fully engaged. Where it becomes necessary to employ flexible connectors at motor or fixture terminals (Fig. 5-25), flexible fittings approved for the particular class location shall be used.

Seal-off fittings (Fig. 5-26) are required in conduit systems to prevent the passage of gases, vapors, or flames from one portion of the electrical installation to another through the conduit. For Class I, Division 1 locations, the NE Code [Section 501-5(1)] states:

"In each conduit run entering an enclosure for switches, circuit breakers, fuses, relays, resistors, or other apparatus which may produce arcs, sparks, or high temperatures, seals shall be installed within 18 inches from such enclosures. Explosion-proof unions, couplings, reducers, elbows, capped elbows and con-

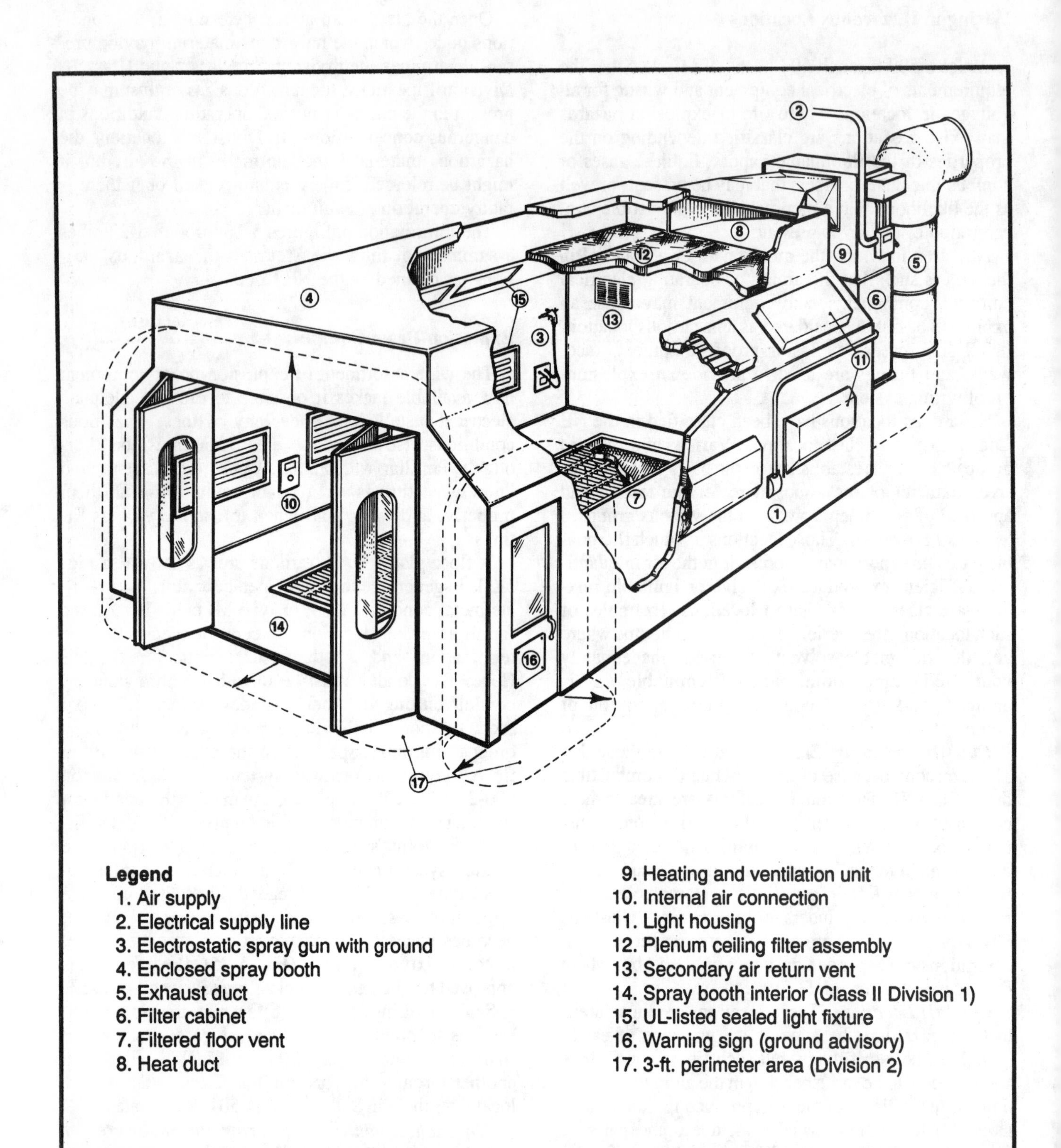

Fig. 5-16: Paint spray booths fall under the classification of hazardous area.

■ INSTALLATION RULES FOR HAZARDOUS LOCATIONS

Class I, Division 1

Components	Characteristics	NE Code Section
Boxes, fitting	Explosion proof	501-4(a)
Seal offs	Approved for purpose	501-5(c)
Wiring methods	Rigid metal conduit, steel intermediate metal conduit, or Type MI cable	501-4(a)
Receptacles	Explosion proof	501-12
Lighting fixtures	Explosion proof	501(9)
Panelboards	Explosion proof	501-6(a)
Circuit breakers	Class I enclosure	501-6(a)
Fuses	Class I	501-6(a)
Switches	Class I enclosure	501-6(a)
Motors	Class I, totally enclosed. or submerged	501-8(a)
Motor controls	Class I, Division 1	501-10(a)
Liquid-filled transformers	Installed in approved vault	501-2(a)
Dry-type transformers	Class I, Division 1 enclosures	501-7(a)
Utilization equipment	Class I, Division 1	501-10(a)
Flexible connections	Class I explosion proof	501-4
Portable lamps	Explosion proof	501-9(a)
Generators	Class I, totally enclosed or submerged	501-8(a)
Alarm systems	Class I	Division 1

Fig. 5-17: Installation rules for Class I, Division 1 locations.

■ INSTALLATION RULES FOR HAZARDOUS LOCATIONS

Class I, Division 2

Components	Characteristics	NE Code Section
Boxes, fittings	Do not have to be explosion proof unless current interrupting contacts are exposed	501-4(b)
Seal offs	Approved for purpose	501-5(c)
Wiring methods	Rigid metal conduit, steel intermediate metal conduit, or Types MI, MC, MV, TC, SMN, PLTC cables, or enclosed gasketed busways or wireways	501-4(b)
Receptacles	Explosion proof	501-12
Lighting fixtures	Protected from physical damage	501-9(b)
Panelboards	General purpose with exceptions	501-6(b)
Circuit breakers	Class I enclosure	501-6(a)
Fuses	Class I	501-6(a)
Switches	Class I enclosure	501-6(b)
Motors	General purpose unless motor has sliding contacts, switching contacts or integral resistance devices; if so, use Class I	501-8(a)
Motor controls	Class I, Division 1	501-7(b)
Liquid-filled transformers	General purpose	501-2(b)
Dry-type transformers	Class I, General Purpose except switching mechanism Division 1 enclosures	501-7(b)
Utilization equipment	Class I, Division 1	501-10(b)
Flexible connections	Class I explosion proof	501-11
Portable lamps	Explosion proof	501-9(a)
Generators	Class I, totally enclosed or submerged	501-8(b)
		501-14(b)

Fig. 5-18: Installation rules for Class I, Division 2 hazardous locations.

■ INSTALLATION RULES FOR HAZARDOUS LOCATIONS

Class II, Division 1

Components	Characteristics	NE Code Section
Boxes, fittings	Class II boxes required when using taps, joints or other connections; otherwise, use dust tight boxes with no openings	502-4(a)
Wiring methods	Rigid metal conduit, steel intermediate metal conduit, or Types MI, MC, MV, TC, SMN, PLTC cables, or enclosed dust-tight busways or wireways	502-4(b)
Receptacles	Class II	502-13(a)
Lighting fixtures	Class II	502-11(a)
Panelboards	Dust-ignition proof	502-6(a)
Circuit breakers	Dust-ignition proof enclosure	502-6(a)
Fuses	Dust-ignition proof enclosure	502-6(a)
Switches	Dust-ignition proof enclosures	502-6(a)
Motors	Class II, Division 1 or totally enclosed	502-8(a)
Motor controls	Dust-ignition proof	502-6(a)
Liquid-filled transformers	Install in vault	502-2(a)
Dry-type transformers	Class II vault	502-2(a)
Utilization equipment	Class II, Division 1	502-10(a)
Flexible connections	Extra-hard usage	502-4(a)
Portable lamps	Class II	502-11(a)
Generators	Class II, Division 1 or totally enclosed	501-8(a)

Fig. 5-19: Installation rules for Class II, Division 1 hazardous locations.

■ INSTALLATION RULES FOR HAZARDOUS LOCATIONS

Class II, Division 2

Components	Characteristics	NE Code Section
Boxes, fittings	Use tight covers to minimize entrance of dust	502-4(b)
Wiring methods	Rigid metal conduit, steel intermediate metal conduit, or Types MI, MC, TC, SMN, PLTC, SI cables, or enclosed dust-tight busways or wireways	502-4(b)
Receptacles	Exposed live parts are not allowed	502-13(b)
Lighting fixtures	Class II	502-11(b)
Panelboards	Dust-tight enclosures	502-6(b)
Circuit breakers	Dust-tight enclosure	502-6(b)
Fuses	Dust-tight enclosure	502-6(b)
Switches	Dust-tight enclosure	502-6(b)
Motors	Class II, Division 1 or totally enclosed	502-8(a)
Motor controls	Dust-tight enclosures	502-6(b)
Liquid-filled transformers	Install in vault	502-2(b)
Dry-type transformers	Class II vault	502-2(b)
Utilization equipment	Class II	502-10(b)
Flexible connections	Extra-hard usage	502-4(b)
Portable lamps	Class II	502-11(b)
Generators	Class II, Division 1 or totally enclosed	501-8(b)

Fig. 5-20: Installation rules for Class II, Division 2 hazardous locations.

■ INSTALLATION RULES FOR HAZARDOUS LOCATIONS

Class III, Divisions 1 and 2

Components	Characteristics	NE Code Section
Boxes, fittings	Use tight covers to minimize entrance of dust	502-4(b)
Wiring methods	Rigid metal conduit, steel intermediate metal conduit, EMT, or Types MI, MC, SMN cable, or enclosed dust-tight busways or wireways	502-4(b)
Receptacles	Exposed live parts are not allowed	502-13(b)
Lighting fixtures	Tight enclosure with no openings	503-9(a)
Panelboards	Dust-tight enclosures	503-4
Circuit breakers	Dust-tight enclosure	503-4
Fuses	Tight metal enclosure with no openings	503-4
Switches	Dust-tight enclosure	503-4
Motors	Totally enclosed	503-6
Motor controls	Dust-tight enclosures	503-4
Liquid-filled transformers	Install in vault	503-2
Dry-type transformers	Class II vault	503-2
Utilization equipment	Class II	503-8
Flexible connections	Extra-hard usage	503-3
Portable lamps	Unswitched, guarded with tight enclosure for lamp	503-9
Generators	Totally enclosed	503-6

Fig. 5-21: Installation rules for Class III, Divisions 1 and 2 hazardous locations.

■ INSTALLATION RULES FOR HAZARDOUS LOCATIONS

Gasoline Dispensing Pumps

Application	NE Code Regulation	NE Code Section
Equipment in hazardous locations	All wiring and components must conform to the rules for Class I locations	514-3
Equipment above hazardous loactions	All wiring must conform to the rules for such equipment in commercial garages	514-4
Gasoline dispenser	A disconnecting means must be provided for each circuit leading to or through a dispensing pump to disconnect all conductors including the grounded neutral An approved seal (seal-off) is required in each conduit entering or leaving a dispenser	514-5 514-6(a)
Grounding	Metal portions of all noncurrent-carrying parts of dispensers	514-7
Underground wiring	Underground wiring must be installed within 2 feet of ground level-in rigid metal or steel intermediate conduit If underground wiring is buried 2 feet or more, rigid nonmetalic conduit may be used along with the types mentioned above, Type MI cable may also be used in some cases	514-8

Fig. 5-22: Installation rules for service stations.

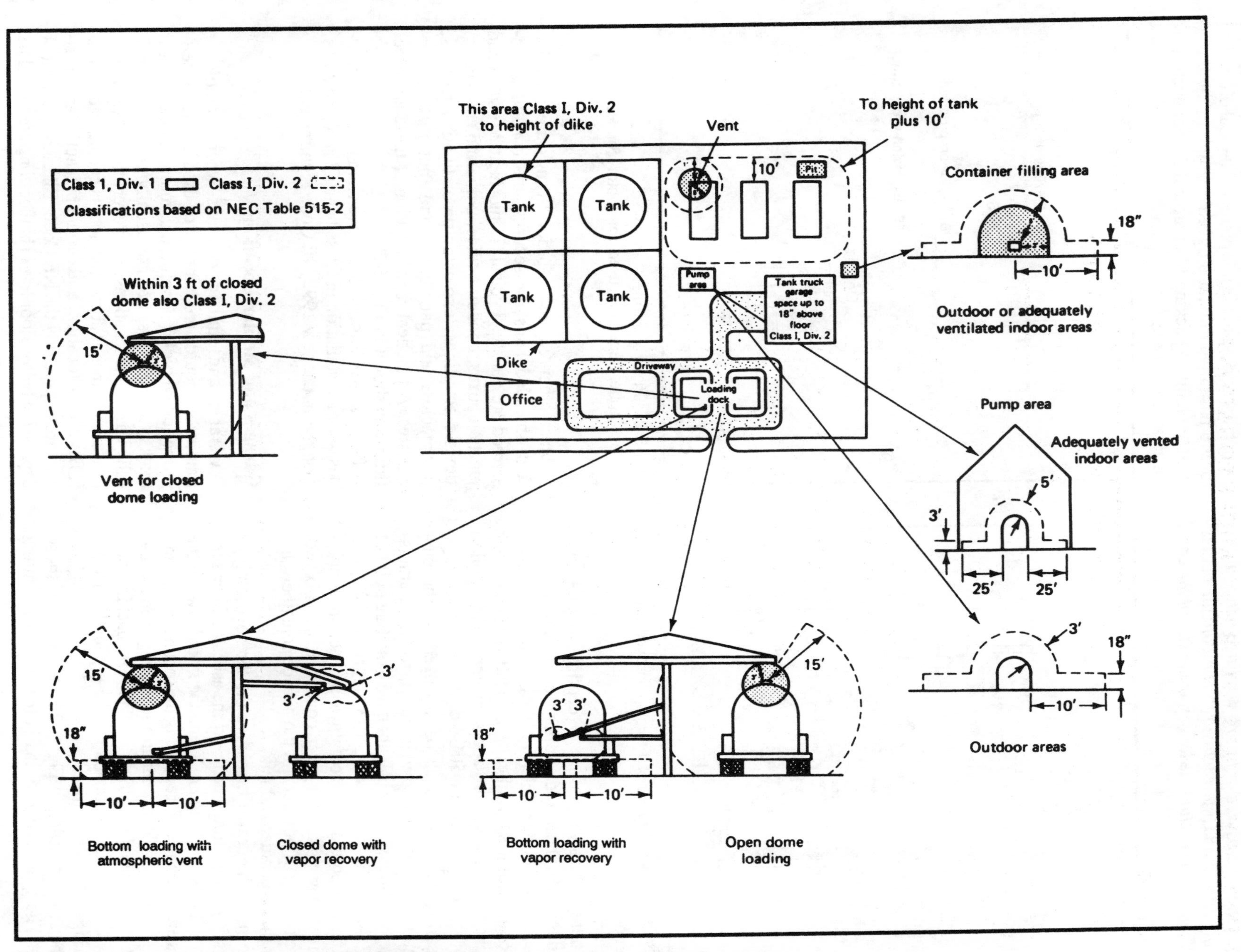

Fig. 5-23: Floor plan for a hazardous area showing the various Divisions as stipulated in the NE Code.

duit bodies similar to "L," "T," and "cross" types shall be the only enclosures or fitting permitted between the sealing fitting and the enclosure. The conduit bodies shall not be larger than the largest trade size of the conduits."

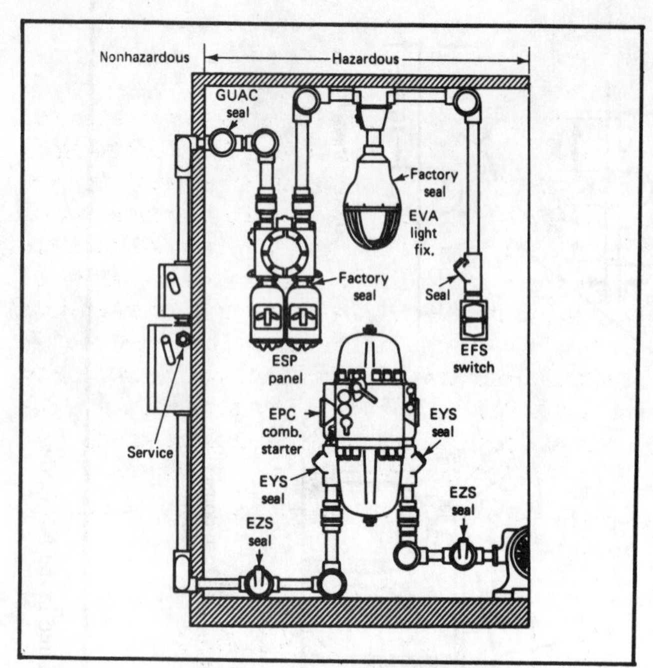

Fig. 5-24: Several types of fittings approved for hazardous areas.

There is, however, one exception to this rule: Conduits 1½ inches and smaller are not required to be sealed if the current-interrupting contacts are either enclosed within a chamber hermetically sealed against the entrance of gases or vapors, or immersed in oil in accordance with Section 501-6 of the NE Code.

Sealing compound shall be approved for the purpose; it shall not be affected by the surrounding atmosphere or liquids; and it shall not have a melting point of less than 200 degrees F. (93 degrees C.) Most sealing-compound kits contain a powder in a polyethylene bag within an outer container. To mix, remove the bag of powder, fill the outside container, and pour in the powder and mix.

To pack the seal off, remove the threaded plug or plugs from the fitting and insert the asbestos fiber supplied with the packing kit. Tamp the fiber between the wires and the hub before pouring the sealing compound into the fitting. Then pour in the sealing cement and reset the threaded plug tightly. The fiber packing prevents the sealing compound (in the liquid state) from entering the conduit lines.

The seal-off fitting in Fig. 5-26 are typical of those used. The type in Fig. 5-26a is for vertical mounting and is provided with a threaded, plugged opening into which the sealing cement is poured, The seal off in Fig. 5-26b has an additional plugged opening in the lower hub to facilitate packing fiber around the conductors to form a dam for the sealing cement.

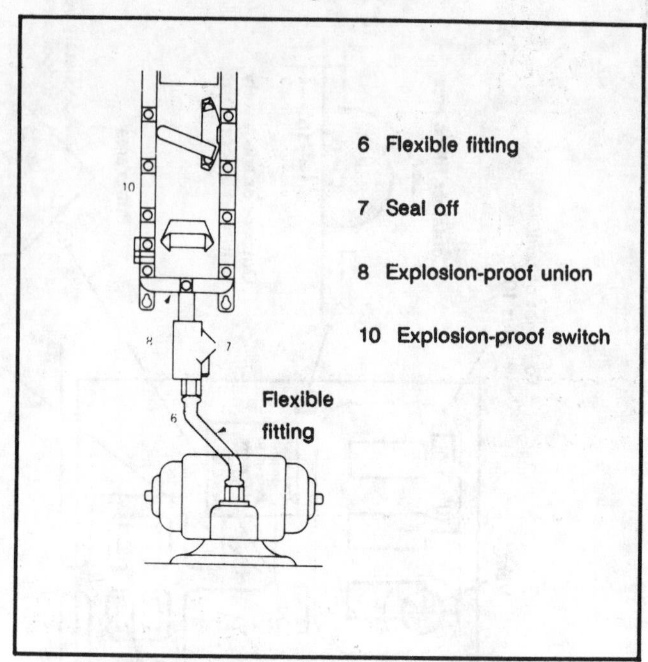

Fig. 5-25: Flexible fittings, approved for hazardous areas are frequently used for motor connections.

Most other explosion-proof fittings are provided with threaded hubs for securing the conduit as described previously. Typical fittings include switch and junction boxes, conduit bodies, union and connectors, flexible couplings, explosion-proof lighting fixtures, receptacles, and panelboard and motor starter enclosures. A practical representation of these and other fittings is shown in Fig. 5-27.

Figures 5-28 through 5-31 give examples of hazardous wiring installation in Class I and Class II locations, along with the new 1993 NE Code requirements.

Garages and Similar Locations

Garages and similar locations where volatile or flammable liquids are handled or used as fuel in self-propelled vehicles (including automobiles, buses, trucks, and tractors) are not usually considered critically hazardous locations. However, the entire area up to a level 18 inches above the floor is considered a Class I, Division 2 location, and certain precautionary measures are required by the NE Code. Likewise, any pit or depression below floor level shall be considered a Class

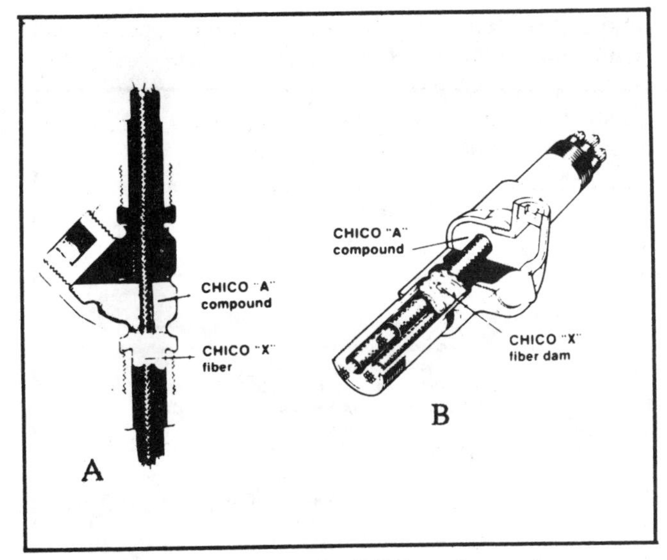

Fig. 5-26: Two types of seal-off fittings for use in hazardous areas.

I, Division 2 location, and the pit or depression may be judged as Class I, Division I location if it is unvented.

Normal raceway (conduit) and wiring may be used for the wiring method above this hazardous level, except where conditions indicate that the area concerned is more hazardous than usual. In this case, the applicable type of explosion-proof wiring may be required.

Approved seal-off fittings should be used on all conduit passing from hazardous areas to nonhazardous areas. The requirements set forth in NE Code Sections 501-5 and 501-5(b)(2) shall apply to horizontal as well as vertical boundaries of the defined hazardous areas. Raceways embedded in a masonry floor or buried beneath a floor are considered to be within the hazardous area above the floor if any connections or extensions lead into or through such an area. However, conduit systems terminating to an open raceway, in an outdoor unclassified area, shall not be required to be sealed between the point at which the conduit leaves the classified location and enters the open raceway.

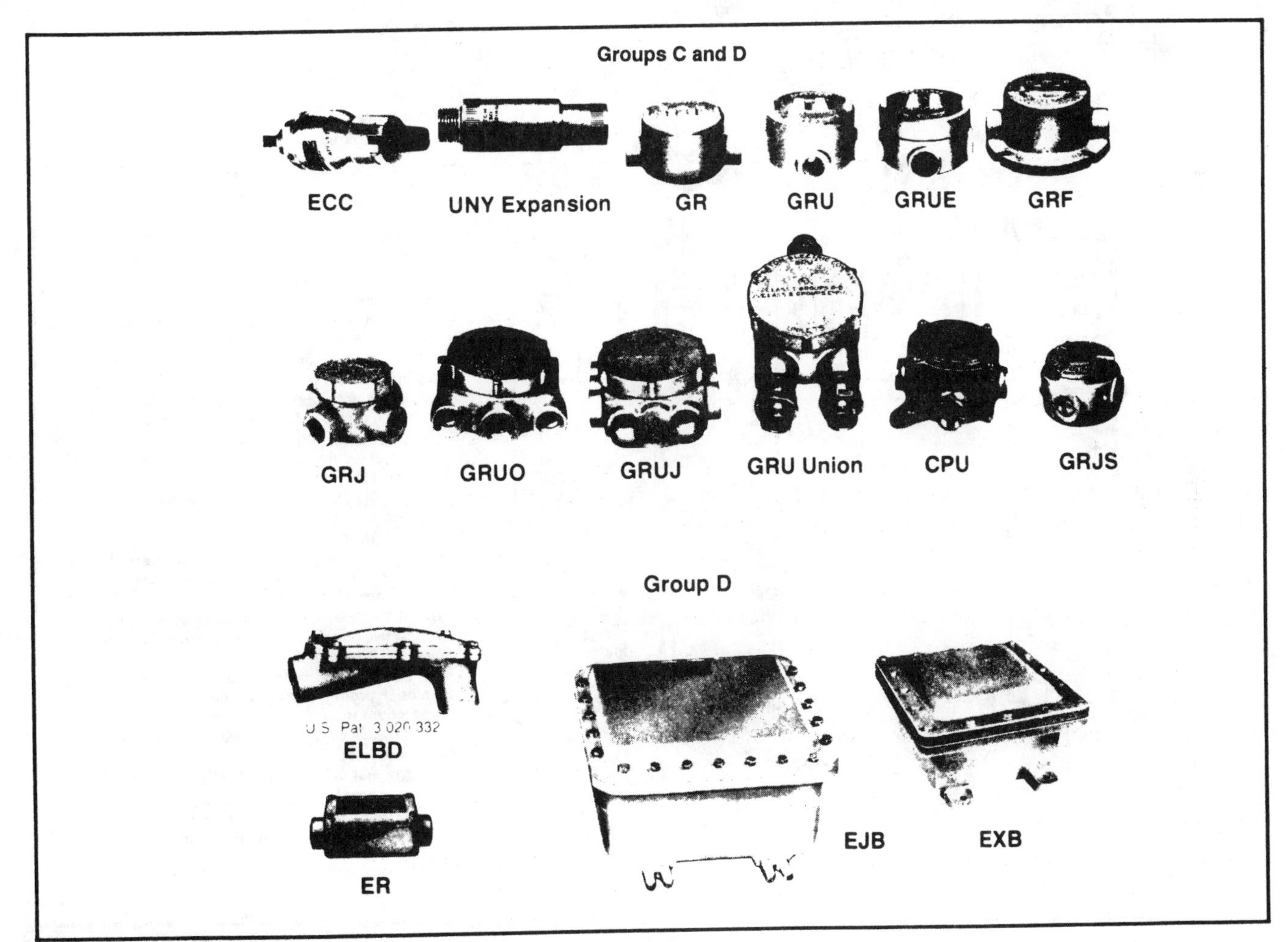

Fig. 5-27: A practical representation of explosion-proof fittings.

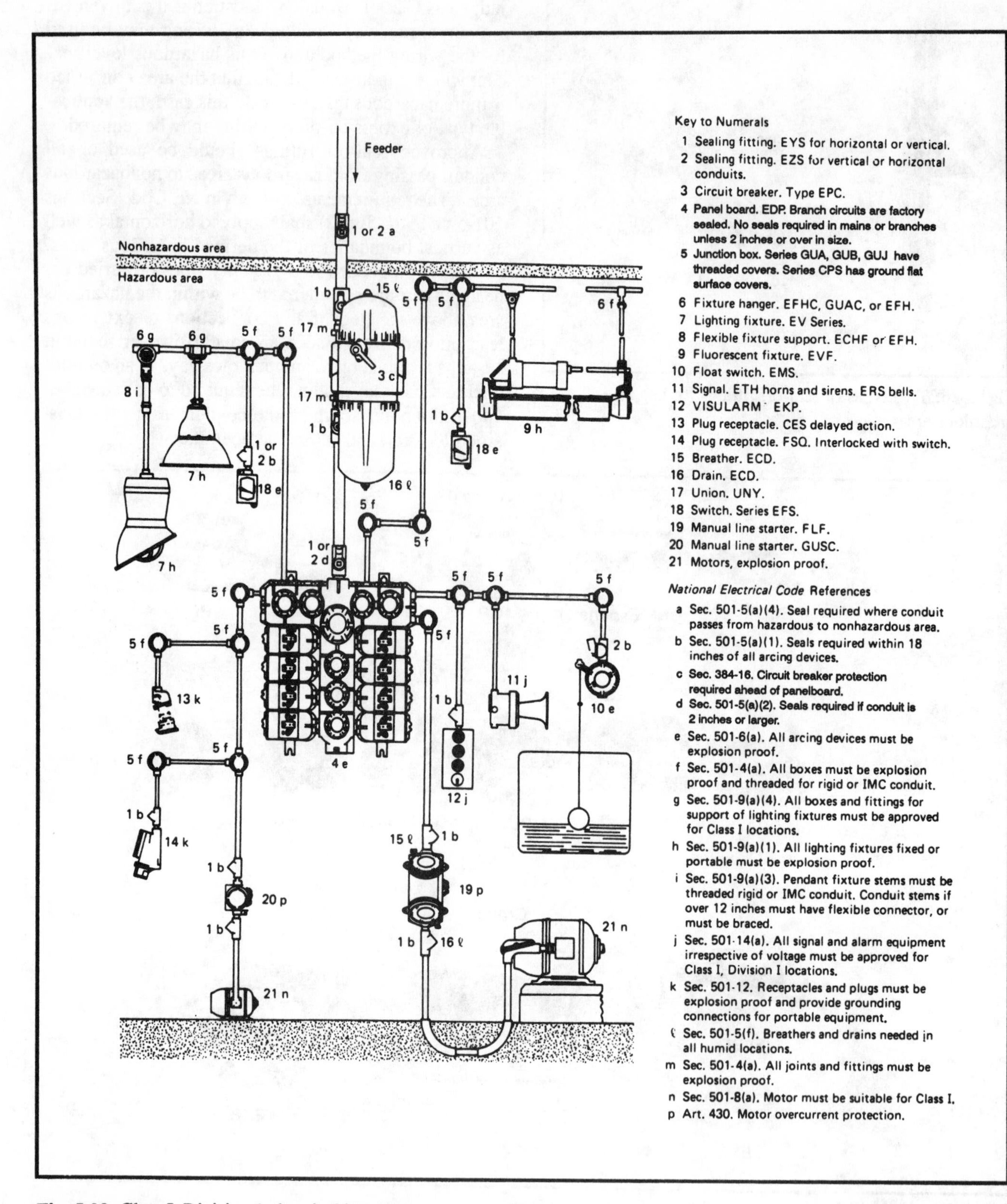

Feeder
1 or 2 a
Nonhazardous area
Hazardous area
Key to Numerals
1 Sealing fitting. EYS for horizontal or vertical.
2 Sealing fitting. EZS for vertical or horizontal conduits.
3 Circuit breaker. Type EPC.
4 Panel board. EDP. Branch circuits are factory sealed. No seals required in mains or branches unless 2 inches or over in size.
5 Junction box. Series GUA, GUB, GUJ have threaded covers. Series CPS has ground flat surface covers.
6 Fixture hanger. EFHC, GUAC, or EFH.
7 Lighting fixture. EV Series.
8 Flexible fixture support. ECHF or EFH.
9 Fluorescent fixture. EVF.
10 Float switch. EMS.
11 Signal. ETH horns and sirens. ERS bells.
12 VISULARM EKP.
13 Plug receptacle. CES delayed action.
14 Plug receptacle. FSQ. Interlocked with switch.
15 Breather. ECD.
16 Drain. ECD.
17 Union. UNY.
18 Switch. Series EFS.
19 Manual line starter. FLF.
20 Manual line starter. GUSC.
21 Motors, explosion proof.
National Electrical Code References
a Sec. 501-5(a)(4). Seal required where conduit passes from hazardous to nonhazardous area.
b Sec. 501-5(a)(1). Seals required within 18 inches of all arcing devices.
c Sec. 384-16. Circuit breaker protection required ahead of panelboard.
d Sec. 501-5(a)(2). Seals required if conduit is 2 inches or larger.
e Sec. 501-6(a). All arcing devices must be explosion proof.
f Sec. 501-4(a). All boxes must be explosion proof and threaded for rigid or IMC conduit.
g Sec. 501-9(a)(4). All boxes and fittings for support of lighting fixtures must be approved for Class I locations.
h Sec. 501-9(a)(1). All lighting fixtures fixed or portable must be explosion proof.
i Sec. 501-9(a)(3). Pendant fixture stems must be threaded rigid or IMC conduit. Conduit stems if over 12 inches must have flexible connector, or must be braced.
j Sec. 501-14(a). All signal and alarm equipment irrespective of voltage must be approved for Class I, Division I locations.
k Sec. 501-12. Receptacles and plugs must be explosion proof and provide grounding connections for portable equipment.
ℓ Sec. 501-5(f). Breathers and drains needed in all humid locations.
m Sec. 501-4(a). All joints and fittings must be explosion proof.
n Sec. 501-8(a). Motor must be suitable for Class I.
p Art. 430. Motor overcurrent protection.

Fig. 5-28: Class I, Division 1 electrical installation.

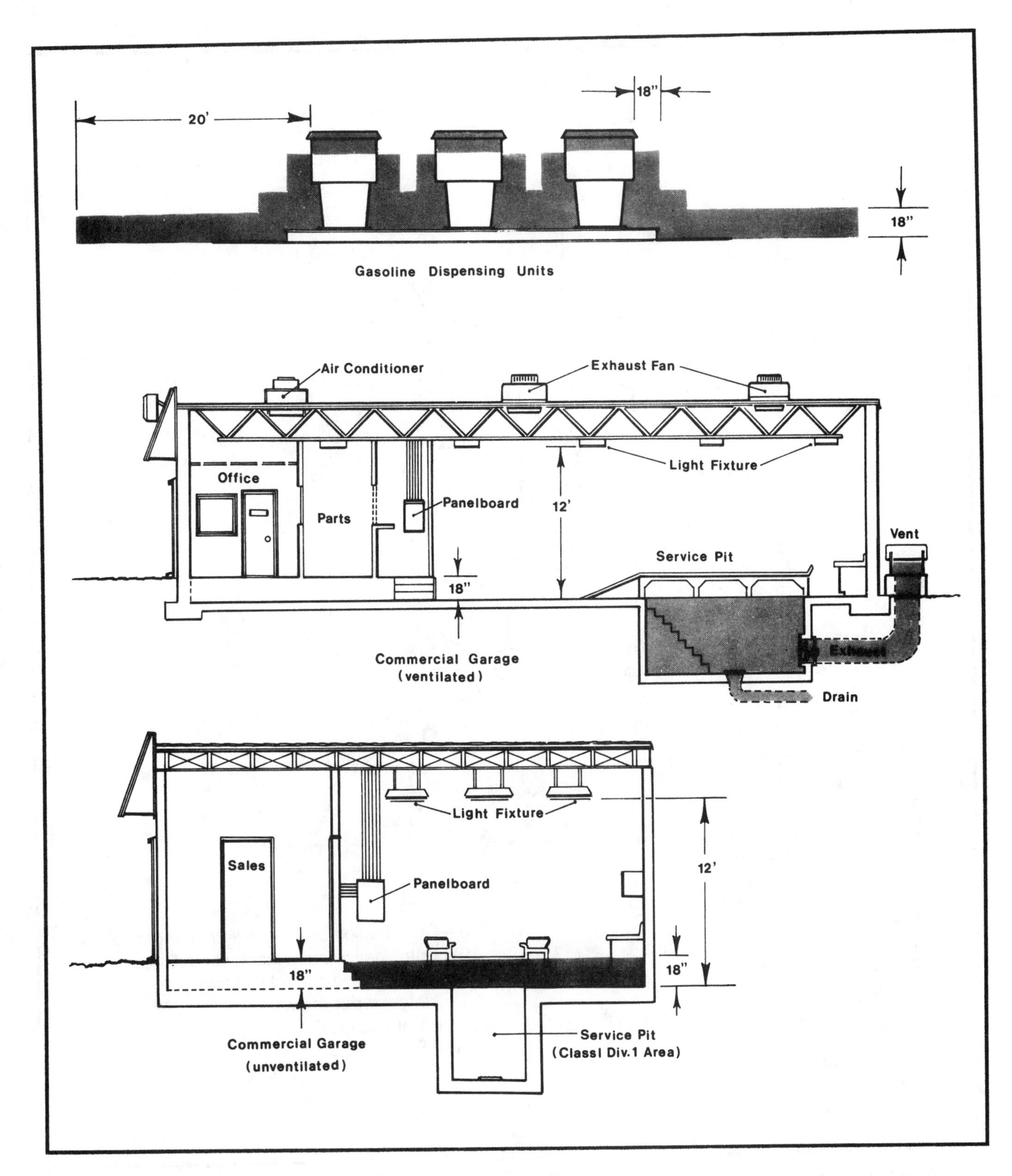

Fig. 5-29: Class I, Division 2, power and lighting wiring diagram.

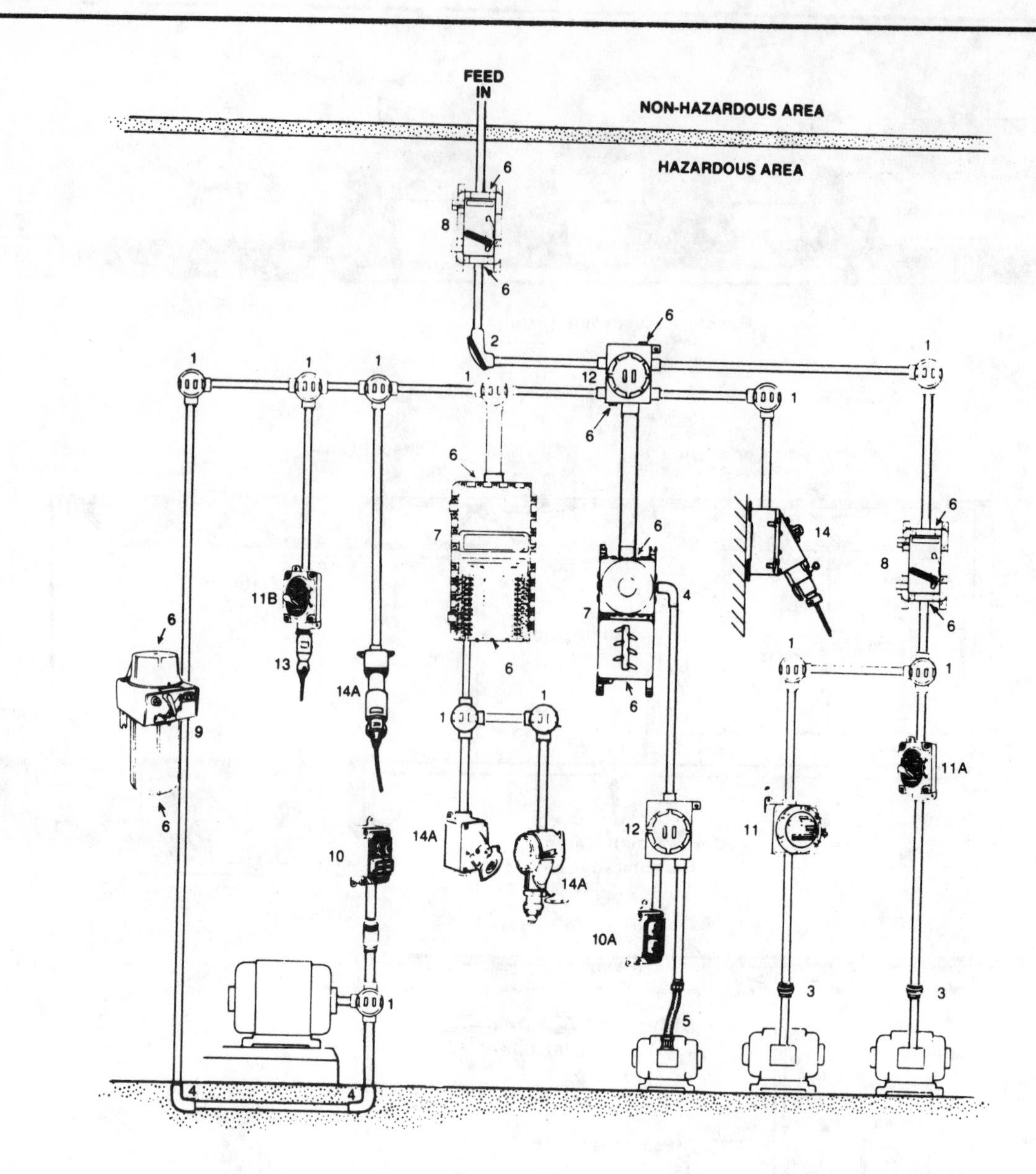

Legend

1 – Cast conduit outlet box unilets
2 – Pulling unilets
3 – Unions
4 – Elbows, plugs and miscellaneous fittings
5 – Flexible couplings
6 – Drain and breathers
7 – Panelboards
8 – Circuit breakers and unilets
9 – Combination circuit breakers/motor starters
10 – Combination pushbutton control station
10A – Pushbutton control station
11 – Motor starter or switch
11A – Motor starter
11B – Switch unilet
12 – Dust-ignition-proof unilet enclosures
13 – Cord connectors
14 – Plugs and receptacles

Fig. 5-30: Power diagram for Class II, Division 1 installations.

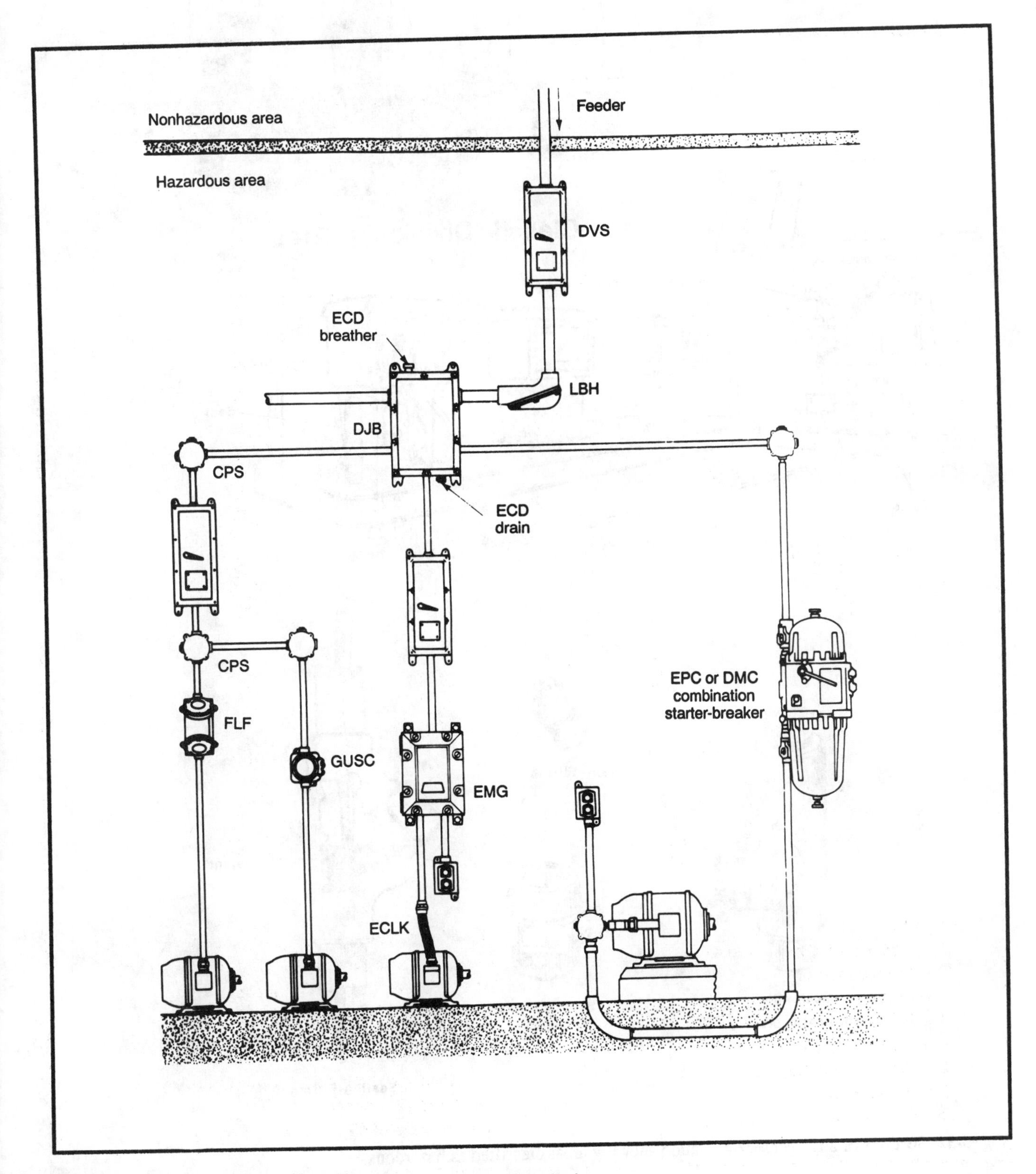

Fig. 5-31: Explosion-proof wiring devices in Class II power installation.

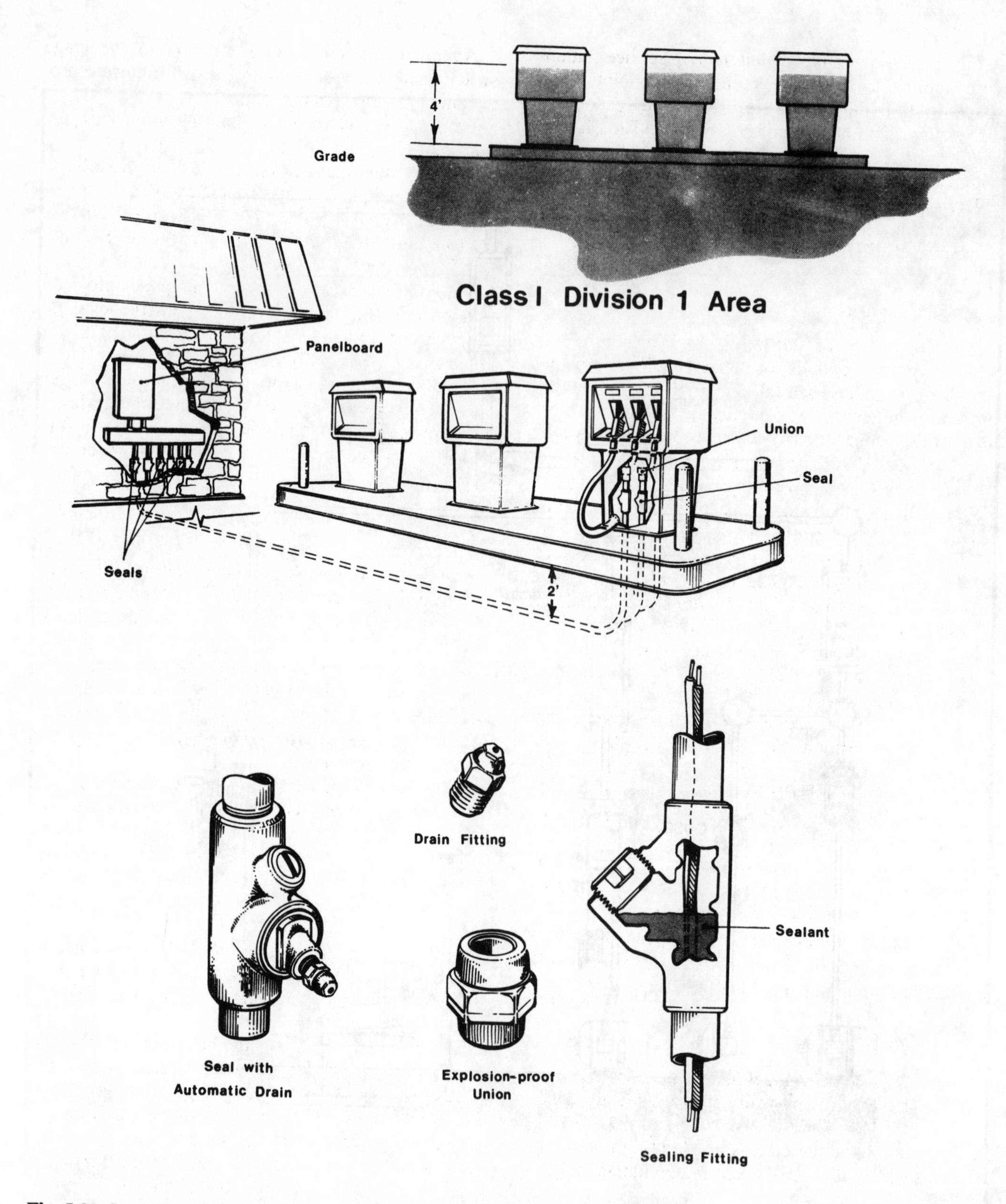

Fig. 5-32: Sections of a typical service station showing areas classified as hazardous.

Fig. 5-32 shows a typical automotive service station with applicable NE Code requirements. Note that space in the immediate vicinity of the gasoline-dispensing island is denoted as Class I, Division I, to a height of 4 feet above grade. The surrounding area, within a radius of 20 feet of the island, falls under Class I, Division 2, to a height of 18 inches above grade. Bulk storage plants for gasoline are subject to comparable restrictions.

Airport Hangars

Buildings used for storing or servicing aircraft in which gasoline, jet fuels, or other volatile flammable liquids or gases are used fall under Article 513 of the NE Code. In general, any pit or depression below the level of the hangar floor is considered to be a Class I, Division I location. The entire area of the hangar including any adjacent and communicating area not suitably cut off from the hangar is considered to be a Class I, Division 2 location up to a level of 18 inches above the floor. The area within 5 feet horizontally from aircraft power plants, fuel tanks, or structures containing fuel is considered to be a Class I, Division 2 hazardous location; this area extends upward from the floor to a level 5 feet above the upper surface of wings and of engine enclosures.

Adjacent areas in which hazardous vapors are not likely to be released, such as stock rooms and electrical control rooms, should not be classed as hazardous when they are adequately ventilated and effectively cut off from the hangar itself by walls or partitions. All fixed wiring in a hangar not within a hazardous area as defined in Section 513-2 must be installed in metallic raceways or shall be Type MI or Type ALS cable; the only exception is wiring in nonhazardous locations as defined in Section 513-2(d), which may be of any type recognized in Chapter 3 (Wiring Methods and Materials) in the NE Code. Figure 5-33 summarizes the NE Code requirements for airport hangars.

Theaters

The NE Code recognizes that hazards to life and property due to fire and panic exist in theaters, cinemas, and the like. The NE Code therefore requires certain precautions in these areas in addition to those for commercial installations. These requirements include the following:

Proper wiring of motion picture projection rooms (Article 540).

Heat-resistant, insulated conductors for certain lighting equipment (Section 520-43(b)).

Adequate guarding and protection of the stage switchboard and proper control and overcurrent protection of circuits (Section 520-22).

Proper type and wiring of lighting dimmers [Sections 520-53(e) and 520-25].

Use of proper types of receptacles and flexible cables for stage lighting equipment (Section 520-45).

Proper stage flue damper control (Section 520-49).

Proper dressing room wiring and control (Sections 520-71, 72, and 73).

Fireproof projection rooms with automatic projector port closures, ventilating equipment, emergency lighting, guarded work lights, and proper location of related equipment (Article 540).

Outdoor or drive-in motion picture theaters do not present the inherent hazards of enclosed auditoriums. However, the projection rooms must be properly ventilated and wired for the protection of the operating personnel.

Hospitals

Hospitals and other health-care facilities fall under Article 517 of the NE Code. Part B of Article 517 covers the general wiring of health-care facilities. Part C covers essential electrical systems for hospitals. Part D gives the performance criteria and wiring methods to minimize shock hazards to patients in electrically susceptible patient areas. Part E covers the requirements for electrical wiring and equipment used in inhalation anesthetizing locations.

With the widespread use of x-ray equipment of varying types in health-care facilities, electricians are often required to wire and connect equipment such as discussed in Article 660 of the NE Code. Conventional wiring methods are used, but provisions should be made for 50- and 60-ampere receptacles for medical x-ray equipment (Section 660-4(a).

Anesthetizing locations of hospitals are deemed to be Class I, Division 1, to a height of 5 feet above floor level. Gas storage rooms are designated as Class I, Division 1, throughout.

The NE Code recommends that wherever possible electrical equipment for hazardous locations be located in less hazardous areas. For example, much time and expense can be saved, in many cases, by installing the service equipment outside the hazardous area; then use feeders and seal-offs to supply loads in the hazardous area. It also suggests that by adequate, positive-pressure ventilation from a clean source of outside air the hazards may be reduced or hazardous locations limited or eliminated. In many cases the installation of dust-collecting systems can greatly reduce the hazards.

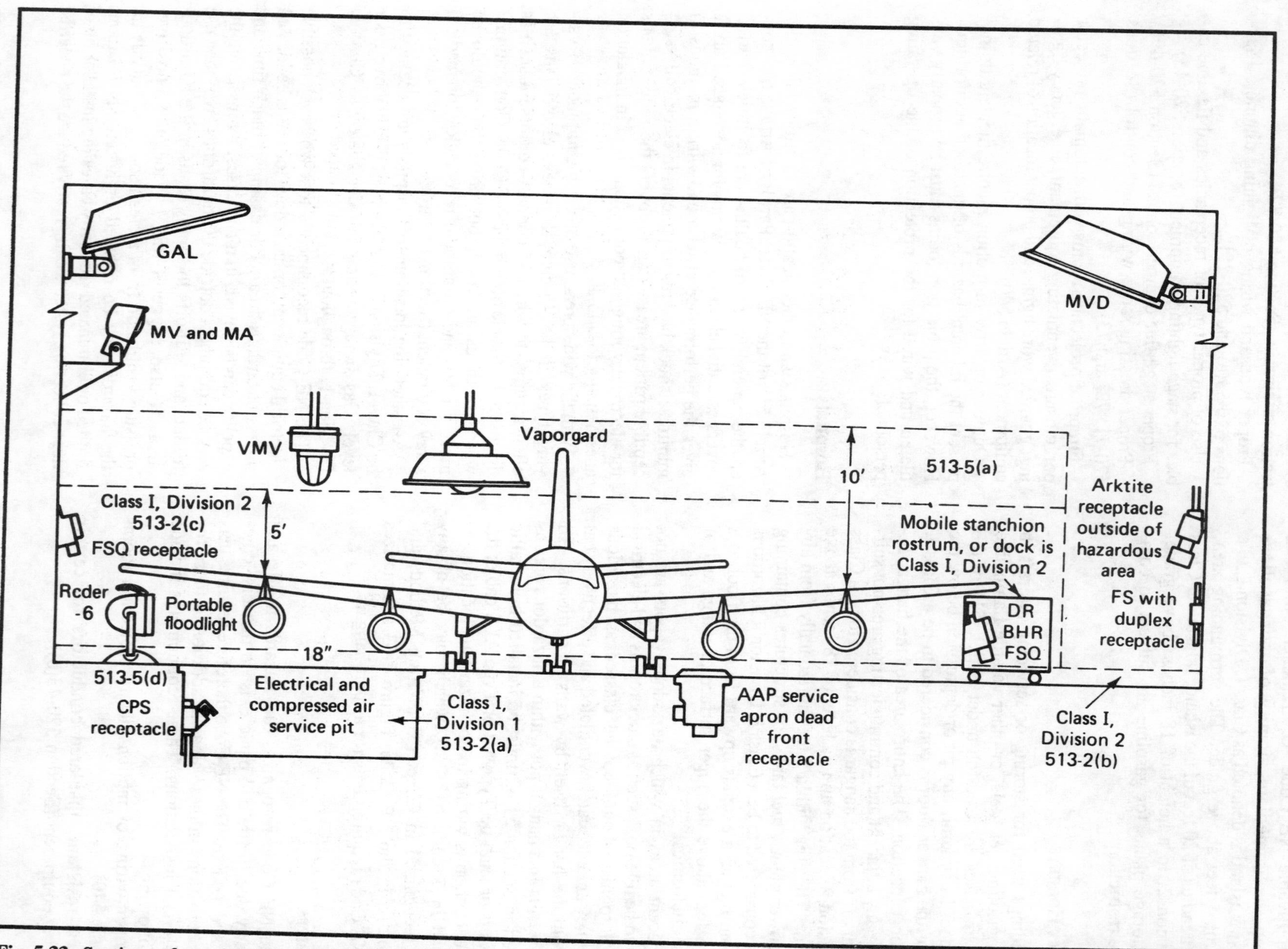

Fig. 5-33: Sections of a typical airport hangar showing hazardous locations.

chapter 6

ELECTRICAL EQUIPMENT—LIGHTING AND ELECTRIC HEAT

Article 410 of the NE Code covers lighting fixtures, lampholders, pendants, receptacles, incandescent filament lamps, arc lamps, electric-discharge lamps, and the related wiring to each. While the rules in this Section provide for a safe installation, guidelines are not given as to good, efficient lighting design.

In general, lighting layouts for any type of building should be designed to provide the highest visual comfort and performance that is consistent with the type of area to be lighted and the budget provided. However, since individual tastes and opinions vary, there can be many solutions to any given lighting situation. Some of these solutions can be commonplace, while others will show imagination and resourcefulness.

General-purpose branch circuits are provided in all occupancies to supply lighting outlets for illumination and receptacle outlets for small appliances and office equipment. When lighting circuits are separate from circuits that supply receptacles the NE Code provides rules for the design of each type of branch circuit. This is usually the case in commercial and industrial occupancies.

Applicable Rules: The lighting load to be used in the branch-circuit calculations for determining the required number of circuits must be the larger of the values obtained by using one of the following:

- The actual load
- A minimum load in volt-amperes or watts per square foot as specified in the NE Code

Table 220-3(b) of the NE Code specifies the minimum unit load in volt-amperes (watts) per square foot of floor area based on outside dimensions for the occupancies listed. If the actual lighting load is known and if it exceeds the minimum determined by the watts per square foot basis, the actual load must be used because the NE Code specifies that branch-circuit conductors shall have an ampacity not less than the maximum load to be served. A store building, for example, with 4000 square feet of floor space (outside dimensions of 40 feet by 100 feet) would have a minimum lighting load of 12 kilovolt-amperes (12,000 watts) based on the 3 watts per square foot unit load specified by the NE Code. If the actual connected load happened to be one hundred 150-watt lamps, or 15 kilowatts, and the NE Code calculation requires only 12 kilowatts of lighting, the actual load that must be used in service-entrance, feeders and branch circuit calculations is 15 kilowatts. Therefore, since 15,000 watts is the larger value, this figure would be used in the calculations.

Branch Circuits for Lighting: The NE Code permits only 15- or 20-ampere branch circuits to supply lighting units with standard lampholders. Branch circuits of

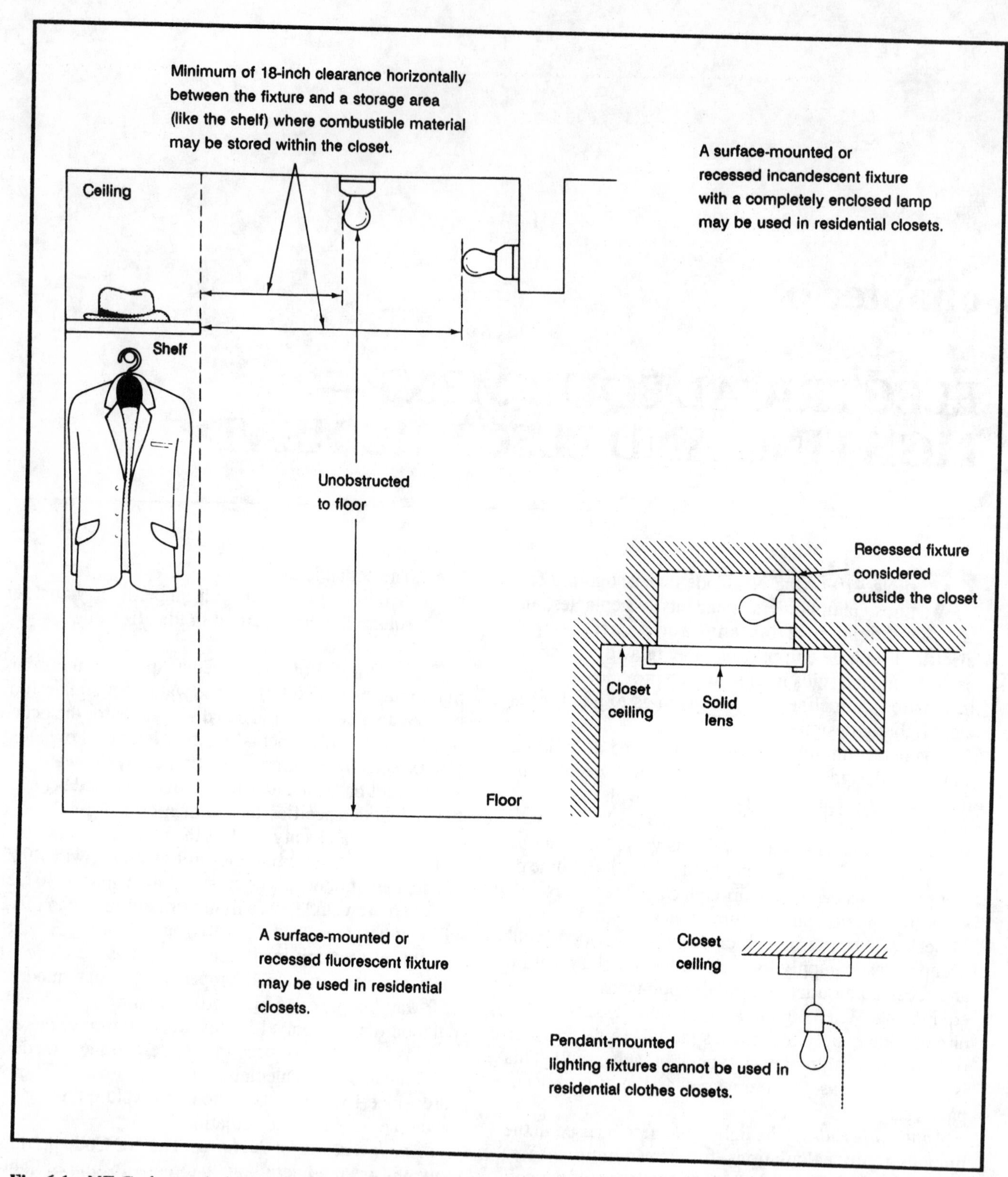
Minimum of 18-inch clearance horizontally
between the fixture and a storage area
(like the shelf) where combustible material
may be stored within the closet.
A surface-mounted or
recessed incandescent fixture
with a completely enclosed lamp
may be used in residential closets.
Ceiling
Shelf
Unobstructed
to floor
Floor
Recessed fixture
considered
outside the closet
Closet
ceiling
Solid
lens
A surface-mounted or
recessed fluorescent fixture
may be used in residential
closets.
Closet
ceiling
Pendant-mounted
lighting fixtures cannot be used in
residential clothes closets.

Fig. 6-1: NE Code regulations governing residential closet lighting.

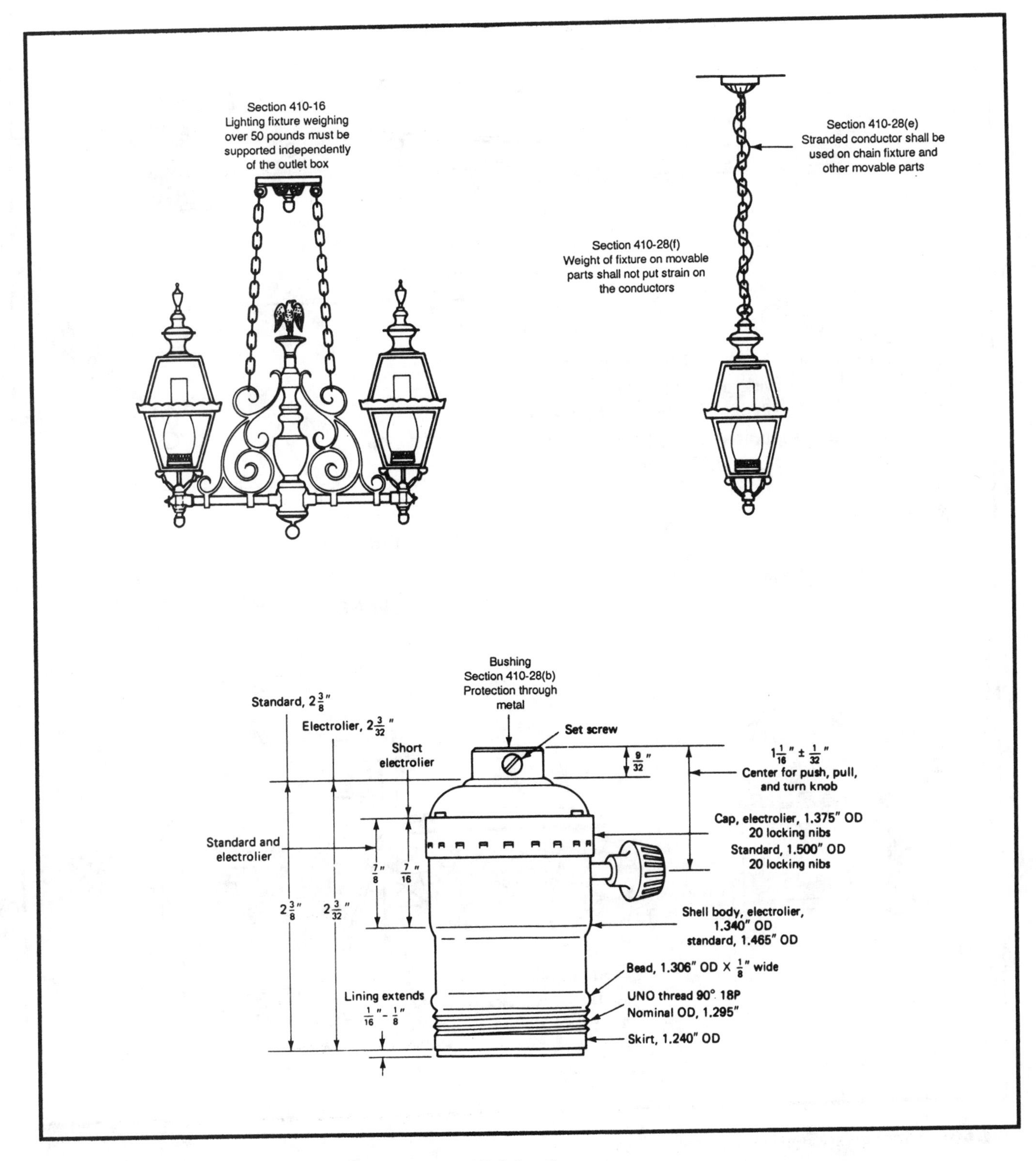

Fig. 6-2: NE Code rules governing pendant-mounted lighting fixtures.

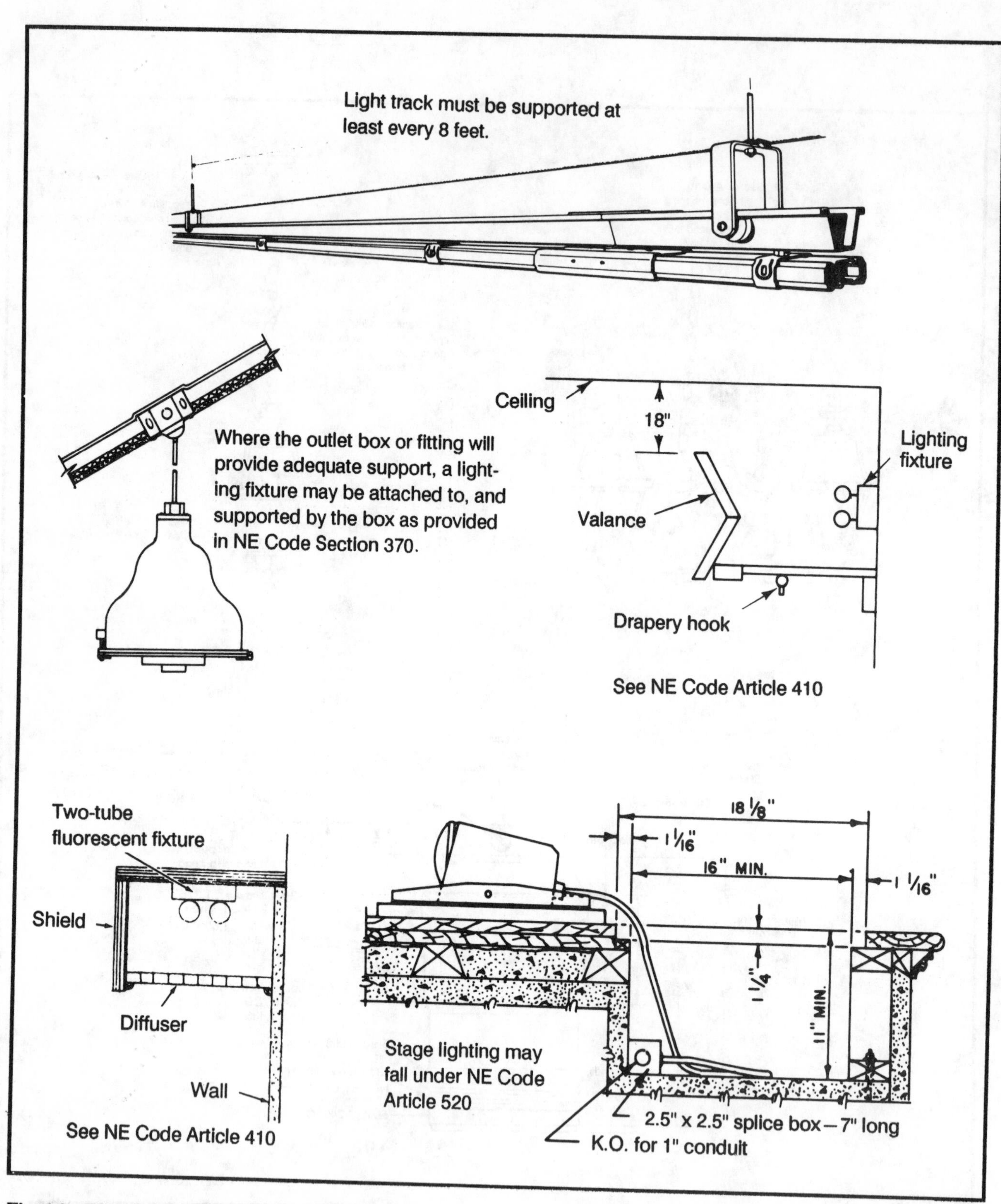
Light track must be supported at least every 8 feet.
Where the outlet box or fitting will provide adequate support, a lighting fixture may be attached to, and supported by the box as provided in NE Code Section 370.
Ceiling
18"
Lighting fixture
Valance
Drapery hook
See NE Code Article 410
Two-tube fluorescent fixture
Shield
Diffuser
Wall
See NE Code Article 410
18 1/8"
1 1/16"
16" MIN.
1 1/16"
1 1/4"
11" MIN.
Stage lighting may fall under NE Code Article 520
2.5" x 2.5" splice box – 7" long
K.O. for 1" conduit

Fig. 6-3: Electric-discharge lighting applications.

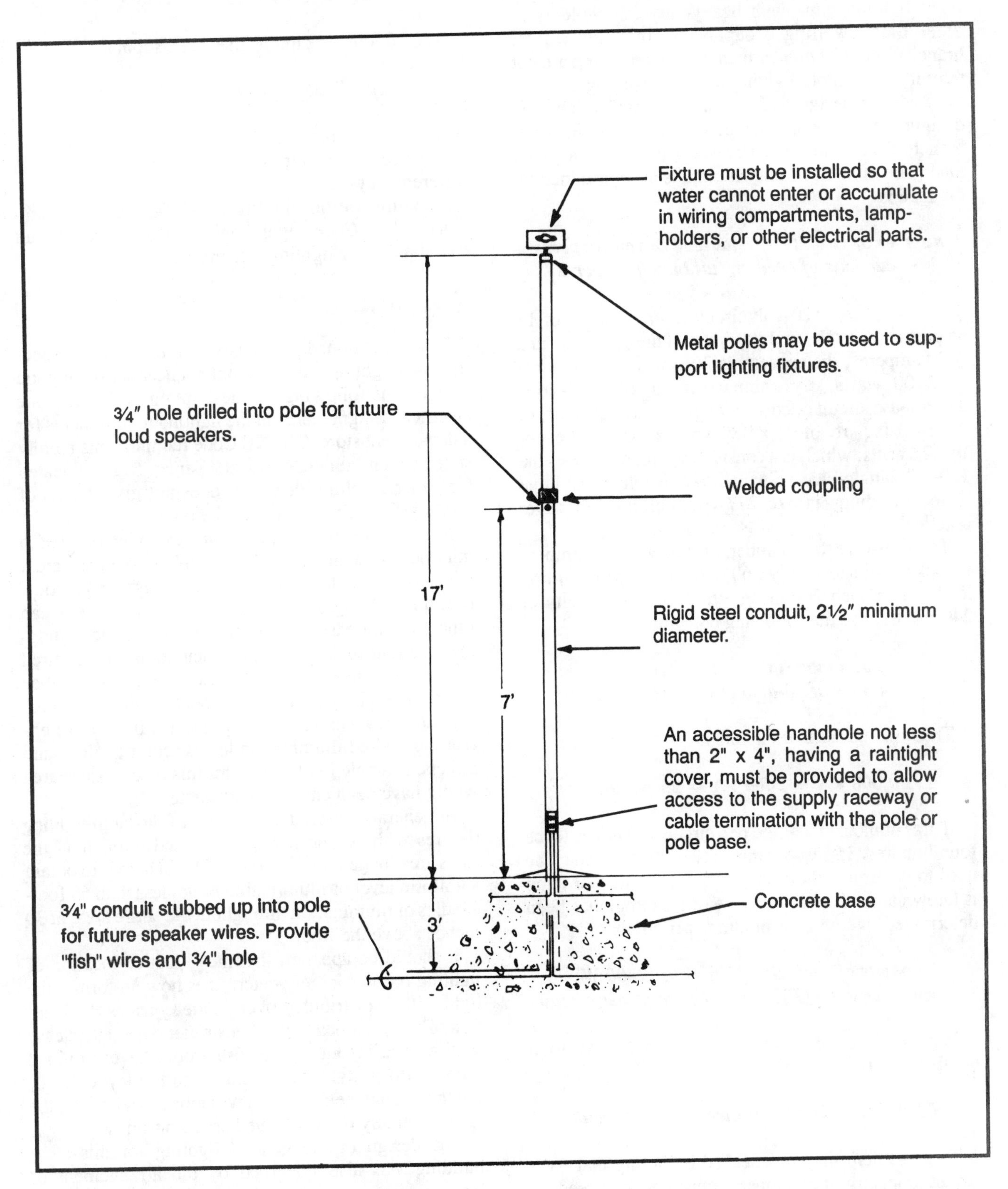

Fixture must be installed so that water cannot enter or accumulate in wiring compartments, lamp-holders, or other electrical parts.
Metal poles may be used to support lighting fixtures.
¾" hole drilled into pole for future loud speakers.
Welded coupling
17'
7'
Rigid steel conduit, 2½" minimum diameter.
An accessible handhole not less than 2" x 4", having a raintight cover, must be provided to allow access to the supply raceway or cable termination with the pole or pole base.
¾" conduit stubbed up into pole for future speaker wires. Provide "fish" wires and ¾" hole
3'
Concrete base

Fig. 6-4: Miscellaneous lighting applications.

greater than 20-ampere rating are permitted to supply fixed lighting units with heavy-duty lampholders in other than dwelling occupancies. In other words, branch circuits of greater than 20-ampere rating are not permitted to supply lighting units in dwellings.

In certain design and installation applications, the designer or worker must determine the number of branch circuits that are necessary to supply a given load. The number of branch circuits as determined by the load is

number of circuits = total load in volt-amperes ÷ capacity of each circuit in volt-amperes

A 15-ampere, 120-volt circuit has a capacity of 15 amperes x 120 volts = 1800 watts. If the circuit is rated at 20 amperes, the capacity is 20 amperes x 120 volts = 2400 watts. By comparison, a 480/277-volt, three-phase circuit rated at 20 amperes has a capacity of the 1.73 [sq. rt. of 3] x 480 volts x 20 amperes = 16,627 watts, which is a considerable increase over the 120-volt single-phase circuits. This is why most large office buildings utilize 277-volt lighting whenever practical.

To determine the number of 120-volt, 20-ampere branch circuits to supply a 60,000 volt-ampere lighting load, when each 20-ampere circuit has a capacity of 2400 watts, use the following equation:

Total load (in watts) ÷ 2400 = number of 20-amp circuits required

Thus, the number of circuits is

60,000 VA ÷ 2400 VA = 25 circuits

If the number of lamps per circuit is known to be four hundred 150-watt lamps, two methods may be used to determine the result. When the watts per lamp is known and when the capacity of the circuit has been determined, the number of lamps per circuit is

capacity of each circuit in watts (2400 W) ÷ watts per lamp (150 W) = 16 lamps per circuit

With each circuit supplying 16 lamps, the total number of circuits required would be

400 lamps/16 lamps per circuit = 25 circuits

This problem may be checked by noting that each circuit ampacity of 20 amperes must not be exceeded. The current drawn by each lamp at 120 volts is

I = 150 W/120 V = 1.25 A

The 20-ampere circuit, then, can supply

20 A/1.25 A per lamp = 16 lamps

or 16 lamps as before. The confidence in the result is high since the answer has been determined by several different ways.

The illustrations in Figs. 6-1 through 6-4 summarize NE Code regulations used in branch-circuit calculations for lighting circuits.

Design Example

It was mentioned previously that the NE Code specifies *safe* lighting practices, but not necessarily of good design. The following is a description of the way one electrical designer laid out the lighting requirements for a department store. The NE Code requirements merely state that a minimum of 3 watts per sq. ft. be allocated for general lighting loads. It does not give details of lighting fixture placement or the type.

Modern practice requires the establishment of a minimum quantity of light throughout a given area, which is usually termed "general lighting." In practical use, this lighting must be arranged so that the eye can function with ease and efficiency. Experienced judgment combined with lighting calculations are required to create the lighting layout shown in Fig. 6-5. Notice that the overall lighting arrangement combines fluorescent and incandescent lighting fixtures in order to obtain the desired illumination levels, color quality, and effects. A single light source for this type of sales area would have been entirely inadequate.

In general, recessed 2- by 4-foot four-lamp lighting fixtures with acrylic lenses were used throughout the sales area for general lighting. The aim here is to obtain a uniform level of illumination of not less than 65 foot-candles of illumination throughout the entire area, at 30 inches above the floor.

It should be apparent that the layout or location of lighting fixtures largely determines how uniformly the light will be distributed over an area, just as the location of sprinkler heads regulates water coverage in case of fire. The layout in Fig. 6-6 would have provided more uniform light distribution than the layout used, but the arrangement selected was more economical; the layout actually used was the best compromise.

The design of the general lighting for this store building was also governed by the arrangement of bays, columns, beams, and other architectural details; they required that the layout be fitted as symmetrically

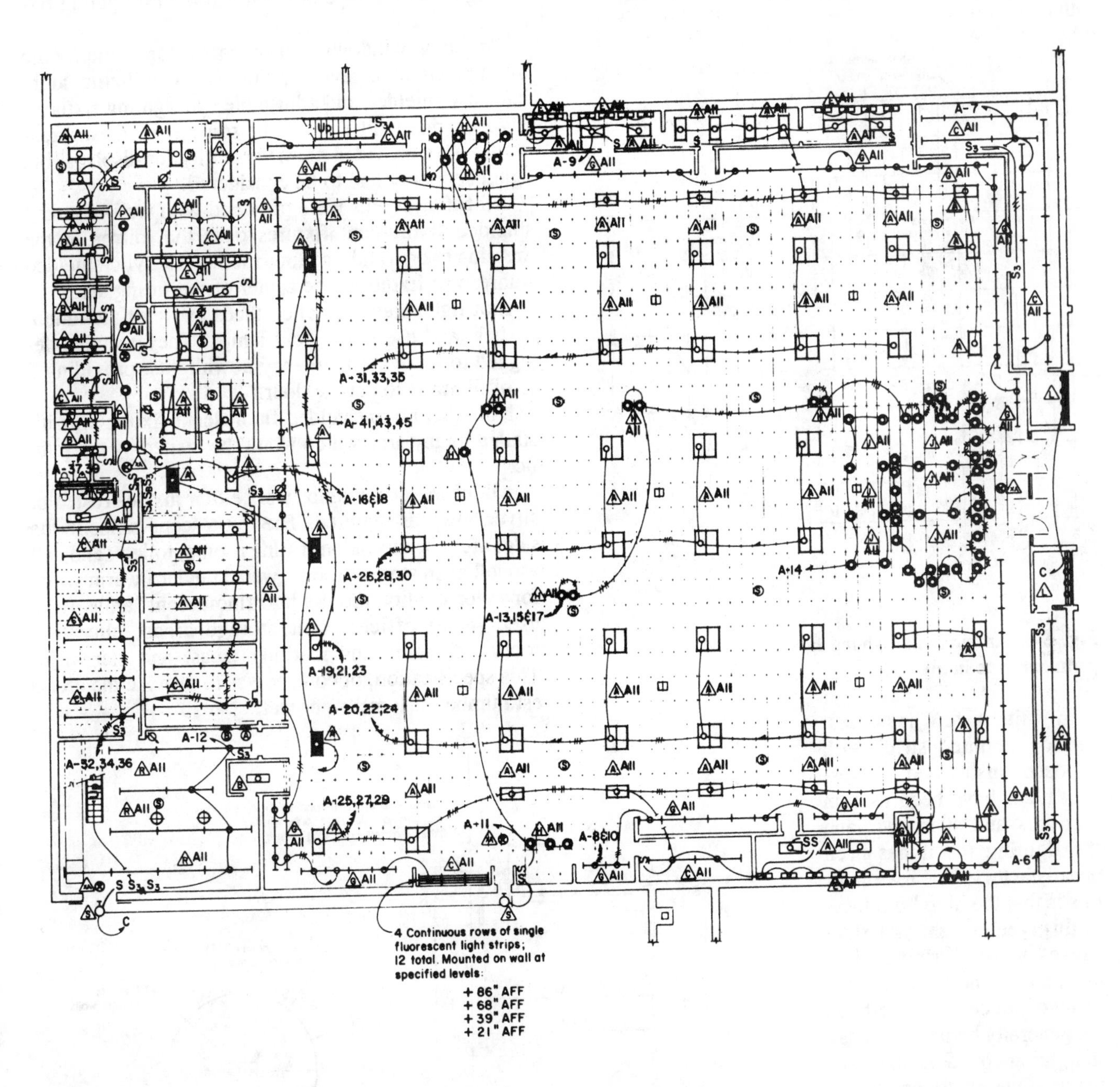

Fig. 6-5: Floor plan showing the lighting layout of a department store.

as possible to the interior but always kept within the limits of spacing directed by ceiling height or the heights at which lighting units could be mounted.

Again, recessed 2- by 4-foot fluorescent lighting fixtures with 100 percent acrylic low-brightness prismatic lenses were used in all office areas. However, in these areas, it was calculated that the maintained illumination level was 150 fc at 30 inches above the floor. The zonal cavity method was used for this calculation.

Lighting in stockrooms and receiving areas was designed to provide an illumination level of 50 fc maintained at 30 inches above the floor. Lighting fixtures in these areas were surface-mounted strips with exposed unshielded fluorescent tubes for economy.

The lighting layout in stairways, as well as in all alteration rooms, was arranged to maintain an illumination level of 100 footcandles (fc). All lighting fixtures in these areas were acrylic-shielded fluorescent—

surface-mounted in some locations and recessed in others.

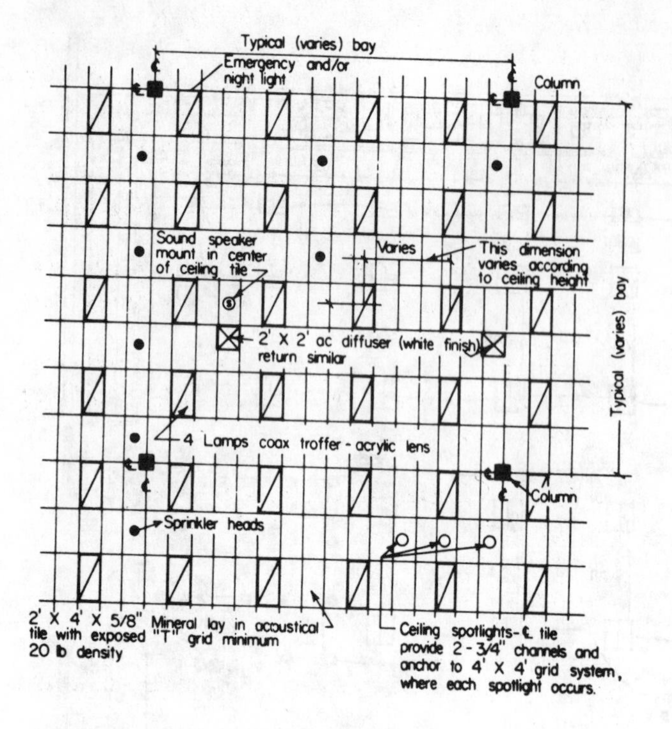

Fig. 6-6: An alternate lighting layout for the department store in Fig. 6-5.

In quite a few instances, surface-mounted strips with exposed unshielded fluorescent tubes have been used in corridors and stairways, and the lighting result has been unsatisfactory. The reason for this is that the high brightness of this type of lighting fixture causes uncomfortable glare and poor, or no, diffusion.

Two successful methods are generally being used today in lighting corridors and stairways. One is to provide a good matte-white ceiling and to use totally indirect lighting (in some, the light sources are completely hidden); the other is to employ a light source of low brightness which in turn provides for high visual comfort. In stairways, the problem of changing lamps should also be considered when selecting either the type or the location of a lighting fixture.

The show windows near the main store entrance are illuminated to a level of 500 fc by utilizing high-intensity shielded and adjustable spotlighting fixtures. Two or more different light sources in these areas would have provided more versatility for displays, but again economy had to be considered.

All fitting rooms were provided with wall-mounted shielded fluorescent fixtures over the mirror in the dressing booths and a ceiling-mounted acrylic-shielded fluorescent fixture in the center of the fitting area.

Exit lights have stenciled faces and cast-aluminum bodies, and use fluorescent lamps. They are also designed to serve as emergency downlights, providing 5 fc of illumination at the door.

Note that no unshielded fluorescent fixture was installed in any area which would be visited by the public.

All incandescent lamps were either inside-frosted or silver bowl, as required. The fluorescent lamps in the sales area, beauty salon, fitting rooms, toilets, and hall as well as in storage rooms were standard warm white for color quality. Cool-white fluorescent lamps were used in all offices, bulk-storage areas, alteration rooms, shipping and marking areas, and stairways. This specification, of course, was to provide the greatest number of lumens per watt.

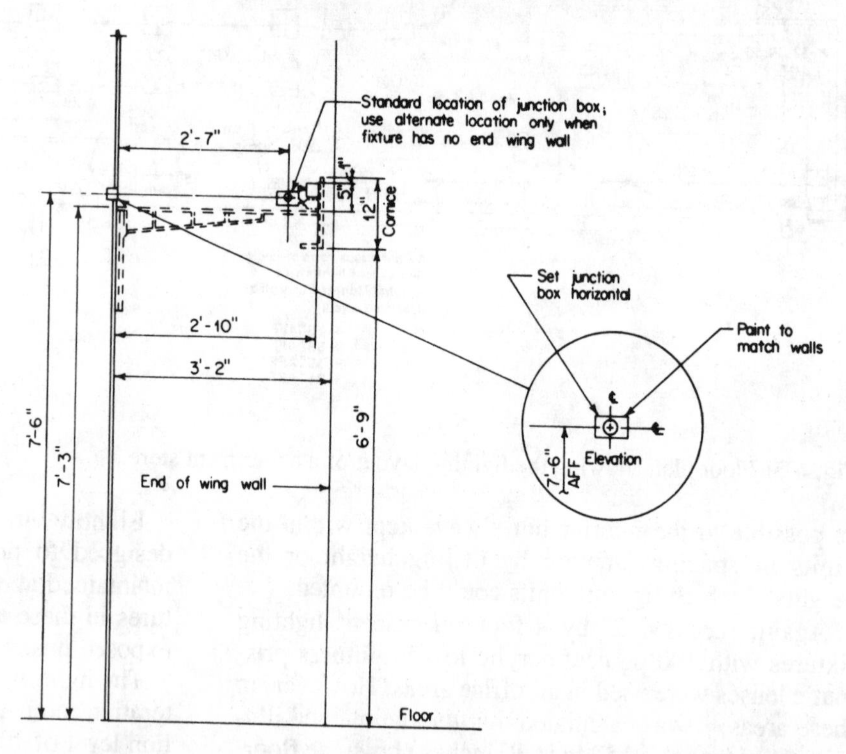

Fig. 6-7: Sectional detail of supplemental lighting.

Supplemental Lighting

The perimeter lighting shown in the floor plan in Fig. 6-5 and also in the detail in Fig. 6-7 serve to highlight the merchandise shelves around the perimeter of the store. Although all these lamps utilized bare tubes, the light sources were not readily visible to the public.

The remaining filament (incandescent) lighting in the sales areas consisted of either recessed downlights or adjustable spots and were used for various merchandise displays as required by the store-fixture layout.

Farm Lighting

Proper lighting is an essential element of modern electrical living and working. The amount and type of lighting required should be determined by the various seeing tasks associated with the management of the home and farm. Types and locations of lighting outlets should conform with the lighting fixtures or equipment to be used. Unless a specified location is stated, lighting outlets can be located anywhere within the area under consideration to produce the desired lighting effects.

For wet and damp locations, such as certain farm buildings, use lamp receptacles with nonmetallic coverings, such as porcelain, plastic, or rubber. Lighting fixtures in feed-grinding rooms, feed-storage rooms, haymows, and other dusty locations should be of the dustproof type.

Yard lighting is a very important consideration for the farmstead. Properly located and installed lights can prevent many accidents on the farm, as well as greatly increase after-hours farm efficiency. The time required to do after-dark chores, such as watering, feeding, and housing livestock and storing machinery, can be greatly reduced with the aid of adequate light.

Outside lights should be so located that the most frequently used paths and work areas are well lighted. Outside feeding and watering areas should be illuminated by lights that can be controlled from the buildings most accessible to the areas. In the absence of buildings to support the lights, poles may be used for attachment.

The control of farm lighting should be given much thought. In general, yard lighting should be controlled from both the house and outbuildings, either by means of three- and four-way switches or by relays. Low-voltage control systems can be installed with a master control panel located in the house to control every lighting circuit on the premises from one location, or at each individual location. Also, these low-voltage systems are advantageous when long distances are encountered between controls. Buck and boost transformers may be used to help with voltage regulation problems.

Whereas yard lighting consisted mainly of incandescent lamps a decade or so ago, these relatively inefficient light sources are now being replaced with quartz and HID lamps, all of which give off more lumens per watt. In any case, yard lighting around the farm is absolutely essential as a protection against thievery and predators and also as a safety factor to prevent accidents.

Many farmers are also using electric lighting systems to start seedlings indoors in the early spring. The results have been promising.

The practice of growing plants in a controlled environment (growth rooms) where natural light is replaced by artificial light permits:

- More predictable plant response
- More precise timing of crops
- Easier management
- The saving of labor

Lighting in greenhouses before sunrise and after sunset extends the light period and increases the normal growth time.

Lighting under benches can double growing areas within a greenhouse.

For example, it has been determined that a winter crop of pot chrysanthemums develops improved growing and flowering qualities where grown under GE Lucalox lamps the last three weeks of their growing cycle. During this three-week period, they are exposed to the lights 12 hours a day. The layout in Fig. 6-8 has been successfully used with 400-watt Lucalox lamps. They are mounted 4 feet above the bed and produce about 500 footcandles—16 watts per square foot.

The table in Fig. 6-9 will serve as a guide in determining a proper lighting layout for plant growth.

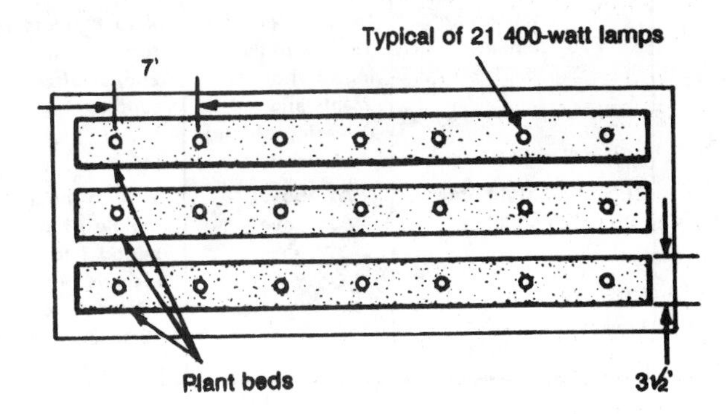

Fig. 6-8: A lighting system using 400-watt Lucalox lamps to speed the growth of seedlings.

Object of Lighting	Applications	Time Applied	Total Effective Light Period (hours)	Lamp (W/ft^2)	Light Sources	Luminaires
I. Photosynthetic						
A. Supplementary						
1. Daylength extension	a. Seed germination, seedlings, bulb forcing b. Mature plants	4 to 10 hours before sunrise and/or after sunset	a. 12 to continuous b. 10 to continuous	a. 5 to 20 b. 10 to 40	Fluorescent, mercury & fluorescent-mercury lamps of various wattages with & without internal reflectors & used with or without 10 to 30% of installed watts of incandescent	Moisture-resistant luminaires of industrial or custom-made designs with mountings fixed or adjustable providing minimum interference with greenhouse routine & uniform light distribution
2. Dark day	As above	Total light period	As above 16	As above		
3. Night	As above	4 to 6 hours in middle of dark period		Same		
4. Under bench	As above	Total light period	10 to continuous	Same	Fluorescent lamps	Moisture-resistant, direct reflector units with mounting for uniform light distribution
B. Growth room						
1. Professional horticulture	Seed germination, seedlings, cuttings, bulb forcing	Total light period	12 to continuous	5 to 30	Fluorescent lamps with or without 10 to 30% incandescent or combination of plant growth lamps	Industrial direct reflector luminaires which are moisture resistant & are mounted in a shelf arrangement
2. Amateur horticulture	Seed germination, seedlings, cuttings, bulb forcing, mature plants, etc.	Total light period	10 to continuous	5 to 30	Fluorescent lamps (plant growth lamps) with and without incandescent	As above
3. Experimental horticulture	All types of plant responses	Total light period	0 to continuous	0 to 140 & higher	Many types used to fit the requirements of tests. Generally fluorescent with 10 to 30% incandescent lamps or combination of plant growth and wide-spectrum lamps.	Custom built with minimum spacing for maximum light output of lamps with uniform light distribution
II. Photoperiodic						
A. Supplementary						
1. Daylength extension	Long-day effect to prevent flowering of short-day plants and induce flowering of long-day plants	4 to 8 hours before sunrise and/or after sunset	14 to 16	.5 to 5	Fluorescent, fluorescent-mercury, and incandescent lamps	As for photosynthetic supplementary lighting
2. Night break	As above	2 to 5 hours in middle of dark period	14 to 16	.5 to 5	As above	As above
3. Cyclic	As above	1 to 4 seconds per minute, 1 to 4 or 10 to 30 minutes per hour	14 to 15	1 to 5	Mostly incandescent, or fluorescent lamps with flashing ballasts	As above

Fig. 6-9: The requirements for photosynthetic and photoperiodic lighting.

Residential Lighting

The basic requirements for residential lighting are that it provide adequate light of the right quality and give proper attention to the appearance and artistic features of the lighting fixtures and the effect they produce.

Properly designed lighting is one of the greatest comforts and conveniences that any home owner can enjoy, and lighting should be considered as important as furniture placement, choice of draperies, color schemes, etc., because it is one of the most important features of decorating the home.

Unlike large commercial applications, residential lighting does not require elaborate calculations, but some guide should be used for the initial planning. The following method has been recommended by General Electric. To understand this procedure, the reader should be familiar with the following definitions:

Lumen: a unit of light quality (luminous flux) produced by a light source.
Footcandle: the amount of direct illumination on a surface 1 foot from the flame of a standard candle.
One lumen per square foot equals one *Footcandle.*

It is important to remember, when calculating the total lumen requirements by this method, that lighter colors reflect light while darker colors absorb it. Therefore, to achieve the same lighting level, a room with dark surfaces will require a greater number of lumens than a room with lighter surfaces.

The following table lists the required lumens per square foot for various areas in the home. These are recommended lumens when portable lamps, surface-mounted fixtures, and structural-lighting techniques are used. When a high percentage of the light for the area comes from recessed lighting fixtures, the figures in the table should be approximately doubled.

Lumens Required Per Square Foot

Area	Lumens	Area	Lumens
Living room	80	Bedroom	70
Dining room	45	Hallway	45
Kitchen	80	Laundry	70
Bathroom	65	Work bench area	70

Now let's take a typical residential living room to see how one solution to a lighting problem is solved. Since the above table recommends 80 lumens per square foot, determine the total lumens (lm) required to achieve this amount of illumination in a living room 15 feet wide and 20 feet long.

Multiply room width by room length to find square footage; that is,

15 ft. x 20 ft. = 300 sq. ft.

Multiply square footage by desired lumens per square foot to get total lumens required

300 sq. ft. x 80 lm per sq. ft. = 24,000 lm

We now know that a total of 24,000 lm are required to provide the recommended illumination for this residential living room. However, there are dozens of solutions and the one to choose will depend upon several factors; that is, the home owner's preferences, the allotted budget for lighting fixtures, and so on. One solution is as follows:

- Use four 40-W WWX (warm-white deluxe) fluorescents, at 2,080 lm each, installed in a drapery cornice.
- Use one 200-W inside-frosted bulb, at 3,940 lm, installed in a study lamp.
- Use two 75-W recessed R-30 spotlights, at 860 lm each, over the piano.
- Use two table lamps, each containing three-way (100-200-300-W) soft-white bulbs.
- Use one (50-100-150-W) soft-white three-way bulb, at 2,190 lm in a chairside table lamp.

The total of all lamps used yields 25,630 lm, or 85 lm per sq. ft. While this is slightly over the "desired" illumination level, it should be pointed out that this lumen method should be used only as a guide and not as an absolute rule. A slight differential between "desired" and "actual" lumens is permitted in residential lighting.

This lumen method of calculating residential lighting can be an important aid in planning the lighting for a single room or for an entire home or apartment complex. With this guide, it is possible to determine the total light output of all light sources in a given area. This makes it possible to determine quickly the number of lumens required in a specific lighting design to achieve the desired illumination level. The table in Fig. 6-10 gives lumen output of most popular lamps suited for residential applications.

Practical Residential Application

The floor plan of a small residence is illustrated in Fig. 6-11 and serves to demonstrate how the lighting system of this residence was laid out.

LAMP LUMEN OUTPUT TABLE

Incandescent

Watts	Bulb	Designation and finish	Approx. initial lumens
25	A-19	Inside Frosted	232
		Soft-White	222
40	A-19	Inside Frosted	450
		Soft-White	435
		Dawn Pink	340
50	A-19	Inside Frosted	680
60	A-19	Inside Frosted	855
		Soft-White	840
		Dawn Pink	650
75	A-19	Inside Frosted	1170
		Soft-White	1140
		Dawn Pink	870
		Sky Blue	450
100	A-19	Inside Frosted	1750
		Soft-White	1710
100	A-21	Dawn Pink	1200
		Sky Blue	610
150	A-21	Inside Frosted	2830
		Soft-White	2710
150	R-40	Soft-White	2300
200	A-23	Inside Frosted	3940
		Soft-White	3840

3-Way bulbs

Watts	Bulb	Designation and finish	Approx. initial lumens
30/70/100	A-21	Soft-White	275/1010/1285
50/100/150	A-23	Soft-White	560/1630/2190
		Dawn Pink (med. base)	435/1253/1688
50/100/150	R-40	Soft-White Indirect	560/1630/2190
50/200/250	PS-25	Soft-White	550/3560/4110
100/200/300	PS-25	Soft-White	1290/3440/4730
		Dawn Pink (mogul base)	968/2580/3548

Clear spots and floodlights

Watts	Finish and beam type	Bulb designation	Approx. initial lumens
30	Spot	R-20	200
50	Spot	R-20	430
75	Spot or Flood	PAR-38	745
75	Spot or Flood	R-30	860
150	Spot or Flood	PAR-38	1730
150	Spot or Flood	R-40	1950

Colored spots and floodlights

Percent initial lumen output—colored sources as relates to corresponding clear floodlights.

Watts and bulb designation	Amber	Blue	Blue white	Green	Pink	Red	Yellow
50wR20	—	—	55	—	74	—	—
75wR30/ 150wR40	35	10	30	15	60	15	95
100wPAR38	57	5	39	17	52	7	77
150wPAR38 DICHRO SP/FL	52	6	—	18	—	27	78

NOTE: Flair Chandelight as well as colored light sources are generally for decorative lighting and would not be included in a total lumen count.

Fluorescent

Deluxe warm- or cool-white fluorescent

Watts	Identification	Tube designation, thickness and length	Approx. initial lumens
14	WWX or CWX	T-12 (1½" × 15")	460
15	WWX or CWX	T-8 (1" × 18")	600
15	WWX or CWX	T-12 (1½" × 18")	505
20	WWX or CWX	T-12 (1½" × 24)	820
30	WWX or CWX	T-8 (1" × 36")	1520
30	WWX or CWX	T-12 (1½" × 36")	1480
40	WWX or CWX	T-12 (1½" × 48")	2080

Deluxe warm- or cool-white circline fluorescent

Watts	Identification	Tube designation, thickness and diameter	Approx. initial lumens
22	WWX or CWX	T-9 (1⅛" × 8¼")	745
32	WWX or CWX	T-10 (1¼" × 12")	1240
40	WWX or CWX	T-10 (1¼" × 16")	1760

NOTE: Deluxe white fluorescents render colors as they really are rather than distorting. Deluxe warm-white nearly duplicates the color of incandescent light while deluxe cool-white closely resembles daylight.

Fig. 6-10: Lamp lumen output table.

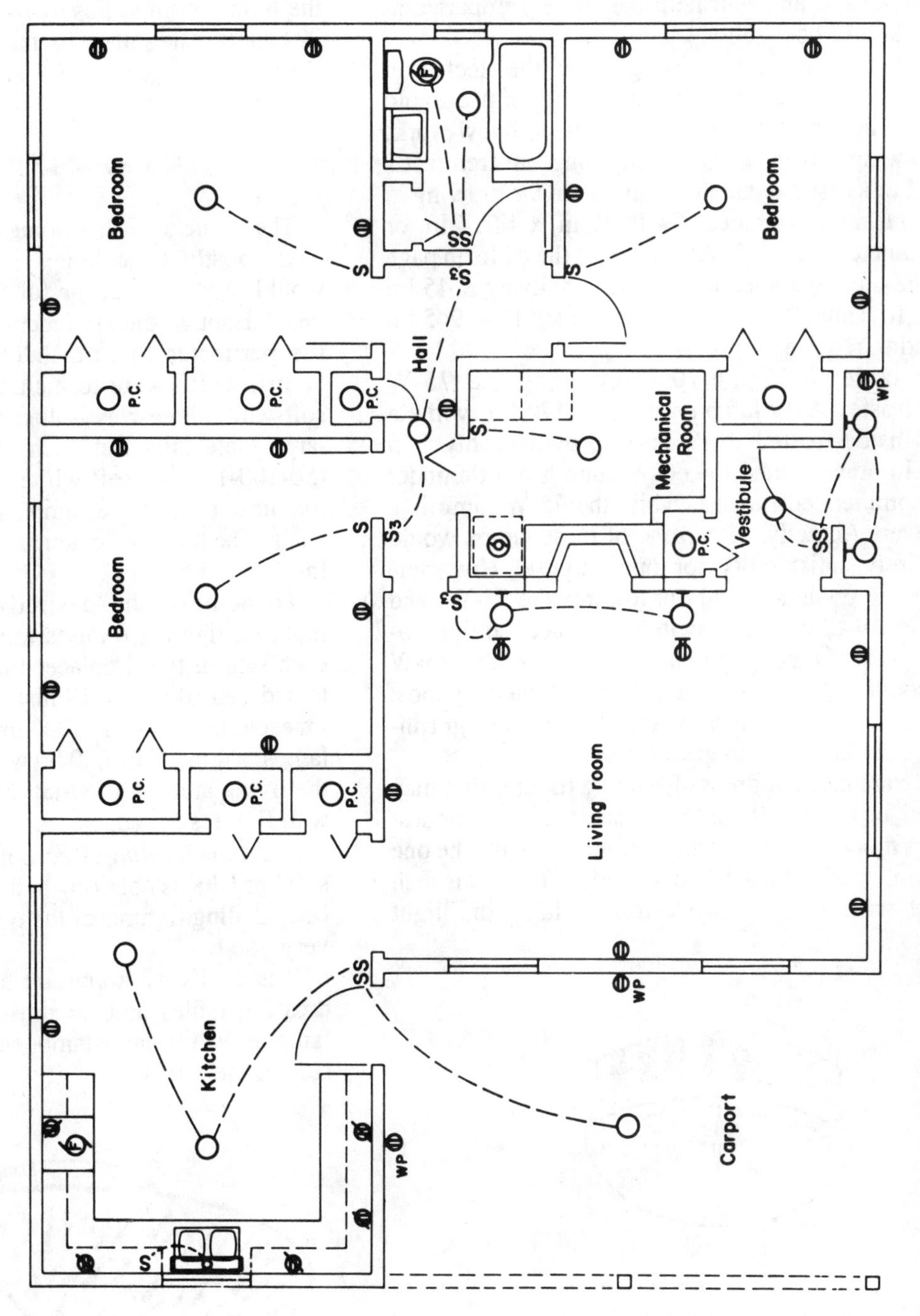

Fig. 6-11: Floor plan of a small residence showing a simple, yet adequate, lighting layout.

Vestibule: The vestibule or entrance hall is the area which gives visitors their first impression of the house. Here the owners greet their guests and help to remove their coats. Carefully planned lighting will show it to best advantage and will help the owners impart graciousness and hospitality.

Where renovating an existing home, the electrician may take actual measurements of the area or use an old set of architectural drawings. In the case of new work, the drawings may be measured using an architect's scale. Let's assume that the vestibule in the drawing in Fig. 6-11 has a floor area of 4 ft. 10 in. x 4 ft. 4 in. or approximately 21 sq. ft. According to the table on page 118, the recommended lumens for a hallway is 45 lm per sq. ft. Thus, 21 sq. ft. x 45 lm per sq. ft.= 945 lm of lighting required in this area.

The table in Fig. 6-10 shows that one 75-W inside-frosted A-19 lamp produces 1,170 lm, while a 60-W inside-frosted A-19 lamp provides only 855 watts. In other words, the 60-W lamp is a little under the recommended lumens, while the 75-W lamp is a little over. Actually, either one of these lamps would be entirely satisfactory for our purpose. However, most home owners would opt towards the 75-W, and then use a wall dimming switch to reduce the light intensity to suit their requirements. That is, the 75-W lamp could be dimmed to, say, half of its intensity most of the time. Then, when visitors arrive, the light intensity may be raised to greet guests.

There are also hundreds of lighting fixtures that may be used in the vestibule, but since this is a modest home, a modest lighting fixture (luminaire) like the one shown in Fig. 6-12 would work nicely. This unit is then lamped with a 75-W inside-frosted lamp or "light bulb."

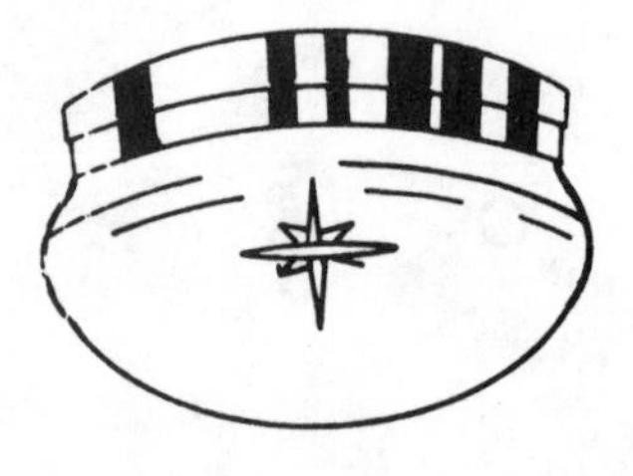

Fig. 6-12: A modest surface-mounted lighting fixtures suited for vestibules and hallways.

Living room: The living room is, of course, one of the most important rooms to have well lighted, since in the average home, like the one under consideration, this room is the one in which members of the family spend much of their time, and also one that the owners wish to have most attractive when guests are present.

Referring again to the chart containing the lumens required and to the floor plan in Fig. 6-11, we find that the living room scales to 14 ft. 0 in. x 16 ft. 0 in., or 224 sq. ft. The square footage is then multiplied by 80, the number of lumens per square foot recommended in the table.

224 x 80 = 17,920 lm required

The owners of this home expressed their desire to have mostly table lamps in the living areas which would provide local or functional lighting for visual tasks. Enough duplex receptacles were placed around the perimeter to furnish electricity for these table lamps, which were to include: two (100-200-300-W) soft-white three-way bulbs, at 4,730 lm each, in two large table lamps at each end of the sofa; and one (50-100-150-W) soft-white three-way bulb, at 2,190 lm, in another table lamp placed next to a reclining chair. The three table lamps produced a total of 11,630 lm.

To achieve the desired 17,920 lm, two wall-mounted lighting fixtures were chosen to be placed on each side of the fireplace. Each of these fixtures contained two 100-W A-19 inside-frosted lamps, at 1,750 lm each, or a total of 7,000 lm. Together with the table lamps, a total of 18,650 lm was available for lighting in the living area. A rheostat dimmer controlled the two wall fixtures for added versatility.

Bedroom lighting: Bedrooms should be lighted with soft light that is not tiring to the eyes of a person lying in bed. Ceiling fixtures of the type shown in Fig. 6-13 are very good.

It is also very important to have sufficient light at dressing tables and on mirrors; either wall-mounted lamps or vanity-table lamps should be provided on each side of the mirror.

Fig. 6-13: One type of lighting fixture suited to bedroom lighting.

Portable lamps on small tables by the beds or clamp-on lights on the headboards are ideal for reading. A lamp at one side of the bed should direct its light so as to provide adequate reading light for the person on that side of the bed, yet not greatly disturb the person on the other side who might be trying to sleep.

A switch controlling one of the lights in the room should be located near enough to a bed to be within easy reach of a person either in the bed or right at its edge. A table lamp with a built-in switch will, of course, accomplish this.

All the bedrooms in the floor plan in Fig. 6-11 are approximately the same size; that is, 158 sq. ft. Referring to the chart, we find that bedrooms require 70 lm per sq. ft. Therefore,

70 x 158 = 11,060 lm required

The surface-mounted ceiling fixtures installed in the center of each room contain three 75-W inside-frosted lamps, at 1,170 lm each, for a total of 3,510 lm. The remaining lumens (7,550) desired were obtained by using table lamps at each side of the bed, each containing one (50-100-150-W) soft-white three-way bulb at 2,190 lm, and two more table lamps on the dresser with a 100-W inside-frosted lamp in each (1,750 lm apiece). This gave a total of 11,390 lm of illumination in the bedroom.

Hall: Since the hallway in the residence in Fig. 6-11 has a total of 36 sq. ft. the total amount of lumens required is 36 x 45 = 1,620. A lighting fixture similar to the one in the vestibule is used, except that this time one 100-W A-19 inside-frosted lamp is used, and this produces 1,750 lm.

Mechanical room: The mechanical room is this residence contains the furnace and water heater, also a washer and dryer. Therefore, this room is treated as a laundry area requiring 70 lm per sq. ft.

60 x 70 lm = 4,200 lm

A lighting fixture was chosen that contained four 75-W inside-frosted lamps at 1,170 lm each; one with a diffusing shade that gives a relatively soft and comfortable light for such a high intensity of illumination. The fixture produced 4,680 lm of light.

Kitchen: The kitchen in any residence or apartment complex should always receive careful attention, because it is an area where the homemaker spends a great deal of time.

Good kitchen lighting begins with a light source located in the center of the room and close to the ceiling. Since this is the fixture to furnish general illumination, it should be a glarefree source which will direct light to every corner in the kitchen. The homemaker will find work a lot easier if certain "task" areas are specially lighted—for example, counter lights beneath cabinets, or, as shown in the floor plan, a recessed fixture over the sink to provide good supplementary lighting and eliminate shadows.

The kitchen-dinette shown in the floor plan has a pulldown fixture directly over the table for functional, yet decorative illumination. A combination hood/fan/light is also installed over the kitchen range, producing supplementary lighting for cooking. The two rooms are treated as one area needing 80 lm per sq. ft.

Bathroom: Bathrooms should have either two wall-mounted fixtures, one on each side of the mirror, or one fixture mounted above the mirror for general light inside the room as well as for grooming. The light should be sufficient to illuminate one's face and the underside of the chin for shaving.

In this home, one 20-W fluorescent fixture was used over the mirror to produce 820 lm. Since this was somewhat short of the 2,275-lm requirement determined previously, another lighting fixture was installed in the ceiling; this fixture contained one 100-W inside-frosted lamp at 1,750 lm. A similar fixture, containing one 60-W inside-frosted lamp, was placed in the area containing the linen closet.

Outdoor lighting: Outdoor lighting is a partner in modern living. It welcomes guests and lights their paths, creating a hospitable look. As a safety factor, it protects the home from prowlers and reduces the incidence of outdoor accidents.

A pair of outdoor wall brackets, such as those shown at the front entrance on the floor plan is basic. If a long walk is necessary, a decorative yet practical post light should be included.

The carport is a separate lighting area, and, as shown in the floor plan, an all-purpose lighting fixture centered in the ceiling is adequate to light the owner's way to the back door.

The varieties are almost endless, and any number of designs are possible as long as they comply with the latest edition of the NE Code. Contractors, designers, and electricians should use their experience and judgment to give the very best lighting system to fit the home owner's needs and budget.

Electric Heating

Fixed electric space heating equipment is covered in Article 424 of the 1993 NE Code. Such equipment includes heating cable, unit heaters, boilers, central systems, and other approved fixed electric space heating equipment. This Article, however, does not apply to process heating nor room air conditioning. Equipment used in hazardous locations must comply with Articles 500 through 517 and that equipment incorporating a hermetic refrigerant motor-compressor must also comply with Article 440.

In selecting heating and cooling equipment for the home, there are many factors that must be considered. The living habits of the family who resides in the home is one big factor. How much time is actually spent at home? How much unoccupied space is in the house? Does one member of the household—like an elderly relative—like the temperature in his or her room to be at least 78 degrees F, while the rest of the family prefers 70 degrees F? How many bedrooms are vacant most of the time? All of these are factors to consider when selecting an heating, ventilating, air-conditioning (HVAC) system for the home.

To some extent, the construction features of the house to be conditioned will dictate the type of heating system to use. Other questions to consider are:

- Does the house have a basement?
- Is it feasible to install ductwork in the existing structure, or would this type of installation require too much cutting and patching?
- Does the owner want (and can afford) the best, or should the system be of fundamental design—one to which the owners can add later?
- Will the home require maintaining a constant temperature 24 hours a day, or do family members need heat quickly when they return home in the evening?

Another consideration is the type of fuel to use. The selection will mostly be determined by the location of the house, as the distance from the house to fuel sources affects the fuel cost. But initial fuel cost is not always a good indication of the type to select. A fuel that costs more initially but provides clean and automatically controlled heat could be the cheapest in the long run. This may not always be the case, but should be considered.

Less than 25 years ago, electric heating units were used only for supplemental heat in small, seldom-used areas of the home, such as a laundry room or workshop, or in vacation homes on chilly autumn nights. Today, however, electric heat is used extensively in both new and renovated homes.

In addition to the fact that electricity is the cleanest fuel available, electric heat is usually the least expensive to install and maintain. Individual room heaters are very inexpensive compared to furnaces and ductwork required in oil and gas forced-air systems, no chimney is required, no utility room is necessary since there is no furnace or boiler, and the installation time and labor are less. Combine all these features and we have a heating system that ranks with the best.

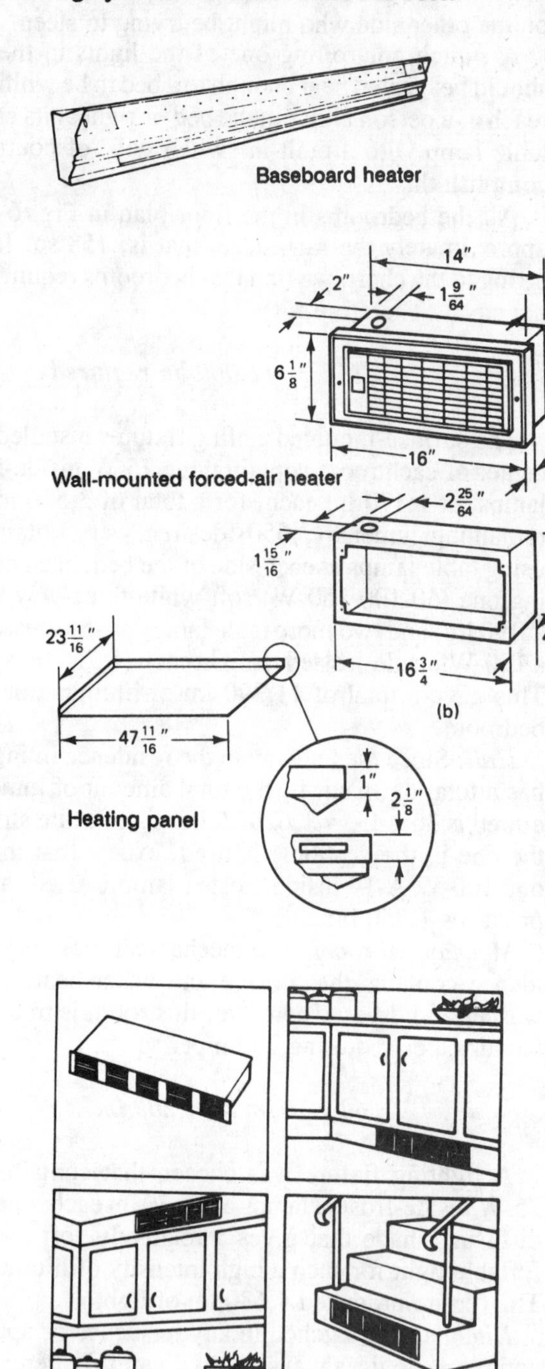

Fig. 6-14: Several types of electric space heaters.

Several types of electric heating units are available (see Fig. 6-14) and a description of each will help you decide which will best suit your needs, and also help in repair and maintenance of such units.

Electric baseboard heaters: Electric baseboard heaters are mounted on the floor along the baseboard, preferably on outside walls under windows for the most efficient operation. It is absolutely noiseless in operation and is the type most often used for heating residential occupancies. The ease with which each room or area may be controlled separately is another great advantage of this heater. Living areas can be heated to, say, 70 degrees F; bedroom heat lowered to, say 55 degrees F, for sleeping comfort; and unused areas may be turned off completely.

Electric baseboard units may be mounted on practically any surface (wood, plaster, drywall, and so on), but if polystyrene foam insulation is used near the unit, a ¾-inch (minimum) ventilated spacer strip must be used between the heater and the wall. In such cases, the heater should also be elevated above the floor or rug to allow ventilation to flow from the floor upward over the total heater space.

One complaint received over the years about this type of heater had been wall discoloration directly above the heating units. When this occurred, the reason was almost always traced to one or more of the following:

- High wattage per square foot of heating element.
- Heavy smoking by occupants.
- Poor housekeeping.

Radiant ceiling heaters: Radiant ceiling heaters are often used in bathrooms and similar areas so that the entire room does not have to be overheated to meet the need for extra warmth after a bath or shower. They are also used in larger areas, such as a garage or basement, or for spot-warming a person standing at a workbench.

Most of these units are rated from 800 to 1,500 watts (W) and normally operate on 120-volt circuits. As with most electric units, they may be controlled by a remote thermostat, but since they are usually used for supplemental heat, a conventional wall switch is often used. They are quickly and easily mounted on an outlet box in much the same way as conventional lighting fixtures. In fact, where very low wattage is used, ceiling heaters may often be installed by merely replacing the ceiling lighting fixture with a heater.

Radiant heating panels: Radiant heating panels are commonly manufactured in 2′ x 4′ sizes and are rated at 500 watts. They may be located on ceiling or walls to provide radiant heat that spreads evenly through the room. Each room may be controlled by its own thermostat. Since this type of heater may be mounted on the ceiling, its use allows complete freedom for room decor, furniture placement, and drapery arrangement. Most are finished in beige to blend in with nearly any room or furniture color.

Units mounted on the ceiling give the best results when located parallel to, and approximately 2 feet from, the outside wall. However, this type of unit may also be mounted on the walls.

Electric infrared heaters: Rays from infrared heaters do not heat the air through which they travel. Rather, they heat only persons and certain objects that they strike. Therefore, infrared heaters are designed to deliver heat into controlled areas for the efficient warming of people and surfaces both indoors and outdoors (such as to heat persons on a patio on a chilly night or around the perimeter of an outdoor swimming pool). This type of heater is excellent for heating a person standing at a workbench without heating the entire room, melting snow from steps or porches, sunlike heat over outdoor areas, and similar applications.

Some of the major advantages of infrared heat include:

- No warm-up period is required. Heat is immediate.
- Heat rays are confined to the desired areas.
- They are easy to install, as no ducts, vents, and so on, are required.
- The infrared quartz lamps provide some light in addition to heat.

Forced-air wall heaters: Forced-air wall heaters are designed to bring quick heat into an area where the sound of a quiet fan will not be disturbing. Some are very noisy. Most of these units are equipped with a built-in thermostat with a sensor mounted in the intake air stream. Some types are available for mounting on high walls or even ceilings, but the additional force required to move the air to a usable area produces even more noise.

Floor insert convection heaters: Floor insert convection heaters require no wall space, they fit into the floor. They are best suited for placement beneath conventional or sliding glass doors to form an effective draft barrier. All are equipped with safety devices, such as a thermal cutout to disconnect the heating element automatically in the event that normal operating temperatures are exceeded.

Floor insert convector heaters may be installed in both old and new homes by cutting through the floor, inserting the metal housing and wiring, according to the manufacturer's instructions. A heavy-gauge floor grille then fits over the entire unit.

Electric kick-space heaters: Modern kitchens contain so many appliances and so much cabinet space for the convenience of the owner that there often is no room to install electric heaters except on the ceiling. Therefore, a kick-space heater was added to the lines of electric heating manufacturers to overcome this problem.

For the most comfort, kick-space heaters should not be installed in such a manner that warm air blows directly on the occupant's feet. Ideally, the air discharge should be directed along the outside wall adjacent to normal working areas, not directly under the sink.

Radiant heating cable: Radiant heating cable provides an enormous heating surface over the ceiling or concrete floor so that the system need not be raised to a high temperature. Rather, gentle warmth radiates downward (in the case of ceiling-mounted cable) or upward (in the case of floor-mounted cable), heating the entire room or area evenly.

There is virtually no maintenance with a radiant heating system, as there are no moving parts and the entire heating system is invisible—except for the thermostat.

Combination heating and cooling units: One way to have individual control of each room or area in the home, as far as heating and cooling are concerned, is to install through-wall heating and cooling units. Such a system gives the occupants complete control of their environment with a room-by-room choice of either heating or cooling at any time of year at any temperature they desire. Operating costs are lower than for many other systems due to the high efficiency of room-by-room control. Another advantage is that if a unit should fail, the defective chassis can be replaced immediately or taken to a shop for repair without shutting down the remaining units in the home.

When selecting any electric heating units, obtain plenty of literature from suppliers and manufacturers before settling on any one type. In most cases you are going to get what you pay for, but shop around at different suppliers before ordering the equipment.

Delivery of any of these units may take some time, so once the brand, size, and supplier have been selected, place your order well before the unit is actually needed.

Electric furnaces: Electric furnaces are becoming more popular, although they are somewhat surpassed by the all-electric heat pump. Most are very compact, versatile units designed for either wall, ceiling, or closet mounting. The vertical model can be flush mounted in a wall or shelf mounted in a closet; the horizontal design can be fitted into a ceiling (flush or recessed).

Central heating systems of the electrically energized type distribute heat from a centrally located source by means of circulating air or water. Compact electric boilers can be mounted on the wall of a basement, utility room, or closet with the necessary control and circuit protection, and will furnish hot water to convectors or to embedded pipes. Immersion heaters may be stepped in one at a time to provide heat capacity to match heat loss.

The majority of electric furnaces are commonly available in sizes up to 24 kilowatts (kw) for residential use. The larger boilers with proper controls can take advantage of lower off-peak electricity rates, where they prevail, by heating water during off-peak periods, storing it in insulated tanks, and circulating it to convectors or radiators to provide heat as needed.

Hot-Water Systems

A zone hydronic (hot-water) heating system permits selection of different temperatures in each zone of the home. Baseboard heaters located along the outer walls of rooms provide a blanket of warmth from floor to ceiling; the heating unit also supplies domestic hot water simultaneously, through separate circuits. A special attachment coupled to the hot-water unit can be used to melt snow and ice on walkways and driveways in winter, and a similar attachment can be used to heat your swimming pool during the spring and fall seasons.

A typical hot-water system operating diagram is shown in Fig. 6-15, and is explained as follows: When a zone thermostat calls for heat, the appropriate zone valve motor begins to run, opening the valve slowly; when the valve is fully opened, the valve motor stops. At that time, the operating relay in the hydrostat is energized, closing contacts to the burner and the circulator circuits. The high-limit control contacts (a safety device) are normally closed so the burner will now fire and operate. If the boiler water temperature exceeds the high-limit setting, the high-limit contacts will open and the burner will stop, but the circulator will continue to run as long as the thermostat continues to call for heat. If the call for heat continues, the resultant drop in boiler water temperature—below the high-limit setting—will bring the burner back on. Thus, the burner will cycle until the thermostat is satisfied; then both the burner and circulator will shut off.

Hot-water boilers for the home are normally manufactured for use with oil, gas, or electricity. While a zoned hot-water system is comparatively costly to install, the cost is still competitive with the better hot-air systems. The chief disadvantage of hot-water systems is that they don't use ducts. If you wish to install central air conditioning, you must install a complete duct system along with the central unit. Chillers, or refrigerated water, may be run through the same pipes, but such systems are usually too costly for most applications.

Practical Applications

When possible, all electric heating units should be installed on an outside wall. However, there are two areas in the residence in Fig. 6-16 where such installations are not practical; the kitchen and utility room. The kitchen, for example, has kitchen cabinets along most of the wall. Therefore, a 2000-kVA kick-space heater is utilized in the kick-space directly below the kitchen sink. Another 1300-kVA baseboard heater is installed on the opposite wall. Since the utility room is centrally located in the house, there are no outside walls. Consequently, a 750-kVA forced-air wall-mounted heater is utilized on one wall. A 750-kVA wall-mounted heater is also used in the bathroom due to limited wall space.

Electric baseboard heaters are utilized in the living room, foyer and all three bedrooms. Note that most of the heaters are controlled by wall-mounted thermostats; the exceptions being the kick-space heater in the kitchen and the wall-mounted heater in the utility room. Both of these heaters have integral thermostats contained within the heating units. Wall-mounted thermostats should be mounted on an inside wall and 48 to 54 inches above the floor. They should always be kept away from heat producing equipment. All line-voltage thermostats should be the two-pole type with an OFF position. Low-voltage thermostats may be used in conjunction with relays.

Circuit requirements for electric heating equipment do not greatly differ from those for any other circuit, but the manufacturer's wiring diagrams must be thoroughly studied for the more complicated systems.

For example, an electric furnace may have a total rating of 24 kVA. It would be natural to calculate the circuit size for the unit as follows:

24,000/240 x 1.25 (NEC factor) = 125 amperes

However, resistance-type heaters found in electric furnaces usually will be connected in steps, that is, four elements of 6 kVA each. If one feeder circuit is run to the furnace, a subpanel would have to be installed to separately feed each of the four heaters; that is, four 40-ampere circuits feeding each element. See Figs. 6-17 and 6-18 for NE Code regulations.

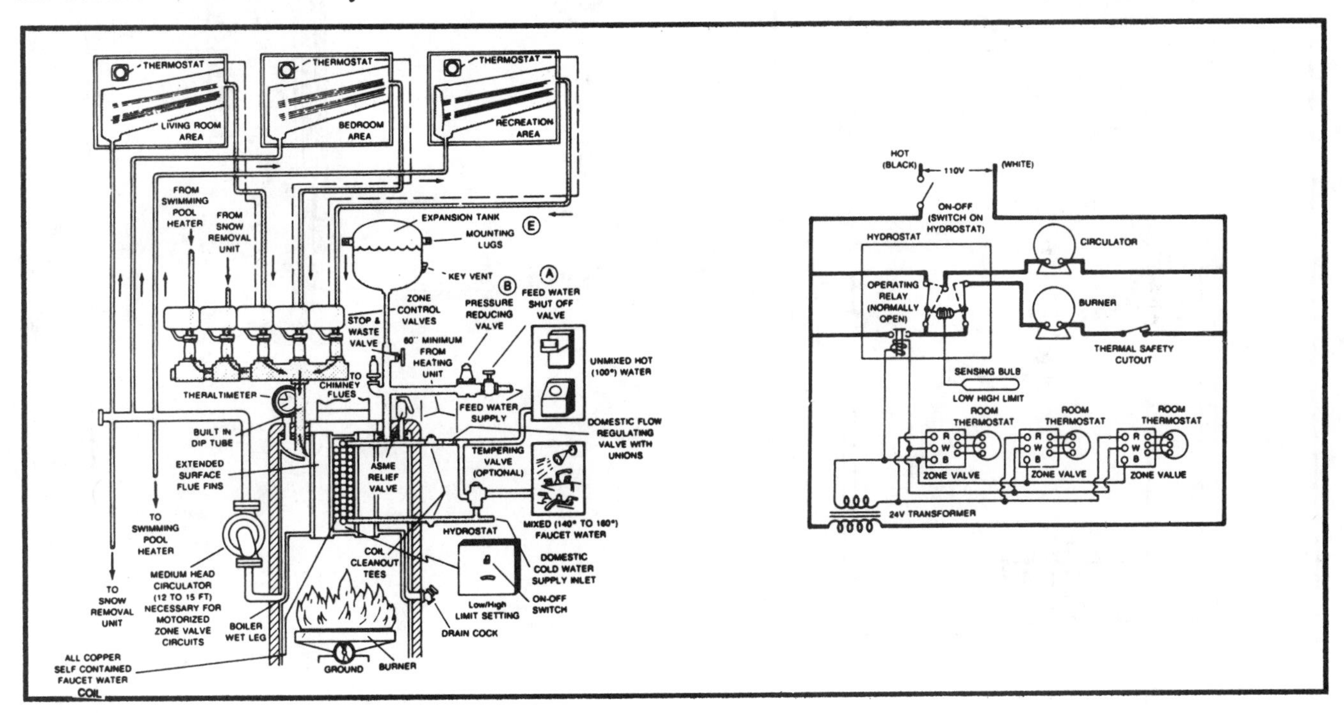

Fig. 6-15: A typical hot-water system operating diagram.

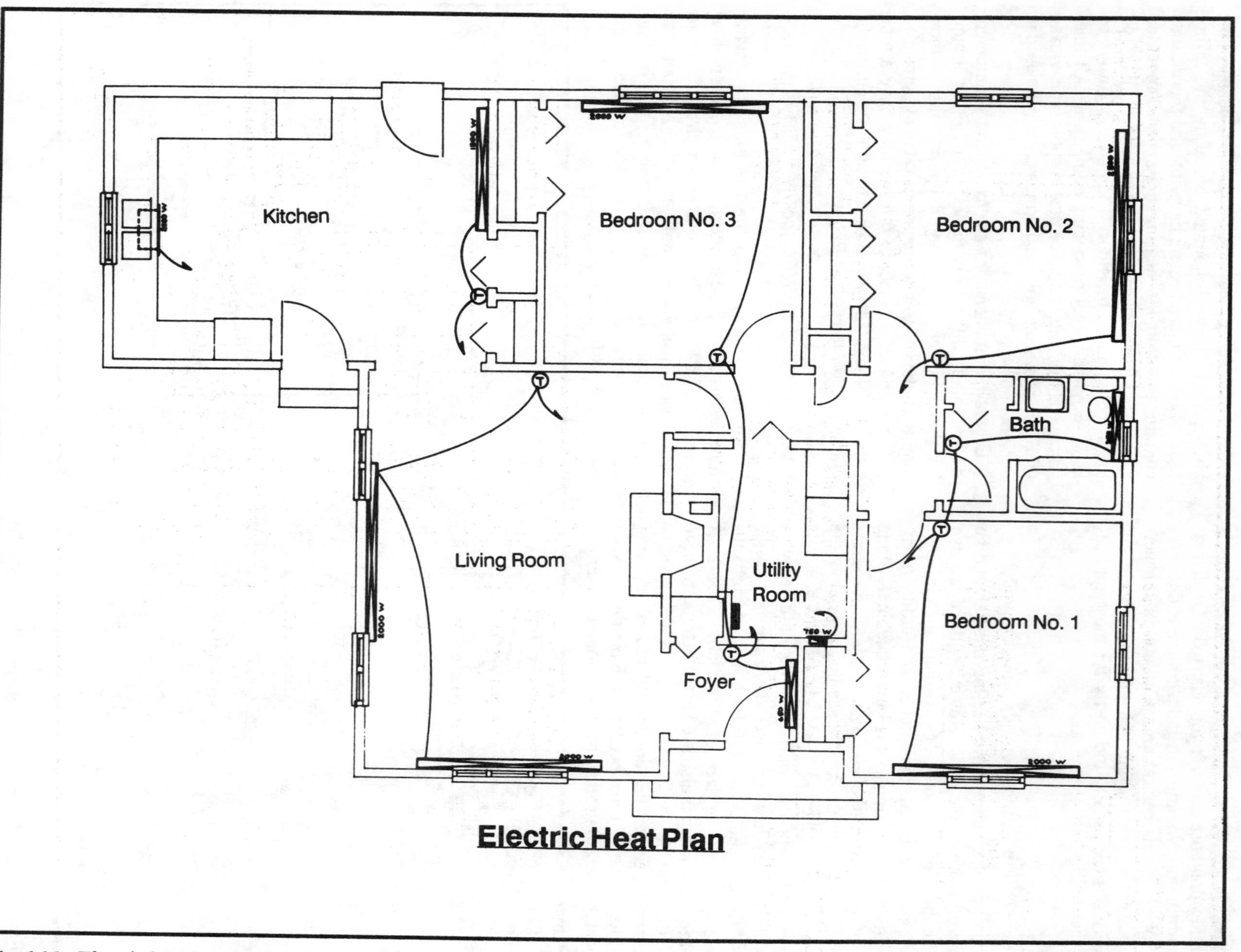

Fig. 6-16: Electric heat layout in a small residence.

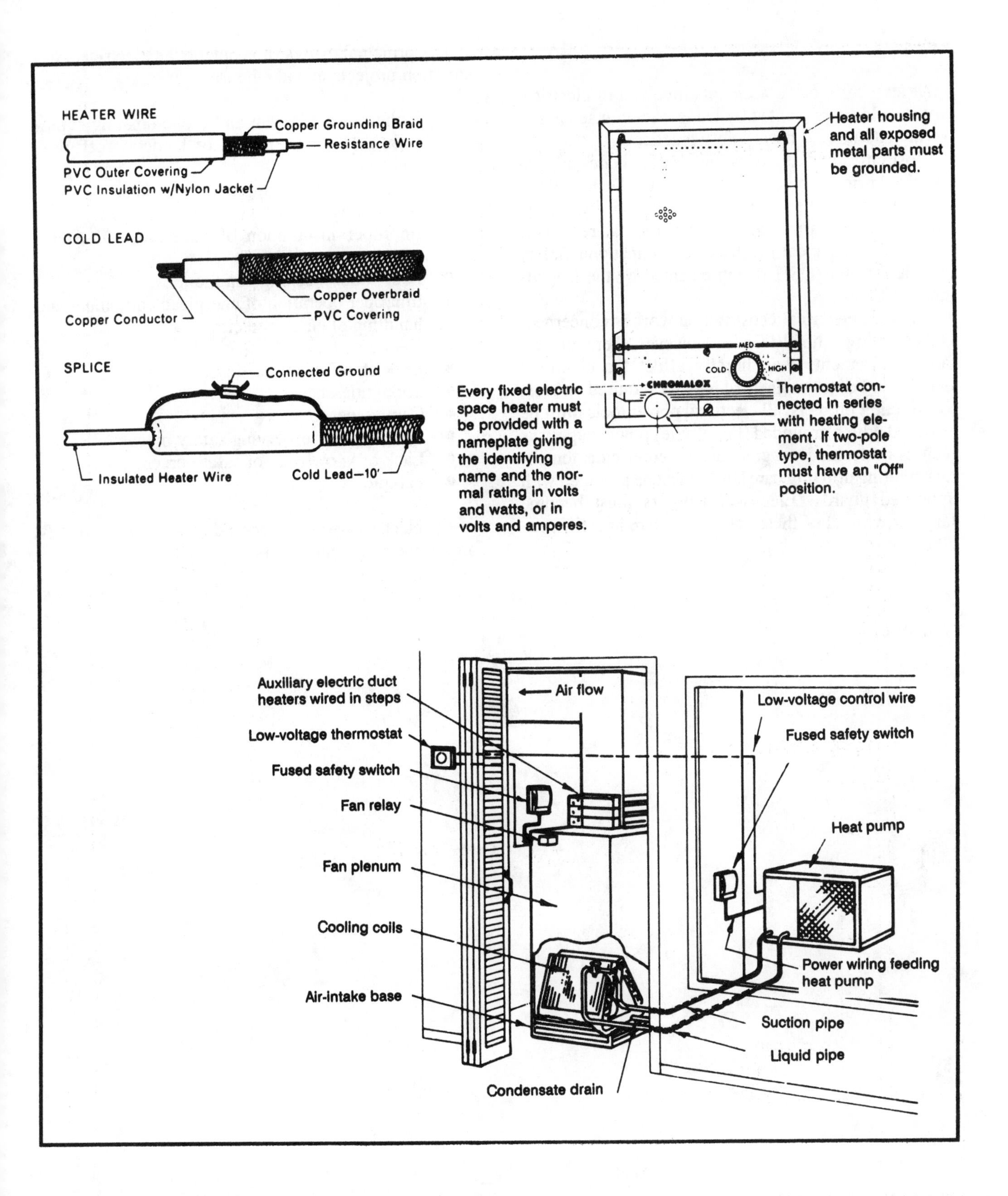
HEATER WIRE
Copper Grounding Braid
Resistance Wire
PVC Outer Covering
PVC Insulation w/Nylon Jacket
COLD LEAD
Copper Overbraid
PVC Covering
Copper Conductor
SPLICE
Connected Ground
Insulated Heater Wire
Cold Lead—10′
Heater housing and all exposed metal parts must be grounded.
MED
COLD
HIGH
CHROMALOX
Every fixed electric space heater must be provided with a nameplate giving the identifying name and the normal rating in volts and watts, or in volts and amperes.
Thermostat connected in series with heating element. If two-pole type, thermostat must have an "Off" position.
Auxiliary electric duct heaters wired in steps
Air flow
Low-voltage thermostat
Low-voltage control wire
Fused safety switch
Fused safety switch
Fan relay
Heat pump
Fan plenum
Cooling coils
Power wiring feeding heat pump
Air-intake base
Suction pipe
Liquid pipe
Condensate drain

Fig. 6-17: NE Code requirements governing electric heating equipment.

Safety

At least 50% of all accidents involved in electrical work could be prevented by using common sense and paying attention to basic safety practice.

Most cities and all states have safety codes and inspection officers to enforce these codes. Details of the codes are available from county and state inspection officers. Most areas now utilize the far-reaching, industry-wide standard such as the Occupational Safety and Health Act (OSHA) implemented by the Department of Labor.

As far as electrical construction work is concerned, OSHA follows the NE Code. However, there are certain requirements that directly affect the electrical technician. For example, hard hats are required on most construction projects; all electrical power tools must be provided with a grounded (three-wire) power supply as well as protected by a ground-fault connector; temporary lighting utilizing bare lamps must be provided with approved guards. Electrical workers must become familiar with all of these activities; there are stiff penalties for not complying.

The principal causes of accidents on electrical construction projects include the following:

- Failure of equipment and tools; hoisting equipment and slings, failure of ladders, scaffolds, and so on.
- Improper use of equipment and tools.
- Attempting to handle excessive weights.
- Improper installation of materials and equipment.
- Poor "housekeeping" on the jobs.
- Improper operation of energized equipment and handling of energized circuits.
- Undue haste.
- Lack of sufficient personnel.
- Thoughtlessness.
- Carelessness.
- Contempt for observing safety precautions.
- Lack of knowledge of safety precautions.
- Fatigue.

All electrical workers should take steps to rectify any of the above inefficiencies.

chapter 7

ELECTRIC MOTORS AND CONTROLLERS

The principal means of changing electrical energy into mechanical energy or power is the electric motor, ranging in size from small fractional-horsepower, low-voltage motors to the very large high-voltage synchronous motors. Electric motors are classified according to size (horsepower); type of application; electrical characteristics; speed, starting, speed control and torque characteristics; and mechanical protection and method of cooling.

Article 430 of the NE Code covers rules governing motors, motor circuits, and controllers, including phase converters. All such equipment must be installed in a location that allows adequate ventilation to cool the motors. Furthermore, the motors should be located so that maintenance, troubleshooting and repairs can be readily performed. Such work could consist of lubricating the motor's bearings, or perhaps replacing worn brushes. Testing the motor for open circuits and ground faults is also necessary from time to time.

When motors must be installed in locations where combustible material, dust, or similar material may be present, special precautions must be taken in selecting and installing motors. See Chapter 4—Wiring in Hazardous Locations.

Exposed live parts of motors operating at 50 volts or more between terminals must be guarded; that is, they must be installed in a room, enclosure, or location so as to allow access by only qualified persons (electrical maintenance personnel). If such a room, enclosure, or location is not feasible, an alternative is to elevate the motors not less than 8 feet above the floor. In all cases, adequate space must be provided around motors with exposed live parts—even when properly guarded—to allow for maintenance, troubleshooting, and repairs.

The illustrations in Fig. 7-1 and 7-2 summarize installation rules for the 1993 NE Code. Further, detailed information may be found in the NE Code book, under the Articles or Sections indicated in the drawings.

Single-Phase Motors

Split-Phase Motors: Split-phase motors are fractional-horsepower units that use an auxiliary winding on the stator to aid in starting the motor until it reaches its proper rotation speed (see Fig. 7-3). This type of motor finds use in small pumps, oil burners, and other residential applications.

In general, the split-phase motor consists of a housing, a laminated iron-core stator with embedded windings forming the inside of the cylindrical housing; a rotor made up of copper bars set in slots in an iron core and connected to each other by copper rings; plates that are bolted to the housing and contain the bearings that support the rotor shaft, and a centrifugal switch inside the housing. This type of rotor is often called a *squirrel cage* rotor since the configuration of the copper bars resembles an actual cage. These motors have no windings as such, and a centrifugal switch is provided to open the circuit to the starting winding when the motor reaches running speed.

■ INSTALLATION RULES FOR TRANSFORMERS

Application	NE Code Regulation	NE Code Section
Location	Motors must be installed in areas with adequate ventilation. They must also be arranged so that sufficient work space is provided for replacement and maintenance.	430-14
	Open motors must be located or protected so that sparks cannot reach combustible materials.	430-16
	In locations where dust or flying material will collect on or in motors in such quantities as to seriously interfere with the ventilation or cooling of motors and thereby cause dangerous temperatures, suitable types of enclosed motors that will not overheat under the prevailing conditions must be used.	
Disconnecting means	A motor disconnecting means must be within sight from the controller location (with exceptions) and disconnect both the motor and controller. The disconnect must be readily accessible and clearly indicate the *Off/On* positions (open/closed).	Article 430 (I)
	Motor-control circuits require a disconnecting means to disconnect them from all supply sources.	430-74
	The service switch may serve to disconnect a single motor if it complies with other rules for a disconnecting means	430-106
	The disconnecting means must be a motor-circuit safety switch rated in horsepower or a circuit breaker.	430-109
Wiring methods	Flexible connections such as Type AC cable, "Greenfield", flexible metallic tubing, etc. are standard for motor connections	Articles 310 and 430

Fig. 7-1: NE Code regulations governing motor installations.

■ **INSTALLATION RULES FOR MOTORS**

Application	NE Code Regulation	NE Code Section
Motor-control circuits	All conductors of a remote motor control circuit outside of the control device must be installed in a raceway or otherwise protected. The circuit must be wired so that an accidental ground in the control device will not start the motor.	430-73
Guards	Exposed live parts of motors and controllers operating at 50 volts or more must be guarded by installation in a room, enclosure, or location so as to allow access by only qualified persons, or elevated 8 feet or more above the floor.	Article 430
Motors operating over 600 volts	Special installation rules apply to motors operating at over 600 volts.	Article 30 (J)
Controller grounding	Motor controllers must have their enclosures grounded.	430-144
Grounding	The NE Code requires that all stationary and portable motors operating at move than 150 volts to ground, must have the frame grounded. Frames of stationary motors either supplied by metal-enclosed wiring or where in a wet or hazardous location must be grounded or suitably insulated from ground.	Article 430 (L)
	Where required, motors must be grounded in a manner specified in NE Code Article 250. In addition, where Type AC cable or specified raceways are used, junction boxes to house motor terminals are required to be connected to armor or raceway according the provisions set forth in the NE Code.	430-145
	A junction box may be placed up to 6 feet from a motor if the leads to the motor are Type AC cable, armored cord, or are stranded leads in metal conduit or tubing not smaller than ⅜ inch.	

Fig. 7-2: NE Code regulations governing motor installations.

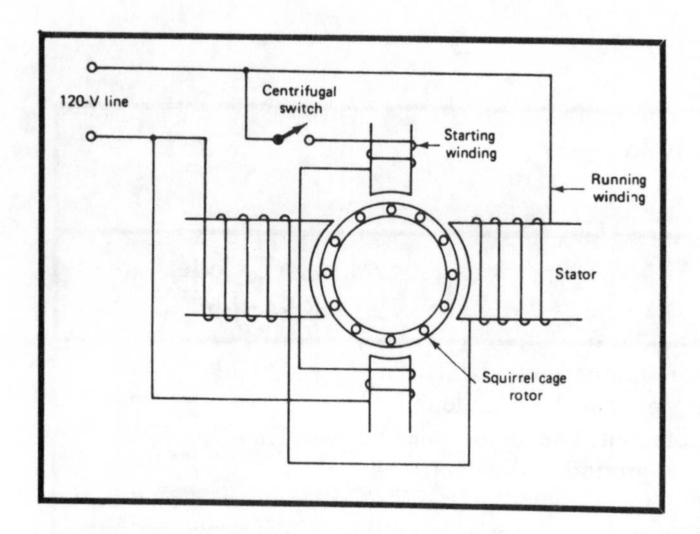

Fig. 7-3: Wiring diagram of a single-phase, split-phase motor.

To understand the operation of a split-phase motor, look at the wiring diagram in Fig. 7-3. Current is applied to the stator windings, both the main winding and the starting winding, which is in parallel with it through the centrifugal switch. The two windings set up a rotating magnetic field, and this field sets up a voltage in the copper bars of the squirrel-cage rotor. Because these bars are shortened at the ends of the rotor, current flows through the rotor bars. The current-carrying rotor bars then react with the magnetic field to produce motor action. When the rotor is turning at the proper speed, the centrifugal switch cuts out the starting winding since it is no longer needed.

Capacitor Motors

Capacitor motors are single-phase ac motors ranging in size from fractional horsepower (hp) to perhaps as high as 15 hp. This type of motor is widely used in all types of single-phase applications such as powering air compressors, refrigerator compressors, and the like. This type of motor is similar in construction to the split-phase motor, except a capacitor is wired in series with the starting winding, as shown in Fig. 7-4.

The capacitor provides higher starting torque, with lower starting current, than does the split-phase motor, and although the capacitor is sometimes mounted inside the motor housing, it is more often mounted on top of the motor, encased in a metal compartment.

In general, two types of capacitor motors are in use: the capacitor-start motor and the capacitor start-and-run motor. As the name implies, the former utilizes the capacitor only for starting; it is disconnected from the circuit once the motor reaches running speed, or at about 75 percent of the motor's full speed. Then the centrifugal switch opens to cut the capacitor out of the circuit.

The capacitor start-and-run motor keeps the capacitor and starting winding in parallel with the running winding, providing a quiet and smooth operation at all times.

Capacitor split-phase motors require the least maintenance of all single-phase motors, but they have a very low starting torque, making them unsuitable for many applications. Its high maximum torque, however, makes it especially useful in HVAC systems to power slow-speed direct-connected fans.

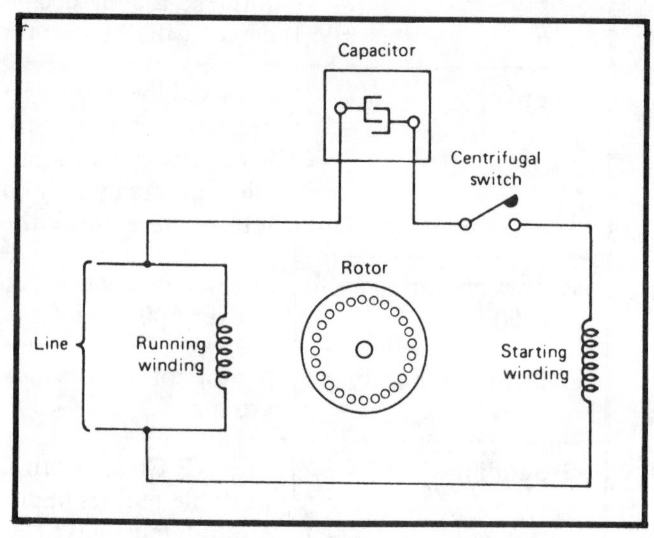

Fig. 7-4: Wiring diagram of a capacitor motor.

Repulsion-Type Motors

Repulsion-type motors are divided into several groups, including (1) repulsion-start, induction-run motors, (2) repulsion motors, and (3) repulsion-induction motors. The repulsion-start, induction-run motor is of the single-phase type, ranging in size from about 1/10 hp to as high as 20 hp. It has high starting torque and a constant-speed characteristic, which makes it suitable for such applications as commercial refrigerators, compressors, pumps, and similar applications requiring high starting torque.

The repulsion motor is distinguished from the repulsion-start, induction-run motor by the fact that it is made exclusively as a brush-riding type and does not have any centrifugal mechanism. Therefore, this motor both starts and runs on the repulsion principle. This type of motor has high starting torque and a variable-speed characteristic. It is reversed by shifting the brush holder to either side of the neutral position. Its speed can be decreased by moving the brush holder farther away from the neutral position.

The repulsion-induction motor combines the high starting torque of the repulsion-type and the good speed regulation of the induction motor. The stator of this motor is provided with a regular single-phase winding, while the rotor winding is similar to that used on a dc motor. When starting, the changing single-phase stator flux cuts across the rotor windings and induces currents in them; thus, when flowing through the commutator, a continuous repulsive action on the stator poles is present.

This motor starts as a straight repulsion-type and accelerates to about 75 percent of normal full speed when a centrifugally operated device connects all the commutator bars together and converts the winding to an equivalent squirrel-cage type. The same mechanism usually raises the brushes to reduce noise and wear. Note that, when the machine is operating as a repulsion-type, the rotor and stator poles reverse at the same instant, and that the current in the commutator and brushes is ac.

This type of motor will develop four to five times normal full-load torque and will draw about three times normal full-load current when starting with full-line voltage applied. The speed variation from no load to full load will not exceed 5 percent of normal full-load speed.

The repulsion-induction motor is used to power air compressors, refrigeration (compressor and fans), pumps, stokers, and the like. In general, this type of motor is suitable for any load that requires a high starting torque and constant-speed operation. Most motors of this type are less than 5 hp.

Universal Motor

This type of motor is a special adaptation of the series-connected dc motor, and it gets its name *universal* from the fact that it can be connected on either ac or dc and operate the same. All are single-phase motors for use on 120 or 240 volts.

In general, the universal motor contains field windings on the stator within the frame, an armature with the ends of its windings brought out to a commutator at one end, and carbon brushes that are held in place by the motor's end plate, allowing them to have a proper contact with the commutator. When current is applied to a universal motor, either ac or dc, the current flows through the field coils and the armature windings in series. The magnetic field set up by the field coils in the stator reacts with the current-carrying wires on the armature to produce rotation.

Universal motors are frequently used on small fans, electric razors, small kitchen appliances and similar applications.

Shaded-Pole Motor

A shaded-pole motor is a single-phase induction motor provided with an uninsulated and permanently short-circuited auxiliary winding displaced in magnetic position from the main winding. The auxiliary winding is known as the shading coil and usually surrounds from one-third to one-half of the pole (see Fig. 7-5). The main winding surrounds the entire pole and may consist of one or more coils per pole.

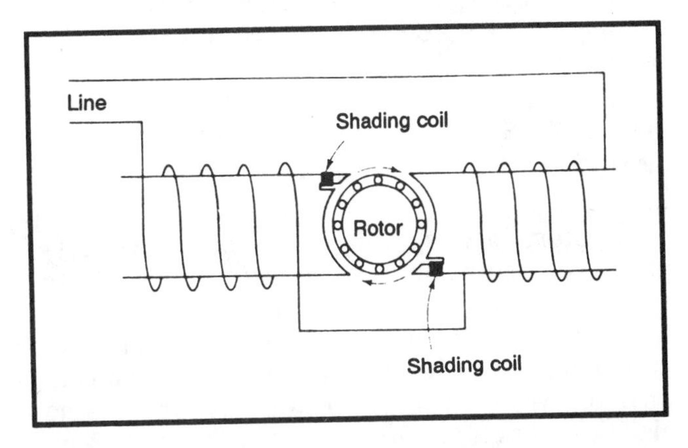

Fig. 7-5: Wiring diagram of a shaded-pole motor.

Applications for the shaded-pole motor include small fans, timing devices, relays, instrument dials, or any constant-speed load not requiring high starting torque.

Polyphase Motors

Three-phase motors offer extremely efficient and economical application and are usually the preferred type for commercial and industrial applications when three-phase service is available. In fact, the great bulk of motors sold are standard ac three-phase motors. These motors are available in ratings from fractional horsepower up to thousands of horsepower in practically every standard voltage and frequency. In fact, there are few applications for which the three-phase motor cannot be put to use.

Three-phase motors are noted for their relatively constant speed characteristic and are available in designs giving a variety of torque characteristics; that is, some have a high starting torque and others a low starting torque. Some are designed to draw a normal starting current, others a high starting current.

A typical three-phase motor is shown in Fig. 7-6. Note that the three main parts are the stator, rotor, and end plates. It is very similar in construction to conventional split-phase motors except that the three-phase motor has no centrifugal switch.

The stator consists of a steel frame and a laminated iron core and winding formed of individual coils placed in slots. The rotor may be a squirrel-cage or wound-rotor type. Both types contain a laminated core pressed onto a shaft. The squirrel-cage rotor is similar to a split-phase motor. The wound rotor has a winding on the core that is connected to three slip rings mounted on the shaft.

The end plates or brackets are bolted to each side of the stator frame and contain the bearings in which the shaft revolves. Either ball bearings or sleeve bearings are used.

Induction motors, both single-phase and polyphase, get their name from the fact that they utilize the principle of electromagnetic induction. An induction motor has a stationary part, or stator, with windings connected to the ac supply, and a rotation part, or rotor, which contains coils or bars. There is no electrical connection between the stator and rotor. The magnetic field produced in the stator windings induces a voltage in the rotor coils or bars.

Since the stator windings act in the same way as the primary winding of a transformer, the stator of an induction motor is sometimes called the *primary*. Similarly, the rotor is called the *secondary* because it carries the induced voltage in the same way as the secondary of a transformer.

The magnetic field necessary for induction to take place is produced by the stator windings. Therefore, the induction-motor stator is often called the *field* and its windings are called *field windings*.

The terms primary and secondary relate to the electrical characteristics and the terms stator and rotor to the mechanical features of induction motors.

The rotor transfers the rotating motion to its shaft, and the revolving shaft drives a mechanical load or a machine, such as a pump, spindle, or clock.

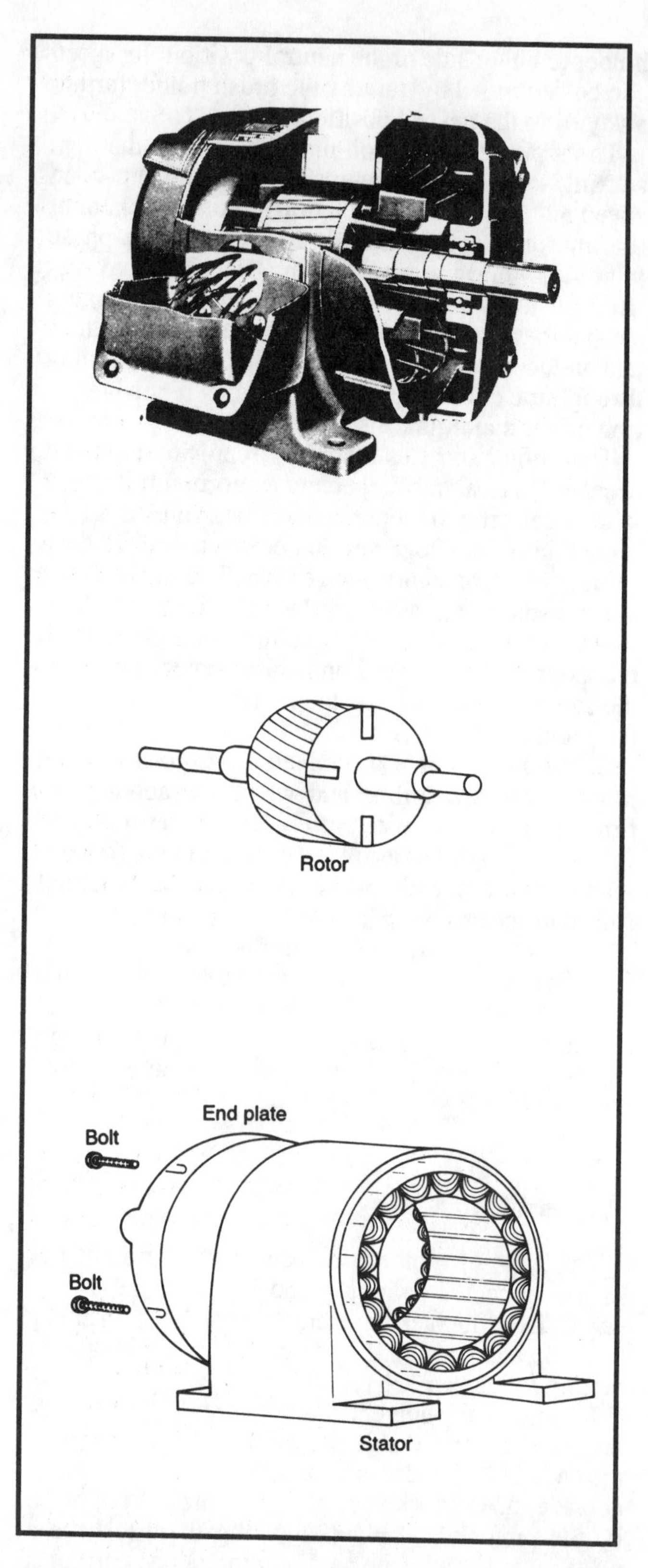

Fig. 7-6: Basic parts of a three-phase induction motor.

Commutator segments, which are essential parts of dc motors, are not needed on induction motors. This simplifies greatly the design and the maintenance of induction motors as compared to dc motors.

The turning of the rotor in an induction motor is due to induction. The rotor, or secondary, is not connected to any source of voltage. If the magnetic field of the stator, or primary, revolves, it will induce a voltage in the rotor, or secondary. The magnetic field produced by the induced voltage acts in such a way that it makes the secondary follow the movement of the primary field.

The stator, or primary, of the induction motor does not move physically. The movement of the primary magnetic field must thus be achieved electrically. A rotating magnetic field is made possible by a combination of two or more ac voltages that are out of phase with each other and applied to the stator coils. Direct current will not produce a rotating magnetic field. In three-phase induction motors, the rotating magnetic field is obtained by applying a three-phase system to the stator windings.

The direction of rotation of the rotor in an ac motor is the same as that of its rotating magnetic field. In a three-phase motor the direction can be reversed by interchanging the connections of any two supply leads. This interchange will reverse the sequence of phases in the stator, the direction of the field rotation, and therefore the direction of rotor rotation.

Motor Enclosures

Electric motors differ in construction and appearance, depending on the type of service for which they are to be used. Open and closed frames are quite common. In the former enclosure, the motor's parts are covered for protection, but the air can freely enter the enclosure. Further designations for this type of enclosure include drip-proof, weather-protected, and splash-proof.

Totally enclosed motors have an airtight enclosure. They may be fan cooled or self-ventilated. An enclosed motor equipped with a fan has the fan as an integral part of the machine, but external to the enclosed parts. In the self-ventilated enclosure, no external means of cooling is provided.

The type of enclosure to use will depend on the ambient and surrounding conditions. In a drip-proof machine, for example, all ventilating openings are so constructed that drops of liquid or solid particles falling on the machine at an angle of not greater than 15 degrees from the vertical cannot enter the machine, even directly or by striking and running along a horizontal or inclined surface of the machine. The application of this machine would lend itself to areas where liquids are processed.

An open motor having all air openings that give direct access to live or rotating parts, other than the shaft, limited in size by the design of the parts or by screen to prevent accidental contact with such parts is classified as a drip-proof, fully guarded machine. In such enclosures, openings shall not permit the passage of a cylindrical rod ½ inch in diameter, except where the distance from the guard to the live rotating parts is more than 4 inches, in which case the openings shall not permit the passage of a cylindrical rod ¾ inch in diameter.

The NE Code requires the enclosures of motors installed in Class I, Division 1 hazardous locations to be of the totally enclosed type. Furthermore, the enclosure must be supplied with positive-pressure ventilation from a source of clean air with its discharge to a safe area. This set-up must be arranged to prevent energizing of the machine until ventilation has been established and the enclosure has been purged with at least 10 volumes of air, and also arranged to automatically deenergize the equipment when the air supply fails.

An alternative to the above is a totally enclosed inert gas-filled type supplied with a suitable reliable source of inert gas for pressuring the enclosure, with devices provided to ensure a positive pressure in the enclosure and arranged to automatically deenergize the equipment when the gas supply fails.

Still another solution is a motor enclosure designed to be submerged in a liquid which is flammable only when vaporized and mixed with air, or in a gas or vapor at a pressure greater than atmospheric and which is flammable only when mixed with air. In this case, the machine must be so arranged to prevent energizing it until it has been purged with the liquid or gas to exclude air, and also arranged to automatically deenergize the equipment when the supply of liquid, or gas or vapor fails or the pressure is reduced to atmospheric.

The external surfaces of the first two types of enclosures shall not exceed 80 percent of the ignition temperature of the gas or vapor involved.

There are other types of drip-proof machines for special applications such as externally ventilated and pipe ventilated, which as the names imply are either ventilated by a separate motor-driven blower or cooled by ventilating air from inlet ducts or pipes.

An enclosed motor whose enclosure is designed and constructed to withstand an explosion of a specified gas or vapor that may occur within the motor and to prevent the ignition of this gas or vapor surrounding the machine is designated *explosionproof* (XP) motors.

Hazardous atmospheres (requiring XP enclosures) of both a gaseous and dusty nature are classified by the NE Code as follows:

- Class I, Group A: atmospheres containing acetylene.
- Class I, Group B: atmospheres containing hydrogen gases or vapors of equivalent hazards such as manufactured gas.
- Class I, Group C: atmospheres containing ethyl ether vapor.
- Class I, Group D: atmospheres containing gasoline, petroleum, naphtha, alcohols, acetone, lacquer-solvent vapors, and natural gas.
- Class II, Group E: atmospheres containing metal dust.
- Class II, Group F: atmospheres containing carbon-black, coal, or coke dust.
- Class II, Group G: atmospheres containing grain dust.

The proper motor enclosure must be selected to fit the particular atmospheres. However, explosionproof equipment is not generally available for Class I, Groups A and B, and it is therefore necessary to isolate motors from the hazardous area.

Motor Type

The type of motor will determine the electrical characteristics of the design. NEMA-designated designs for polyphase motors are given in Fig. 7-7.

An A motor is a three-phase, squirrel-cage motor designed to withstand full-voltage starting with locked rotor current higher than the values for a B motor and having a slip at rated load of less than 5 percent.

A B motor is a three-phase, squirrel-cage motor designed to withstand full-voltage starting and developing locked rotor and breakdown torques adequate for general application.

A C motor is a three-phase, squirrel-cage motor designed to withstand full-voltage starting, developing locked rotor torque for special high-torque applications, and having a slip at rated load of less than 5 percent.

Design D is also a three-phase, squirrel-cage motor designed to withstand full-voltage starting, developing 275 percent locked rotor torque; slip at 5 percent.

Selection of Electric Motors

Each type of motor has its particular field of usefulness. Because of its simplicity, economy, and durability, the induction motor is more widely used for industrial purposes than any other type of ac motor, especially if a high-speed drive is desired.

If ac power is available, all drives requiring constant speed should use squirrel-cage induction or synchronous motors on account of their ruggedness and lower cost. Drives requiring varying speed, such as fans, blowers, or pumps may be driven by wound-rotor induction motors. However, if there are applications requiring adjustable speed or a wide range of speed control, it will probably be desirable to install dc motors on such equipment and supply them from the ac system by motor-generator sets of electronic rectifiers.

Practically all constant-speed machines may be driven by ac squirrel-cage motors because they are made with a variety of speed and torque characteristics. When large motors are required or when power supply is limited, the wound-rotor is used even for driving constant-speed machines. A wound-rotor motor, with its controller and resistance, can develop full-load torque at starting with not more than full-load torque at

NEMA Design	Starting Torque	Starting Current	Breakdown Torque	Full-Load Slip
A	Normal	Normal	High	Low
B	Normal	Low	Medium	Low
C	High	Low	Normal	Low
D	Very high	Low	—	High

Fig. 7-7: NEMA-designated motor designs.

starting, depending on the type of motor and the starter used.

For varying-speed service, wound-rotor motors with resistance control are used for fans, blowers, and other apparatus for continuous duty, and for other intermittent duty applications. The controller and resistors must be properly chosen for the particular application.

Cost is an important consideration where more than one type of ac motor is applicable. The squirrel-cage motor is the least expensive ac motor of the three types considered and requires very little control equipment. The wound-rotor is more expensive and requires additional secondary control.

Motor Controls

Every motor in use must be controlled, if only to start and stop it, before it becomes of any value. Motor controllers cover a wide range of types and sizes, from a simple toggle switch to a complex system with such components as relays, timers, and switches. The common function, however, is the same in any case, that is, to control some operation of an electric motor. A motor controller will include some or all of the following functions:

- Starting and stopping
- Overload protection
- Overcurrent protection
- Reversing
- Changing speed
- Jogging
- Plugging
- Sequence control
- Pilot light indication

The controller can also provide the control for auxiliary equipment such as brakes, clutches, solenoids, heaters, and signals, and may be used to control a single motor or a group of motors.

A suitable disconnecting means must be available to assure that the motor and its controller are disconnected from all supply circuits. Furthermore, the disconnecting means must be within sight of the controller location and should also be within sight of the motor.

The term *motor starter* is often used in the electrical field and means practically the same thing as a *controller*. Strictly, a motor starter is the simplest form of controller and is capable of starting and stopping the motor and providing it with overload protection. This type of switch is housed in a small metal enclosure—about the size of a *utility* or *handy* box—like the one appearing in Fig. 7-8. The switch operates like the conventional toggle switch, but has overload protection connected in series with the load.

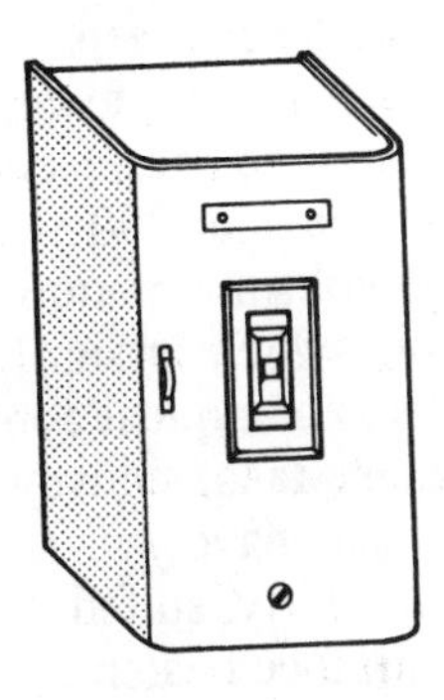

Fig. 7-8: Typical manual motor starter.

Manual Starter

A manual starter is a motor controller whose contact mechanism is operated by a mechanical linkage from a toggle handle or push button, which is in turn operated by hand. A thermal unit and direct-acting overload mechanism provide motor running overload protection. Basically, a manual starter is an ON-OFF switch with overload relays.

Manual starters are used mostly on small machine tools, fans and blowers, pumps, compressors, and conveyors. They have the lowest cost of all motor starters, have a simple mechanism, and provide quiet operation with no ac magnet hum. The contacts, however, remain closed and the lever stays in the ON position in the event of a power failure, causing the motor to automatically restart when the power returns. Therefore, low voltage protection and low-voltage release are not possible with these manually operated starters. However, this action is an advantage when the starter is applied to motors that run continuously.

Fractional-horsepower manual starters are designed to control and provide overload protection for motors of 1 hp or less on 120-or 240-volt single-phase circuits. They are available in single- and two-pole versions and are operated by a toggle handle on the front. When a serious overload occurs, the thermal unit trips to open the starter contacts, disconnecting the motor from the line. The contacts cannot be reclosed until the overload relay has been reset by moving the handle to the full OFF position, after allowing about 2 minutes for the thermal unit to cool. The open-type starter will fit into a standard outlet box and can be used with a standard flush plate. The compact construction of this type of device makes it possible to mount it directly on the driven machinery and in various other places where the available space is small.

Manual motor starting switches provide ON-OFF control of single- or three-phase ac motors where overload protection is not required or is separately provided. Two- or three-pole switches are available with ratings up to 10 hp, 600 volts, three phase. The continuous current rating is 30 amperes at 250 volts maximum and 20 amperes at 600 volts maximum. The toggle operation of the manual switch is similar to the fractional-horsepower starter, and typical applications of the switch include pumps, fans, conveyors, and other electrical machinery that have separate motor protection. They are particularly suited to switch nonmotor loads, such as resistance heaters.

The integral horsepower manual starter is available in two- and three-pole versions to control single-phase motors up to 5 hp and polyphase motors up to 10 hp, respectively.

Two-pole starters have one overload relay and three-pole starters usually have three overload relays. When an overload relay trips, the starter mechanism unlatches, opening the contacts to stop the motor. The contacts cannot be reclosed until the starter mechanism has been reset by pressing the STOP button or moving the handle to the RESET position, after allowing time for the thermal unit to cool.

Integral horsepower manual starters with low-voltage protection prevent automatic start-up of motors after a power loss. This is accomplished with a continuous-duty solenoid, which is energized whenever the line-side voltage is present. If the line voltage is lost or disconnected, the solenoid de-energizes, opening the starter contacts. The contacts will not automatically close when the voltage is restored to the line. To close the contacts, the device must be manually reset. This manual starter will not function unless the line terminals are energized. This is a safety feature that can protect personnel or equipment from damage and is used on such equipment as conveyors, grinders, metal-working machines, mixers, woodworking, etc.

Magnetic Controllers

Magnetic motor controllers use electromagnetic energy for closing switches. The electromagnet consists of a coil of wire placed on an iron core. When current flows through the coil, the iron of the magnet becomes magnetized and attracts the iron bar, called the *armature*. An interruption of the current flow through the coil of wire causes the armature to drop out due to the presence of an air gap in the magnetic circuit.

Line-voltage magnetic motor starters are electromechanical devices that provide a safe, convenient, and economic means for starting and stopping motors, and they have the advantage of being controlled remotely. The great bulk of motor controllers are of this type. Therefore, the operating principles and applications of magnetic motor controllers should be fully understood.

In the construction of a magnetic controller, the armature is mechanically connected to a set of contacts so that, when the armature moves to its closed position, the contacts also close. When the coil has been energized and the armature has moved to the closed position, the controller is said to be *picked up* and the armature is seated or sealed-in. Some of the magnet and armature assemblies in current use are as follows:

1. *Clapper type*: In this type, the armature is hinged. As it pivots to seal in, the movable contacts close against the stationary contacts.
2. *Vertical action*: The action is a straight line motion with the armature and contacts being guided so that they move in a vertical plane.
3. *Horizontal action*: Both armature and contacts move in a straight line through a horizontal plane.
4. *Bell crank*: A bell crank lever transforms the vertical action of the armature into a horizontal contact motion. The shock of armature pickup is not transmitted to the contacts, resulting in minimum contact bounce and longer contact life.

The magnetic circuit of a controller consists of the magnet assembly, the coil, and the armature. It is so named from a comparison with an electrical circuit. The coil and the current flowing in it cause magnetic flux to be set up through the iron in a similar manner to a voltage causing current to flow through a system of conductors. The changing magnetic flux produced by alternating currents results in a temperature rise in the magnetic circuit. The heating effect is reduced by laminating the magnet assembly and armature. By placing a coil of many turns of wire around a soft iron core, the magnetic flux set up by the energized coil tends to be concentrated; therefore, the magnetic field effect is strengthened. Since the iron core is the path of least resistance to the flow of the magnetic lines of force, magnetic attraction will concentrate according to the shape of the magnet.

The magnetic assembly is the stationary part of the magnetic circuit. The coil is supported by and surrounds part of the magnet assembly in order to induce magnetic flux into the magnetic circuit.

The armature is the moving part of the magnetic circuit. When it has been attracted into its sealed-in position, it completes the magnetic circuit. To provide maximum pull and to help ensure quietness, the faces of the armature and the magnetic assembly are ground to a very close tolerance.

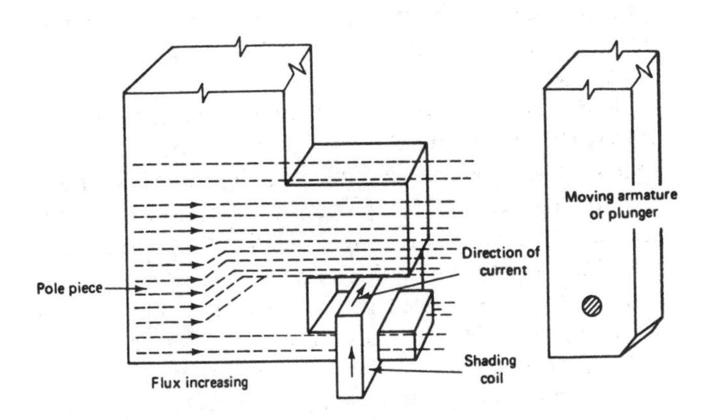

Fig. 7-8: When the flux is increasing in the pole from left to right, the induced current in the coil is in a clockwise direction.

When a controller's armature is sealed-in, it is held closely against the magnet assembly. However, a small gap is always deliberately left in the iron circuit. When the coil becomes de-energized, some magnetic flux (residual magnetism) always remains, and if it were not for the gap in the iron circuit, the residual magnetism might be sufficient to hold the armature in the sealed-in position.

The shaded-pole principle is used to provide a time delay in the decay of flux in dc coils, but it is used more frequently to prevent a chatter and wear in the moving parts of ac magnets. A shading coil is a single turn of conducting material mounted in the face of the magnet assembly or armature. The alternating main magnetic flux induces currents in the shading coil, and these currents set up auxiliary magnetic flux that is out of phase from the pull due to the main flux, and this keeps the armature sealed-in when the main flux falls to zero (which occurs 120 times per second with 60-cycle ac). Without the shading coil, the armature would tend to open each time the main flux goes through zero. Excessive noise, wear on magnet faces, and heat would result.

Figure 7-8 shows an exaggerated view of a pole face with a copper band or short-circuited coil of low resistance connected around a portion of the pole tip. When the flux is increasing in the pole from left to right, the induced current in the coil is in a clockwise direction.

The magnetomotive force produced by the coil opposes the direction of the flux of the main field. Therefore, the flux density in the shaded portion of the iron will be considerably less, and the flux density in the unshaded portion of the iron will be more than would be the case without the shading coil.

Figure 7-9 shows the pole with the flux still moving from left to right but decreasing in value. Now the current in the coil is in a counterclockwise direction. The magnetomotive force produced by the coil is in the same direction as the main unshaded portion but less than it would be without the shading coil. Consequently, if the electric circuit of a coil is opened, the current decreases rapidly to zero, but the flux decreases much more slowly due to the action of the shading coil.

Electrical ratings for ac magnetic contactors and starters are shown in Fig. 7-10.

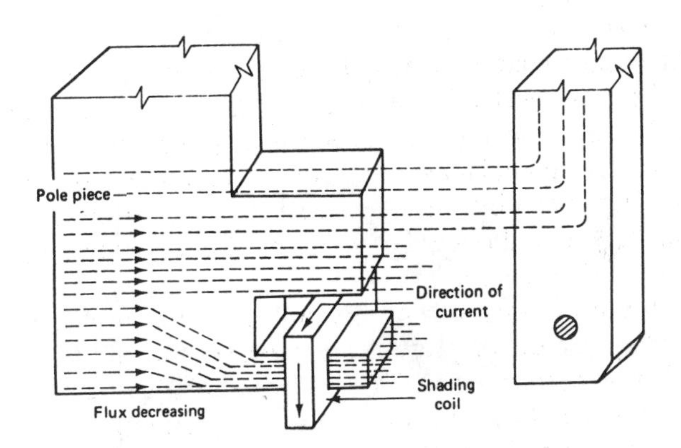

Fig. 7-9: With the flux moving from left to right, the current in the coil is moving in a counterclockwise direction.

Overload Protection

Overload protection for an electric motor is necessary to prevent burnout and to ensure maximum operating life. Electric motors will, if permitted, operate at an output of more than rated capacity. Conditions of motor overload may be caused by an overload on driven machinery, by a low line voltage, or by an open line in a polyphase system, which results in single-phase operation. Under any condition of overload, a motor draws excessive current that causes overheating. Since motor winding insulation deteriorates when subjected to overheating, there are established limits on motor operating temperatures. Overload relays are employed on a motor control to limit the amount of current drawn. This is overload or running protection.

50-60 HERTZ																600 VOLTS MAX.
No. of Poles	NEMA Size	Ratings† Volts	Ratings† Max. HP	Type of Motor	General Purpose Enclosure NEMA Type 1 Type	Price*	Watertight and Dusttight Enclosure NEMA Type 4 Type	Price*	SPIN TOP® For Hazardous Locations Class I Groups C & D Class II Groups E, F & G NEMA Types 7 & 9 Type	Price*	Dusttight and Driptight Industrial Use Enclosure NEMA Type 12 ‡ Type	Price*	Open Type Vertical Type	Open Type Horizontal Type Use For NEMA 1, 4, & 12 Interiors	Price*	Number of Thermal Units Required*
2 Pole Single Phase	00	115 230	⅓ 1	Single Phase 3-Wire	SAG-13	$ 258.	Use NEMA Size 0		Use NEMA Size 0		Use NEMA Size 0		....	SAO-13	$ 242.	1
	0	115 230	1 2		SBG-1	306.	SBW-11▲	$ 490.	SBR-6	$1044.	SBA-1▲	$ 378.	SBO-7	SBO-1	290.	1
	1	115 230	2 3		SCG-2	354.	SCW-11▲	610.	SCR-6	1096.	SCA-1▲	426.	SCO-1	SCO-2	330.	1
3 Pole Single Phase	00	115 230	⅓ 1	4-Wire Rep.-Ind.	SAG-14	266.	Use NEMA Size 0		Use NEMA Size 0		Use NEMA Size 0		....	SAO-14	250.	1
		115 230	⅓ 1	4-Wire Split Ph.	SAG-15	266.	Use NEMA Size 0		Use NEMA Size 0		Use NEMA Size 0		....	SAO-15	250.	1
	0	115 230	1 2	4-Wire Rep.-Ind.	SBG-2	314.	SBW-12▲	498.	SBR-7	1052.	SBA-2▲	386.	SBO-8	SBO-2	298.	1
		115 230	1 2	4-Wire Split Ph.	SBG-3	314.	SBW-13▲	498.	SBR-8	1052.	SBA-3▲	386.	SBO-9	SBO-3	298.	1
	1	115 230	2 3	4-Wire Rep-Ind.	SCG-4	362.	SCW-12▲	618.	SCR-7	1106.	SCA-2▲	434.	SCO-3	SCO-4	338.	1
		115 230	2 3	4-Wire Split Ph.	SCG-6	362.	SCW-13▲	618.	SCR-8	1.06.	SCA-3▲	434.	SCO-5	SCO-6	338.	1
3 Pole Poly-phase	00	200–230 380 460–575	1½ 1½ 2	3 Phase	SAG-16	276.	Use NEMA Size 0		Use NEMA Size 0		Use NEMA Size 0		....	SAO-16	260.	3
	0	200–230 380–575	3 5		SBC 4	324.	SBW-14▲	508.	SBR-9	1064.	SBA-4▲	396.	SBO-10	SBO-4	308.	3
	1	200–230 380–575	7½ 10		SCG-8	372.	SCW-14▲	628.	SCR-9	1116.	SCA-4▲	444.	SCO-7	SCO-8	348.	3
	2	200 230 380–575	10 15 25		SDG-2	724.	SDW-11▲	1140.	SDR-3	1864.	SDA-1▲	844.	SDO-1	SDO-2	660.	3
	3	200 230 380–575	25 30 50		SEG-2	1208.	SEW-11	1824.	SER-3	2996.	SEA-1	1472.	SEO-1	SEO-2	1096.	3
	4	200 230 380 460–575	40 50 75 100		SFG-3	2884.	SFW-11	3972.	SFR-1	5148.	SFA-1	3292.	SFO-1	SFO-3	2676.	3
	5	200 230 380 460–575	75 100 150 200		SGG-3	6346.	SGW-11	7226.	SGR-1	11738.	SGA-1	7226.	SGO-1	SGO-3	5142.	3
	6	200 230 380 460–575	150 200 300 400		SHG-1	14134.	SHW-1	16134.	....	....	SHA-1	15214.	....	SHO-1	12134.	3
	7	230 460–575	300 600		SJG-1	19202.	SJW-1	21202.	....	....	SJA-1	20282.	....	SJO-1	17202.	3
	8	230 460–575	450 900		KG-1	27306.	KW-1	29306.	....	....	KA-1	28386.	KO-1	....	25306.	3
4 Pole Polyphase	0	200–230 380–575	3 5	2 Phase 4-Wire	SBG-5	404.	SBW-15▲	588.	SBR-10	1152.	SBA-5▲	476.	SBO-11	SBO-5	388.	2
	1	200–230 380–575	7½ 10		SCG-10	456.	SCW-15▲	712.	SCR-10	1210.	SCA-5▲	528.	SCO-9	SCO-10	440.	2
	2	200 230 380–575	10 15 25		SDG-4	892.	SDW-12▲	1332.	SDR-4	2076.	SDA-2▲	1012.	....	SDO-4	828.	2
	3	200 230 380–575	25 30 50		SEG-4	1492.	SEW-12	2100.	....	....	SEA-2	1748.	....	SEO-4	1372.	2
	4	200 230 380 460–575	40 50 75 100		SFG-4	3644.	SFW-12	4740.	....	....	SFA-2	4060.	....	SFO-4	3428.	2

*** Prices do not include thermal units.**
▲Separate NEMA Type 4 and 12 enclosures are available; see Class 9991 section.
‡NEMA Type 12 enclosures may be field modified for use in outdoor applications. For details refer to Class 9991 section.
† Voltage ratings in the price table above are those at which motors are rated and are not necessarily identical with the starter (contactor) coil rating. The table below lists the common 60 hz. motor voltage ratings and the corresponding starter (contactor) coil voltage which will be furnished for those ratings. When ordering starters (contactors), always indicate coil voltage.

Fig. 7-10: Electrical ratings for ac magnetic contactors and starters.

The ideal overload protection for a motor is an element with current-sensing properties very similar to the heating curve of the motor (see Fig. 7-11), which would act to open the motor circuit when full-load current is exceeded. The operation of the protective device should be such that the motor is allowed to carry harmless overloads, but is quickly removed from the line when an overload has persisted too long.

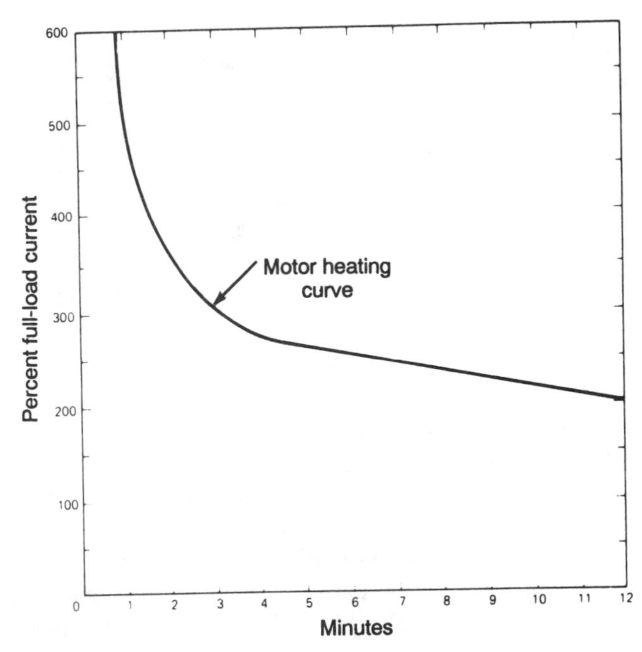

Fig. 7-11: The ideal overload protection for a motor is an element with current-sensing properties very similar to the heating curve of the motor shown here.

Fuses are not designed to provide overload protection. Their basic function is to protect against short circuits (overcurrent protection). Motors draw a high inrush current when starting and conventional single-element fuses have no way of distinguishing between this temporary and harmless inrush current and a damaging overload. Such fuses, chosen on the basis of motor full-load current, would blow every time the motor is started. On the other hand, if a fuse were chosen large enough to pass the starting or inrush current, it would not protect the motor against small, harmful overloads that might occur later.

Dual-element or time-delay fuses can provide motor overload protection, but suffer the disadvantages of being nonrenewable and must be replaced.

The overload relay is the heart of motor protection. It has inverse trip-time characteristics, permitting it to hold in during the accelerating period (when inrush current is drawn), yet providing protection on small overloads above the full-load current when the motor is running. Unlike dual-element fuses, overload relays are renewable and can withstand repeated trip and reset cycles without need of replacement. They cannot, however, take the place of overcurrent protective equipment.

The overload relay consists of a current-sensing unit connected in the line to the motor, plus a mechanism, actuated by the sensing unit, that serves to directly or indirectly break the circuit. In a manual starter, an overload trips a mechanical latch and causes the starter contacts to open and disconnect the motor from the line. In magnetic starters, an overload opens a set of contacts within the overload relay itself. These contacts are wired in series with the starter coil in the control circuit of the magnetic starter. Breaking the coil circuit causes the starter contacts to open, disconnecting the motor from the line.

Overload relays can be classified as being either thermal or magnetic. Magnetic overload relays react only to current excesses and are not affected by temperature. As the name implies, thermal overload relays rely on the rising temperatures caused by the overload current to trip the overload mechanism. Thermal overload relays can be further subdivided into two types, melting alloy and bimetallic.

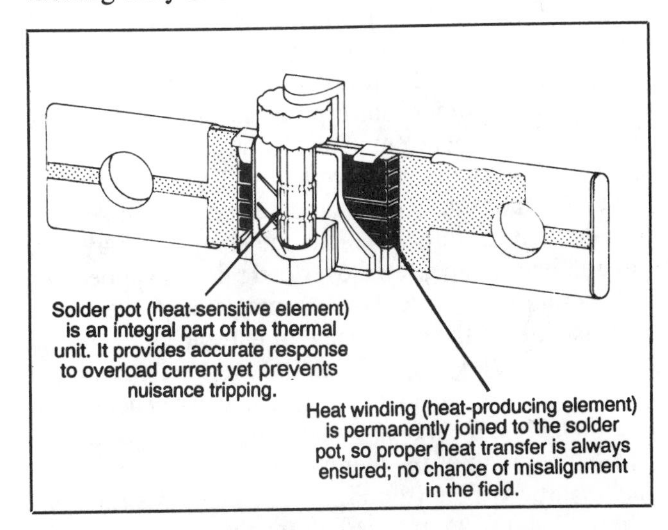

Solder pot (heat-sensitive element) is an integral part of the thermal unit. It provides accurate response to overload current yet prevents nuisance tripping.
Heat winding (heat-producing element) is permanently joined to the solder pot, so proper heat transfer is always ensured; no chance of misalignment in the field.

Fig. 7-12: The melting alloy assembly of a typical heater element overload relay and solder pot.

The melting alloy assembly of the heater element overload relay and solder pot is shown in Fig. 7-12. Excessive overload motor current passes through the heater element, thereby melting a eutectic alloy solder pot. The ratchet wheel will then be allowed to turn in the molten pool, and a tripping action of the starter control circuit results, stopping the motor. A cooling off period is required to allow the solder pot to "freeze" before the overload relay assembly may be reset and motor service restored.

Melting alloy thermal units are interchangeable and of a one-piece construction, which ensures a constant relationship between the heater element and solder pot and allows factory calibration, making them virtually tamper-proof in the field. These important features are not possible with any other type of overload relay construction. A wide selection of these interchangeable thermal units is available to give exact overload protection of any full-load current to a motor.

Bimetallic overload relays are designed specifically for two general types of application: the automatic reset feature is of decided advantage when devices are mounted in locations not easily accessible for manual operation and, second, these relays can easily be adjusted to trip within a range of 85 to 115 percent of the nominal trip rating of the heater unit. This feature is useful when the recommended heater size might result in unnecessary tripping, while the next larger size would not give adequate protection. Ambient temperatures affect overload relays operating on the principle of heat.

Ambient-compensated bimetallic overload relays were designed for one particular situation, that is, when the motor is at a constant temperature and the controller is located separately in a varying temperature. In this case, if a standard thermal overload relay were used, it would not trip consistently at the same level of motor current if the controller temperature changed. This thermal overload relay is always affected by the surrounding temperature. To compensate for the temperature variations the controller may see, an ambient-compensated overload relay is applied. Its trip point is not affected by temperature and it performs consistently at the same value of current.

Melting alloy and bimetallic overload relays are designed to approximate the heat actually generated in the motor. As the motor temperature increases, so does the temperature of the thermal unit. The motor and relay heating curves (see Fig. 7-13) show this relationship. From this graph, we can see that, no matter how high the current drawn, the overload relay will provide protection, yet the relay will not trip out unnecessarily.

When selecting thermal overload relays, the following must be considered:

1. Motor full-load current
2. Type of motor
3. Difference in ambient temperature between motor and controller.

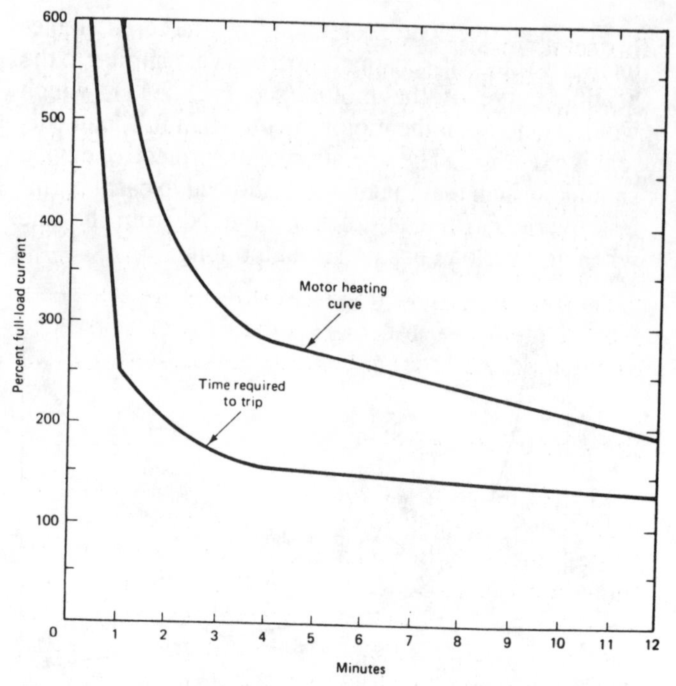

Fig. 7-13: As the motor temperature increases, so does the temperature of the thermal unit.

Motors of the same horsepower and speed do not all have the same full-load current, and the motor nameplate must always be carefully checked to obtain the full-load amperes for a particular motor. Do not use a published table. Thermal unit selection tables are published on the basis of continuous-duty motors, with 1.15 service factor, operating under normal conditions. The tables are shown in the catalog of manufacturers and also appear on the inside of the door or cover of the motor controller. These selections will properly protect the motor and allow the motor to develop its full horsepower, allowing for the service factor, if the ambient temperature is the same at the motor as at the controller. If the temperatures are not the same, or if the motor service factor is less than 1.15, a special procedure is required to select the proper thermal unit.

Standard overload relay contacts are closed under normal conditions and open when the relay trips. An alarm signal is sometimes required to indicate when a motor has stopped due to an overload trip. Also, with some machines, particularly those associated with continuous processing, it may be required to signal an overload condition, rather than have the motor and process stop automatically. This is done by fitting the overload relay with a set of contacts that close when the relay trips, thus completing the alarm circuit. These contacts are appropriately called *alarm* contacts.

A magnetic overload relay has a movable magnetic core inside a coil that carries the motor current. The flux set up inside the coil pulls the core upward. When the core rises far enough, it trips a set of contacts on the top of the relay. The movement of the core is slowed by a piston working in an oil-filled dashpot mounted below the coil. This produces an inverse-time characteristic. The effective tripping current is adjusted by moving the core on a threaded rod. The tripping time is varied by uncovering oil bypass holes in the piston. Because of the time and current adjustments, the magnetic overload relay is sometimes used to protect motors having long accelerating times or unusual duty cycles.

Protective Enclosures

The correct selection and installation of an enclosure for a particular application can contribute considerably to the length of life and trouble-free operation. To shield electrically live parts from accidental contact, some form of enclosure is always necessary. This function is usually filled by a general-purpose, sheet-steel cabinet. Frequently, however, dust, moisture, or explosive gases make it necessary to employ a special enclosure to protect the motor controller from corrosion or the surrounding equipment from explosion. In selecting and installing control apparatus, it is always necessary to consider carefully the conditions under which the apparatus must operate; there are many applications where a general-purpose enclosure does not afford protection.

Underwriters' Laboratories has defined the requirements for protective enclosures according to the hazardous conditions, and the National Electrical Manufacturers Association (NEMA) has standardized enclosures from these requirements:

NEMA 1: general purpose. The general-purpose enclosure is intended primarily to prevent accidental contact with the enclosed apparatus. It is suitable for general-purpose applications indoors where it is not exposed to unusual service conditions. A NEMA 1 enclosure serves as protection against dust and light and indirect splashing, but is not dust-tight.

NEMA 3: dust-tight, raintight. This enclosure is intended to provide suitable protection against specified weather hazards. A NEMA 3 enclosure is suitable for application outdoors, such as construction work. It is also sleet-resistant.

NEMA 3R: rainproof, sleet resistant. This enclosure protects against interference in operation of the contained equipment due to rain, and resists damage from exposure to sleet. It is designed with conduit hubs and external mounting as well as drainage provisions.

NEMA 4: watertight. A watertight enclosure is designed to meet a hose test which consists of a stream of water from a hose with a 1-inch nozzle, delivering at least 65 gallons per minute. The water is directed on the enclosure from a distance of not less than 10 feet and for a period of 5 minutes. During this period, it may be directed in one or more directions, as desired. There should be no leakage of water into the enclosure under these conditions.

NEMA 4X: watertight, corrosion-resistant. These enclosures are generally constructed along the lines of NEMA 4 enclosures except that they are made of a material that is highly resistant to corrosion. For this reason, they are ideal in applications such as meat packing and chemical plants, where contaminants would ordinarily destroy a steel enclosure over a period of time.

NEMA 7: hazardous locations, class I. These enclosures are designed to meet the application requirements of the *National Electrical Code* for Class I hazardous locations: "Class I locations are those in which flammable gases or vapors are or may be present in the air in quantities sufficient to produce explosive or ignitible mixtures." In this type of equipment, the circuit interruption occurs in air.

NEMA 9: hazardous locations, class II. These enclosures are designed to meet the application requirements of the NE Code for Class II hazardous locations. "Class II locations are those which are hazardous because of the presence of combustible dust." The letter or letters following the type number indicates the particular group or groups of hazardous locations (as defined in the NE Code) for which the enclosure is designed. The designation is incomplete without a suffix letter or letters.

NEMA 12: industrial use. This type of enclosure is designed for use in those industries where it is desired to exclude such materials as dust, lint, fibers and flyings, oil seepage, or coolant seepage. There are no conduit openings or knockouts in the enclosure, and mounting is by means of flanges or mounting feet.

NEMA 13: oil-tight, dust-tight. NEMA 13 enclosures are generally made of cast iron, gasketed, or permit use in the same environments as NEMA 12 devices. The essential difference is that due to its cast housing, a conduit entry is provided as an integral part of the NEMA 13 enclosure, and mounting is by means of blind holes rather than mounting brackets.

Controls

Two-wire control: Figure 7-14 shows wiring diagrams for a two-wire control circuit. The control itself could be a thermostat, float switch, limit switch, or other contact device connected to the magnetic starter.

When the contacts of the control device close, they complete the coil circuit of the motor starter, causing it to pick up and connect the motor to the lines. When the control device contacts open, the starter is de-energized, stopping the motor.

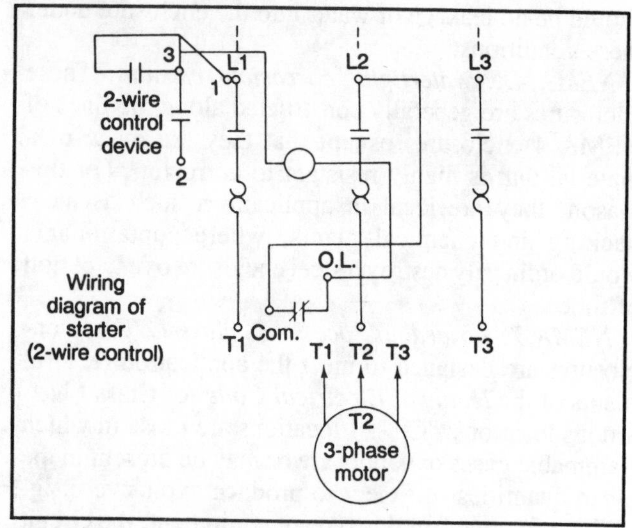

Fig. 7-14: Two-wire control circuit used to operate a flow switch, thermostat, limit switch or similar device.

Two-wire control provides low-voltage release but not low-voltage protection. When wired as illustrated, the starter will function automatically in response to the direction of the control device, without the attention of an operator. In this type of connection, a holding circuit interlock is not necessary.

Three-wire control: A three-wire control circuit uses momentary contact, start-stop buttons, and a holding circuit interlock, wired in parallel with the start button to maintain the circuit. Pressing the normally open (NO) start button completes the circuit to the coil.The power circuit contacts in lines 1, 2, and 3 close, completing the circuit to the motor, and the holding circuit contact also closes. Once the starter has picked up, the start button can be released, as the now-closed interlock contact provided an alternative current path around the reopened start contact.

Pressing the normally closed (NC) stop button will open the circuit to the coil, causing the starter to drop out. An overload condition, which caused the overload contact to open, a power failure, or a drop in voltage to less than the seal-in value would also de-energize the starter. When the starter drops out, the interlock contact reopens, and both current paths to the coil, through the start button and the interlock, are now open.

Since three wires from the pushbutton station are connected into the starter—at points 1, 2, and 3—this wiring scheme is commonly referred to as three-wire control (see Fig. 7-15).

The holding circuit interlock is a normally open auxiliary contact provided on the standard magnetic starters and contactors. It closes when the coil is energized to form a holding circuit for the starter after the start button has been released.

In addition to the main or power contacts which carry the motor current, and the holding circuit interlock, a starter can be provided with externally attached auxiliary contacts, commonly called *electrical interlocks*. Interlocks are rated to carry only control circuit currents, not motor currents. Both NO and NC versions are available. Among a wide variety of applications, interlocks can be used to control other magnetic devices where sequence operation is desired; to electrically prevent another controller from being energized at the same time; and to make and break circuits to indicating or alarm devices such as pilot lights, bells, or other signals.

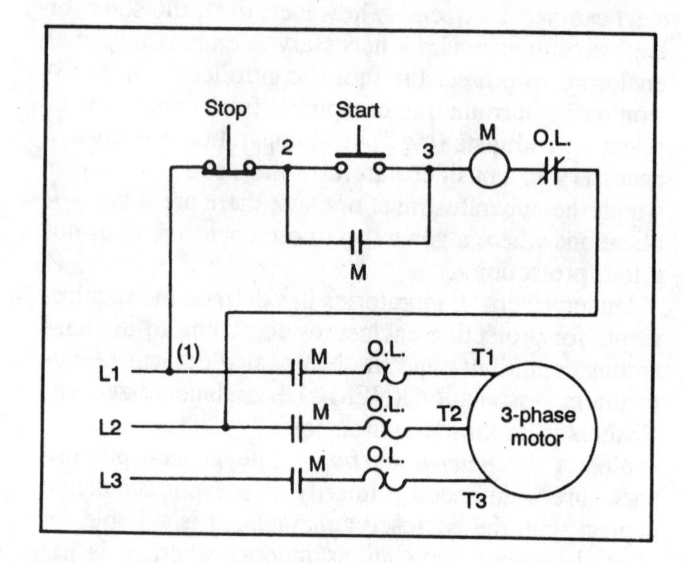

Fig. 7-15: Three-wire motor control circuit.

The circuit in Fig. 7-16 shows a three-pole reversing starter used to control a three-phase motor. Three-phase squirrel-cage motors can be reversed by reconnecting any two of the three line connections to the motor. By interwiring two contactors, an electromagnetic method of making the reconnection can be obtained.

As seen in the power circuit (Fig. 7-16), the contacts (F) of the forward contactor—when closed—connect lines 1, 2, and 3 to the motor terminals T1, T2, and T3, respectively. As long as the forward contacts are closed, mechanical and electrical interlocks prevent the reverse contactor from being energized.

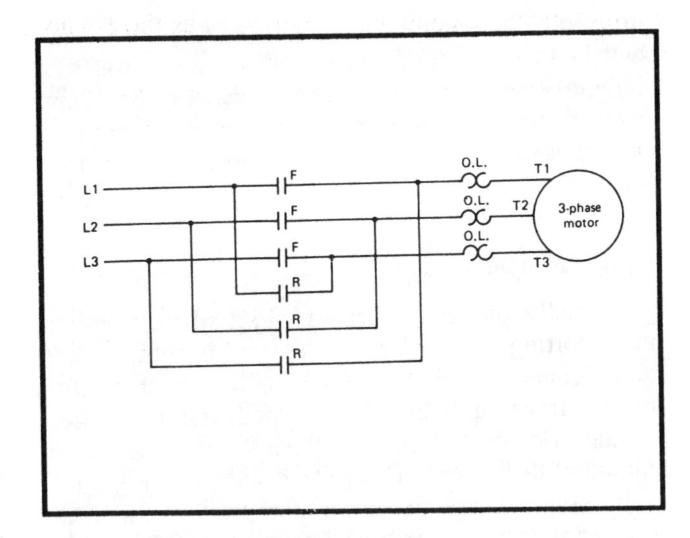

Fig. 7-16: Wiring diagram of a three-pole reversing starter for use with a three-phase motor.

When the forward contactor is de-energized, the second contactor can be picked up, closing its contacts (R), which reconnect the lines to the motor. Note that by running through the reverse contacts, line 1 is connected to motor terminal T3, and line 3 is connected to motor terminal T1. The motor will now run in reverse.

Manual reversing starters (employing two manual starters) are also available. As in the magnetic version, the forward and reverse switching mechanisms are mechanically interlocked, but since coils are not used in the manually operated equipment, electrical interlocks are not furnished.

Control relays: A relay is an electromagnetic device whose contacts are used in control circuits of magnetic starters, contractors, solenoids, timers, and other relays. They are generally used to amplify the contact capability or to multiply the switching functions of a pilot device.

The wiring diagrams in Fig. 7-17 demonstrate how a relay amplifies contact capacity. Fig. 7-17(a) represents a current amplification. Relay and starter coil voltages are the same, but the ampere rating of the temperature switch is too low to handle the current drawn by the starter coil (M). A relay is interposed between the temperature switch and the starter coil. The current drawn by the relay coil (CR) is within the rating of the temperature switch, and the relay contact (CR) has a rating adequate for the current drawn by the starter coil.

Figure 7-17(b) represents a voltage amplification. A condition may exist in which the voltage rating of the temperature switch is too low to permit its direct use in a starter control circuit operating at a higher voltage. In this application, the coil of the interposing relay and the

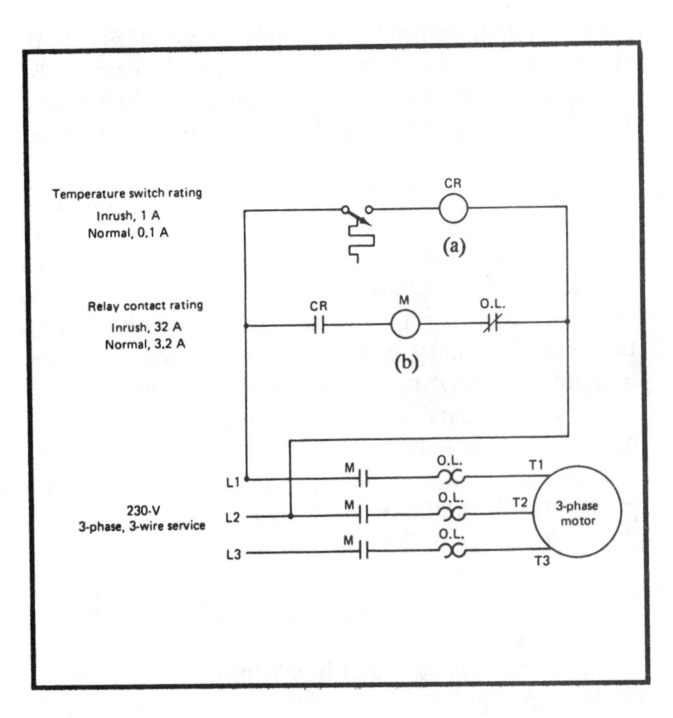

Fig. 7-17: Wiring diagrams demonstrating how a relay amplifies contact capacity.

pilot device are wired to a low-voltage source of power compatible with the rating of the pilot device. The relay contact, with its higher voltage rating, is then used to control the operation of the starter.

Relays are commonly used in complex controllers to provide the logic or "brains" to set up and initiate the proper sequencing and control of a number of interrelated operations. In selecting a relay for a particular application, one of the first steps should be a determination of the control voltage at which the relay will operate. Once the voltage is known, the relays that have the necessary contact rating can be further reviewed, and a selection made, on the basis of the number of contacts and other characteristics needed.

Additional Controlling Equipment

Timers and timing relays: A pneumatic timer or timing relay is similar to a control relay, except that certain kinds of its contacts are designed to operate at a preset time interval after the coil is energized or de-energized. A delay on energization is also referred to as *on delay*. A time delay on de-energization is also called *off delay*.

A timed function is useful in such applications as the lubrication system of a large machine, in which a small oil pump must deliver lubricant to the bearings of the main motor for a set period of time before the main motor starts.

In pneumatic timers, the timing is accomplished by the transfer of air through a restricted orifice. The amount of restriction is controlled by an adjustable needle valve, permitting changes to be made in the timing period.

Drum switch: A drum switch is a manually operated three-position three pole switch which carries a horsepower rating and is used for manual reversing of single- or three-phase motors. Drum switches are available in several sizes and can be spring-return-to-off (momentary contact) or maintained contact. Separate overload protection, by manual or magnetic starters, must usually be provided, as drum switches do not include this feature.

Pushbutton station: A control station may contain pushbuttons, selector switches, and pilot lights. Pushbuttons may be momentary- or maintained-contact. Selector switches are usually maintained-contact, or can be spring-return to give momentary-contact operation.

Stand-duty stations will handle the coil currents of contractors up to size 4. Heavy-duty stations have higher contact ratings and provide greater flexibility through a wider variety of operators and interchangeability of units.

Foot switch: A foot switch is a control device operated by a foot pedal used where the process or machine requires that the operator have both hands free. Foot switches usually have momentary contacts but are available with latches which enable them to be used as maintained-contact devices.

Limit switch: A limit switch is a control device that converts mechanical motion into an electrical control signal. Its main function is to limit movement, usually by opening a control circuit when the limit of travel is reached. Limit switches may be momentary-contact (spring-return) or maintained-contact types. Among other applications, limit switches can be used to start, stop, reverse, slow down, speed up, or recycle machine operation.

Snap switch: Snap switches for motor control purposes are enclosed, precision switches which require low operating forces and have a high repeat accuracy. They are used as interlocks and as the switch mechanism for control devices such as precision limit switches and pressure switches. They are also available with integral operators for use as compact limit switches, door operated interlocks, and so on. Single-pole double-throw and two-pole double-throw versions are available.

Pressure switch: The control of pumps, air compressors, and machine tools requires control devices that respond to the pressure of a medium such as wire, air, or oil. The control device that does this is a pressure switch. It has a set of contacts which are operated by the movement of a piston, bellows, or diaphragm against a set of springs. The spring pressure determines the pressures at which the switch closes and opens its contacts.

Practical Application

A small repair shop, located in back of the sales area in a sporting goods store, is depicted in Fig. 7-18. A quick glance at this cut-away view readily shows two motor-driven power tools: a pedestal two-wheel grinder (lower left in the illustration), and bench-mounted drill press. The grinder uses a 1½ hp, 120-volt, single-phase capacitor-start motor. Since the grinder is supplied with an integral ON-OFF switch, capacitor and thermal protector, this is all the control necessary. The power cord which plugs into a nearby 120-volt wall receptacle acts as a disconnect. It is recommended that the outlet serving motors, such as this one, be fed with a separate 15- or 20-ampere circuit and protected with a fuse or circuit breaker.

The drill press, however, is a different story. This is a 1½ hp, 3-phase, 208-volt induction motor with no built-in controls. Therefore, the following NE Code rules must be followed when installing this machine.

Let's first consider the disconnecting means. The nameplate rating of this motor is 1725 watts (not rpm's) and when operating on a three-wire, three-phase, 208-volt circuit, the motor has a current rating of 4.79. Section 430-110 of the NE Code states that the disconnect switch must be rated at least 115% of the motor nameplate current. So, 4.79 x 1.15 = 5.5 amperes. A standard 30-amp, three-pole safety switch will work fine. Another alternative would be a general-use, three-pole, snap switch rated for at least 11 amperes (must be at least twice the full-current rating of the motor).

The overcurrent protection for this motor cannot be over 300% of the full-load current when using non-time delay fuses, 175% when using time-delay fuses, and 700% when using instantaneous trip circuit breakers. Referring to Table 430-150 of the 1993 NE Code, we find that a 1½ HP induction three-phase motor, operating at 208 volts has a full-load current of 5.7 amperes. Therefore, the absolute maximum fuse size for a non-time delay fuse is 17.1 amperes. A 15-amp cartridge fuse will probably be the one to use. A 7-amp fuse would be the size for fuses of the time-delay type, while a circuit breaker could be rated as high as 40 amperes. The wire size must be rated for the maximum overcurrent protection; that is, if this circuit was pro-

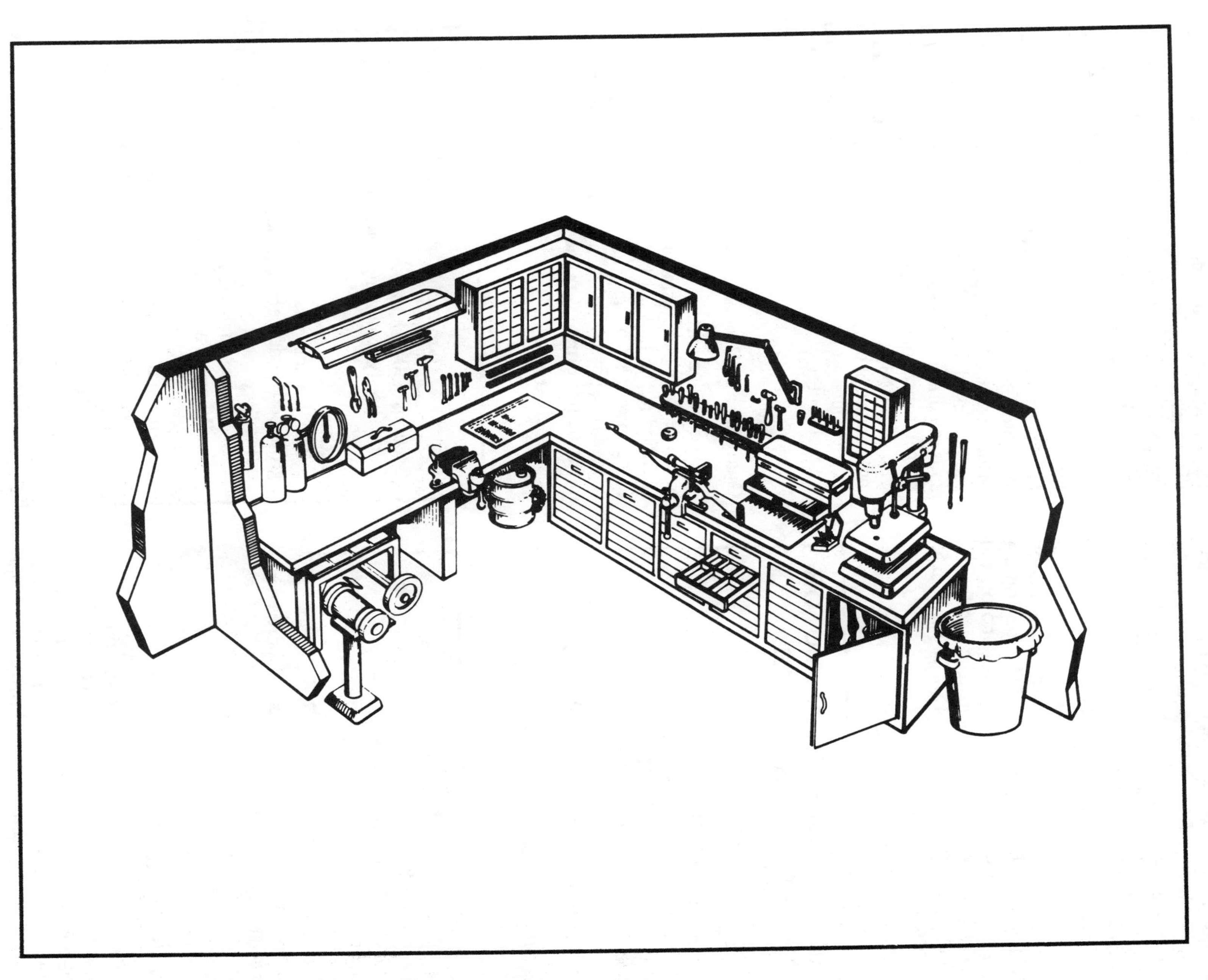

Fig. 7-18: Cut-away view of a repair shop with two motor-driven machines.

tected with a 40-ampere circuit breaker, the circuit conductors must be rated for at least 4 amperes also. However, if this is the only motor on the circuit, the electrician will more than likely use a 15-ampere, 3-pole circuit breaker, and No. 12 AWG wire.

Refer to Section 430-22 of the NE Code for rules governing the selection and installation of the motor controller. Since the motor in question will sometimes be run in reverse, the pushbutton station should have 3 buttons and be wired like the drawing in Fig. 7-19. See Section 430-31 for motor and branch-circuit overload protection. The size for the motor in question should be 115% x 5.7 = 6.55 amperes.

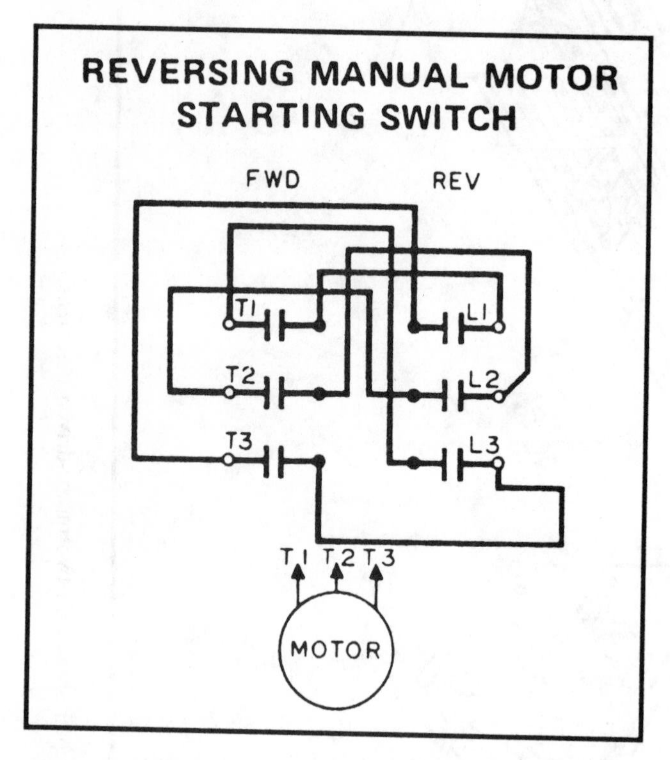

Fig. 7-19: Wiring diagram of motor control for reversing motor rotation.

Disconnecting Means

When motors must be replaced or repaired, a suitable disconnecting means must be available to assure that the motor and its controller are disconnected from all sources of supply. A controller cannot serve as the disconnecting means since the controller must also be disconnected. This does prohibit a single switch or circuit breaker from acting as both controller and disconnecting means as long as it may be disconnected from all ungrounded ("hot") conductors itself and it meets other NE Code requirements.

The disconnecting means must be within sight from the controller location as specified in NE Code Article 100 and Section 430-102 and also in sight of the motor. Consequently, the disconnecting equipment must be within 50 feet and be visible from either location. If it is not practical to locate the disconnect within sight of the motor, a disconnecting means that can be locked open and is within sight of the controller is required. The switch must disconnect the motor from its source of supply, not just stop the motor. This type of installation is typical whenever a central motor control center is used to control a number of motors, many of which are not in sight from the control center.

Guarding and Grounding

Exposed live parts of motors operating at 50 volts or more between terminals must be guarded as specified in the table in Fig. 7-20. Suitable working space must also be provided around motors with exposed live parts even when guarded if maintenance has to be performed with the motor operating.

Grounding of Motors

Sections 430-141 through 430-145 of the NE Code specify grounding regulations for motors.

General:

The frames of stationary motors shall be grounded under any of the following conditions:

1. Where supplied by metal-enclosed wiring.
2. Where installed in a wet location and not isolated or guarded.
3. If installed in a hazardous location as covered in Articles 500 through 517.
4. If the motor operates with any terminal at over 150 volts to ground.

Fig. 7-20: NE Code grounding regulations.

Troubleshooting Control Equipment

Table 7-1 lists troubles encountered with motor controls, their causes, and remedies. This table is of a general nature and covers only the main causes of trouble.

Misapplication of a device can be a cause of serious trouble; however, rather than list this cause repeatedly, it should be noted here that misapplication is a major cause of motor control trouble and should always be questioned when a device is not functioning properly.

Actual physical damage or broken parts can usually be found quickly and replaced. Damage due to water or flood conditions requires special treatment.

Contacts: There are two types of contact wear: electrical and mechanical. The majority of wear to contact tips is due to electrical wear. The mechanical wear is insignificant and requires no further mention.

Arcing causes electrical wear by eroding the contacts. During arcing, a small part of each contact melts, then vaporizes and is blown away.

When a device is new, the contacts are smooth and have a uniform silver color. As the device is used, the contacts become pitted and the color may change to blue, brown, and black. These colors result from the normal formation of metal oxide on the contact's surface and are not detrimental to contact life and performance. Therefore, contacts should not be filed to restore the original color. This practice only shortens contact life and may cause welding.

The contacts should be replaced under the following conditions.

Insufficient contact material: This is when the amount of contact material remaining is inadequate. When less than 1/64 inch remains, replace the contacts.

Irregular surface wear: This type of wear is normal. However, if a corner of the contact material is worn away and a contact may mate with the opposing contact support member, the contacts should be replaced. This condition can result in contact welding.

Pitting: Under normal wear, contact pitting should be uniform. This condition occurs during arcing, as described above. The contacts should be replaced if the pitting becomes excessive and little contact material remains.

Curling of contact surface: This condition results from severe service, producing high contact temperatures, and causes separation of the contact material from the contact support member.

The measurement procedure for checking the contact tip material requires a continuity checker or ohmmeter and a 1/32 inch standard feeler gauge. The procedure is as follows:

1. Place the feeler gauge between the armature and the magnet frame, with the armature held tightly against the magnet frame.

2. Check the continuity of each phase. If there is continuity in all phases, the contacts are in good condition. If not, all contacts should be replaced. Even though the contacts pass condition 1, any of the other conditions would necessitate replacement of the contacts. Contacts should be replaced only when necessary; too frequent replacement is a waste of money and natural resources.

Troubleshooting Motors

The useful life of an induction motor depends largely upon the condition of its insulation. In general, the insulation should be suitable for the operating requirements.

Stator windings: The stator (stationary) windings appear to be so simple and rugged as to cause one to frequently overlook the necessity for certain maintenance procedure. However, a glance into the average motor repair shop will make it apparent that the induction motor stator is after all a vulnerable piece of equipment. Most of the work going on will be replacing or repairing stator windings.

Stator troubles can usually be traced to one or more of the following causes:

- Worn bearings
- Moisture
- Overloading
- Operating single phase
- Poor insulation

Dust and dirt are usually contributing factors. Some forms of dust are highly conductive and contribute materially to insulation breakdown. The effect of dust on the motor temperature through restriction of ventilation is another reason for keeping the machine clean, either by periodically blowing out with compressed air or by dismantling and cleaning. The compressed air must be dry and throttled down to a low pressure which will not endanger the insulation.

Moisture: One of the most subtle enemies of motor insulation is moisture. Needless to say, motor insulation must be kept reasonably dry, although many applications make this practically impossible unless the motor is totally enclosed, or otherwise protected from the direct effects of moisture. If operated in a damp place, special moisture-resisting treatment should be given the windings.

■ TROUBLESHOOTING CHART

Motor Controls

Malfunction	Probable cause	Correction action
Contacts	Poor contact in control circuit.	Replace the contact device or use holding circuit interlock (three-wire control).
	Low voltage.	Check coil terminal voltage and voltage dips during starting.
Welding or freezing	Abnormal inrush of current.	Check for grounds, shorts, or excessive motor load current, or use larger contactor.
	Rapid jogging.	Install larger device rated for jogging service.
	Insufficient tip pressure.	Replace contacts and springs, check contact carrier for deformation or damage.
	Low voltage preventing magnet from sealing.	Check coil terminal voltage and voltage dips during starting.
	Foreign matter preventing contacts from closing.	Clean contacts with Freon. Contacts, starters, and control accessories used with very small current or low voltage, should be cleaned with Freon.
	Short circuit or ground fault.	Remove fault and check to be sure fuse or breaker size is correct.
Short tip life or overheating of tips	Filing or dressing.	Do not file silver tips. Rough spots or discoloration will not harm tips or impair their efficiency.
	Interrupting excessively high currents.	Install larger device or check for grounds, shorts, or excessive motor currents.
	Excessive jogging.	Install larger device rated for jogging service.
	Weak tip pressure.	Replace contacts and springs, check contact carrier for deformation or damage.
	Dirt or foreign matter on contact surface.	Clean contacts with Freon. Take steps to reduce entry of foreign matter into enclosure.
	Short circuits or ground fault.	Remove fault and check to be sure fuse or breaker size is correct.
	Loose connection in power circuit.	Clear and tighten.
	Sustained overload.	Check for excessive motor load current or install larger device.

Fig. 7-21: Troubleshooting chart for motor controls.

Dipping and Baking: The life of a winding depends upon keeping it in its original (or new) condition, as long as possible. In a new machine, the winding is snug in the slots and the insulation is fresh and flexible, being newly treated with varnish, and, therefore, resistant to the deteriorating effects of moisture and other foreign matter. This condition is best maintained by periodic cleaning, followed by varnish and oven treatments.

One condition that frequently hastens winding failure is movement of the coils due to vibration during operation. After insulation dries out it loses its flexibility and the mechanical stresses caused by starting, plugging, stopping, as well as the natural stresses in operation under load, will precipitate short circuits in the coils and possibly failures from coil to ground, this usually at the point where the coil leaves the slot. The effect of periodic varnish and oven treatments properly carried out so as to fill all air spaces caused by drying and shrinkage of the insulation, thereby maintaining a solid winding, will also provide an effective seal against moisture.

Rotor Windings: The rotors of wound rotor motors have many features in common with the stators; therefore, the same comments apply to the care of rotor windings as are given for the care of stator windings. However, the rotor introduces some additional problems because it is a rotating element.

Most wound rotors have a three-phase winding, and are, therefore, susceptible to trouble from single-phase operation. The first symptom of an open-rotor circuit is lack of torque, with slowing down in speed, accompanied by a growling noise, or perhaps failure to start the load. The first place to look for an open secondary circuit is in the resistance bank or the control circuit external to the rotor. Short-circuiting the rotor circuit at the slip rings and then operating the motor will usually determine whether the trouble is in the control circuit or in the rotor itself. It may be one of the stud connections to the slip rings.

If the rotor is wave wound with the windings made up of copper strap coils with clips connecting the top and bottom half of the coil, it will be well to inspect these end connections for possible signs of heating, which would be an indication of a partial open circuit. Faulty or improperly made end connections are a common source of open circuits in rotor windings.

A ground in a rotor circuit will not affect the performance of the motor unless another ground should also develop, which might cause the equivalent of a short-circuit, in which case it would have the effect of unbalancing the rotor electrically. In addition to reduced torque, another symptom of this condition might be excessive vibration of the motor. There might also be sparking and uneven wear of the collector rings.

Another, and reasonably successful manner of checking for short-circuits in the rotor windings, is to raise the brushes off the slip rings and energize the stator. If the rotor winding is free from short-circuits it should have little or no tendency to rotate, even when disconnected from the load. If it does show considerable torque, or have a tendency to come up to speed, the rotor should be removed and the winding opened and examined for a fault. In making this test, note that some rotors having a wide tooth design may show a tendency to rotate even though the windings are in good condition.

Still another check that can be made when the rotor is in place and the stator is energized (also with the brushes raised) is to check the voltage across the rings to see if they are balanced. Judgment will have to be used in making this check to make sure that any inequality in voltage measurements is not due to the relative position of the rotor and stator phases. In other words, the rotor should be moved to several positions in taking these voltage measurements.

Squirrel-Cage Rotors: Squirrel-cage rotors are more rugged and in general require less maintenance than wound rotors, but may also give trouble due to open circuits or high resistance points in the rotor circuit. The symptoms of such conditions are in general the same as with wound rotor motors, that is, slowing down under load and reduced starting torque. Such conditions can usually be detected by looking for evidence of heating at the end ring connections, particularly noticeable when shutting down after operating under load.

In brazed rotors, any fractures in the rotor bars will usually be found either at the point of connection to the end ring or at the point where the bar leaves the laminations. Discolored rotor bars are also evidence of excessive heating.

Brazing broken bars or replacing bars should be done only by a competent person. Considerable technique is required for this kind of work, and it is recommended that the manufacturer's nearest District Sales Office be consulted before attempting such repairs in the shop or plant, unless an experienced operator is available.

With die-cast rotors, look for cracks or other imperfections that may have developed in the end rings. A faulty die-cast rotor can rarely be effectively repaired, and should be replaced if defective.

The Air Gap: A small air gap is characteristic of the induction motor. The size of the air gap has an important bearing on the power factor of the motor and doing anything to affect it, such as grinding the rotor laminations or filing the stator teeth, results in increased magnetizing current, with resultant lower power factor.

Good maintenance procedure calls for periodically checking the air gap with a feeler gauge to insure against a worn sleeve bearing that might permit the rotor to rub the laminations. These measurements should be made on the shaft end of the motor.

On large machines it is desirable to keep a record of these checks. Four measurements should be taken approximately 90 degrees apart, one of these points being the load side; that is, the point on the rotor periphery which corresponds with the load side of the bearing.

A comparison of the new measurements with those previously recorded will permit the early detection of bearing wear.

A very light bearing rub will generate heat sufficient to destroy the coil insulation.

Overloading and Single-Phase Operation: All too frequently a motor of adequate capacity properly applied in the original application is later found to be overloaded or otherwise unsuited for its job, due, for example, to one or more of the following:

- More severe duty imposed on the motor
- A change in equipment
- A change in equipment parts
- A change in operating time

Connecting measuring instruments to the motor circuit will quickly disclose the reason for motor overheating, failure to start the load, or other abnormal symptoms.

Control circuits for many of the older electrical systems were not provided with relay protection, and single-phase operation of polyphase induction motors on such circuits has frequently been responsible for motor burnout. Usually this has resulted from the blowing of one of the fuses, while the motor is up to speed and under load. Under such conditions the portion of winding remaining in the circuit will endeavor to carry the load until it fails due to overheating.

The effect of increasing the load on the motor beyond its rated capacity is simply to increase the operating temperature, which shortens the life of the insulation. Momentary overloads usually do no damage, consequently the tendency to the thermal type of overload protection in present-day control. Obviously the ideal place to measure the thermal effect of overload is on the motor itself.

Without a doubt the polyphase induction motor is the simplest and most fool-proof piece of rotating electrical apparatus. The largest single cause of winding failures is probably due to the rotor rubbing the stator iron, usually because of worn bearing or complete failure of the bearing.

Testing for an Open Motor Coil: A simple test light or continuity tester may be used for this operation. With all power circuits shut off, connect the continuity test leads across each motor coil in turn. If the coil is operational, the light will glow or the dial of the ohmmeter will swing to full scale.

To test for a grounded coil, connect one of the test leads to the motor frame and the other lead to one of the field coil wires. If the light glows or the ohmmeter dial swings towards ZERO, the coil is grounded.

Troubleshooting Charts

The following charts will aid the electrician in troubleshooting and maintenance of electric motors.

■ TROUBLESHOOTING CHART

AC Motors

Malfunction	Probable Cause	Corrective Action
Slow to accelerate	Excess loading.	Reduce load.
	Poor circuit.	Check for high resistance.
	Defective squirrel-cage rotor.	Replace
	Applied voltage too low.	Get power company to increase voltage tap.
Wrong rotation	Wrong sequence of phases.	Reverse connections at motor or at switchboard.
Motor overheats	Check for overload.	Reduce load.
	Wrong blowers or air shields	may be clogged with dirt and prevent proper ventilation of motor.
	Motor may have one phase open.	Check to make sure that all leads are well connected.
	Grounded coil.	Locate and repair.
	Unbalanced terminal voltage.	Check to make sure that all leads are well connected.
	Grounded coil.	Locate and repair.
	Unbalanced terminal voltage.	Check for faulty leads
	Shorted stator coil.	Repair and then check wattmeter reading.
	Faulty connection.	Indicate by high resistance.
	High voltage.	Check terminals of motor with voltmeter
	Low voltage.	Same as above.
Motor stalls	Wrong application	Change type or size. Consult manufacturer.
	Overloaded motor.	Reduce load.
	Low motor voltage.	See that nameplate voltage is maintained.
	Open circuit	Fuses blown
	Incorrect control resistance of wound rotor.	Check control sequence. Replace broken resistors. Repair open circuits.
Motor does not start	One phase open.	See that no phase is open. Reduce load.
	Defective rotor	Look for broken bars or rings.
	Poor stator coil connection.	Remove end bells
Motor runs, then quits	Power failure	Check for loose connections to line, to fuses and to control.
Slow speed	Not applied properly. Voltage too low at motor terminals because of line drop. Open primary circuit.	Consult supplier for proper type. Use higher voltage on transformer terminals or reduce load. Locate fault with testing device and repair.

Continued on next page

■ TROUBLESHOOTING CHART

AC Motors

Malfunction	Probable Cause	Corrective Action
Slow speed	If wound rotor, improper control operation of secondary	Correct secondary control.
	Starting load too high.	Check load motor is supposed to carry at start.
	Low pull-in torque of synchronous motor.	Change rotor starting resistance or change rotor design.
	Check that all brushes are riding on rings.	Check secondary connections. Leave no leads poorly connected.
	Broken rotor bars.	Look for cracks near the rings. A new rotor may be required.
Motor vibrates	Motor misaligned.	Realign.
	Weak foundations.	Strengthen base.
	Coupling out of balance.	Balance coupling.
	Driven equipment unbalanced.	Rebalance driven equipment.
	Defective ball bearing.	Replace bearing.
	Bearing not in line.	Line up properly.
	Balancing weights shifted.	Rebalance rotor.
	Wound rotor coils replaced.	Rebalance rotor.
	Polyphase motor running single phase.	Check for open circuit.
	Excessive end play.	Adjust bearing or add washer.
Unbalanced line current	Unequal terminal volts.	Check leads and connections.
	Single phase operation.	Check for open circuit.
	Poor rotor contacts in control wound rotor resistance.	Check control devices.
	Brushes not in proper position in wound rotor.	See that brushes are properly seated and shunts in good condition.
	Fan rubbing air shield.	Remove interference.
	Fan striking insulation.	Clear fan.
	Loose on bedplate.	Tighten holding bolts.
Magnetic noise	Air gap not uniform.	Check and correct bracket fit or bearing.

chapter 8

MISCELLANEOUS ELECTRICAL SYSTEMS

The NE Code specifies rules for certain special electrical equipment which supplement or modify the general NE Code rules. Some of the items falling under this category include: electric signs and outline lighting, manufactured wiring systems, office furnishings, cranes and hoists, elevators, dumbwaiters, escalators, moving walks, electric welders induction and dielectric heating equipment, industrial machines, and similar applications.

Electric Signs and Outline Lighting

Electric signs and outline lighting are covered in Article 600 of the NE Code. Such equipment is used for decorative and advertising purposes and is usually self-contained and attached to, rather than a part of, the building or wiring system. The NE Code defines an electric sign as illuminated utilization equipment designed to convey information or attract attention. Outline lighting is designed to outline or call attention to certain features of a building or other structure. Various types of lamps are used in both cases: incandescent, electric-discharge, neon tubing, etc.

In general, lighting and the related electrical wiring for outdoor signs must be weatherproof so that exposure to the elements will not adversely affect their operation. External conductors, terminals and devices must be enclosed in metal or other noncombustible material. Controllers known as flashers are often used with electric signs to obtain a blinking effect for further decoration or to gain attention. Any cutouts, flashers, and other control devices must be enclosed in metal boxes with accessible doors.

Signs and outline lighting installations must be clear of open conductors and be elevated at least 16 feet above areas accessible to vehicles, except when the installation is protected from damage at a lower height.

A branch circuit supplying lamps, ballasts, and transformers, or combinations, is limited in its rating to 20 amperes. If the circuit supplies only electric-discharge lighting transformers, however, it may have a rating of up to 30 amperes. While the rating of a branch circuit is limited, no restrictions are placed on the number of branch circuits that may be used to supply an outdoor sign. Each fixed sign and each outline lighting installation must have a disconnecting means that is externally operable, that opens all ungrounded conductors regardless of the number of branch circuits used, and that is within sight of the sign or outline lighting. One type of approved switch is shown in Fig. 8-1.

Switches, flashers, and other devices controlling the primary circuits of transformers may be subject to damage from arcing at the contacts, especially if neon tubes are utilized. Such control devices, therefore, must be a type approved for the purpose or have at least twice the ampere rating of the transformer, except in the case of ac general-use snap switches which may control ac circuits with an inductive load as great as the rating of the switch.

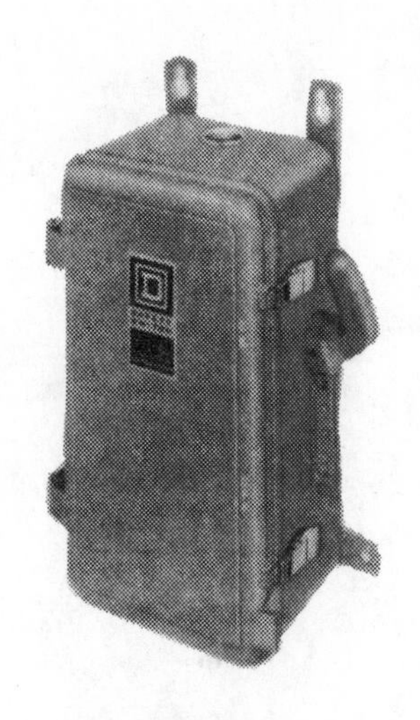

Fig. 8-1: One type of switch approved for use on sign flashers.

The wiring methods, types of lampholders, and type of transformers used in the construction of a sign are defined in detail by the NE Code according to the range of operating voltage of the lamps used. In general, signs operating at 600 volts or less use either incandescent lamps or electric-discharge lamps. In most cases, signs operating at over 600 volts employ electric-discharge lamps only and the NE Code rules are more extensive because of the increased fire and shock hazard associated with high voltages (up to 15,000 volts).

Fig. 8-2 shows important rules for the installation of signs and outline lighting.

Machine Tools

Control wiring and feeder connection terminals on nonportable, electrically driven machines are usually installed at the factory. In most cases, due to the areas in which the equipment is used, the wiring method is restricted to rigid conduit except for short lengths of flexible conduit where necessary for final connection to the equipment. Continuously moving parts of the machine are interconnected with approved type, extra flexible, nonmetallic covered cable. The size of the conductors, type of mounting of control equipment, overcurrent protection, and grounding are covered in Article 670 of the NE Code.

The electric supply for metalworking machines may be from conventional branch circuits or feeders or in the form of bus ducts or wireways. These two latter methods provide a very flexible type of installation allowing the moving of machines from one part of the plant or shop to another. Their reconnection to another part of the bus duct system is almost instantaneous, eliminating changes in the raceway wiring. Bus duct systems are covered in Article 364 of the NE Code. A description of bus duct and wireway systems follows.

Busway

A complete line of plug-in feeder busway, in ratings from 600 amperes through 4000 amperes (aluminum), is constructed in 3-pole or 4-pole full neutral for use on electrical systems up to 600 volts. They may also carry 2-pole ratings with UL listing up to 8650 amperes, dc or single-pole ac.

Many feeder busways and plug-in busways are totally enclosed and do not require ventilation openings in the housing for cooling. Instead, the busway cools by radiation from the housing surface and by convection currents outside the housing. This method of cooling offers several advantages.

The totally enclosed busway needs no derating for different mounting positions, because it cools as efficiently in one position as another. Ventilated busways maintain their maximum operating temperature within allowable limits by utilizing convection currents of air which pass through the housing itself to carry off excess heat. When the busway is mounted so that the perforated housing allows these convection currents to pass freely through the housing and between the conductors, this method of cooling is fairly efficient. However, if the busway is mounted in any position other than this "preferred position," the bus bars themselves interfere with this free passage of cooling air, efficiency is decreased and the operating temperature rises.

Under these conditions, ventilated busway must be "derated" to a substantially lower current carrying capability. Or, if derating is unacceptable, oversized bus bars must be used to reduce overall heating to an acceptable level.

Where totally enclosed construction is used on busways that are normally ventilated, even more stringent derating is required.

Plug-in switches or circuit breakers should be side-mounted for maximum utilization. The "preferred" mounting position of most ventilated busway requires the plug-in units to be mounted on the top and bottom of the run; making those on top hard to get at, and making those on the bottom protrude into available headroom. Totally enclosed busway plug-in units may be side-mounted for maximum utilization, without "derating" the busway.

The NE Code (Section 364-6) requires that busways may extend "vertically through dry floors if total-

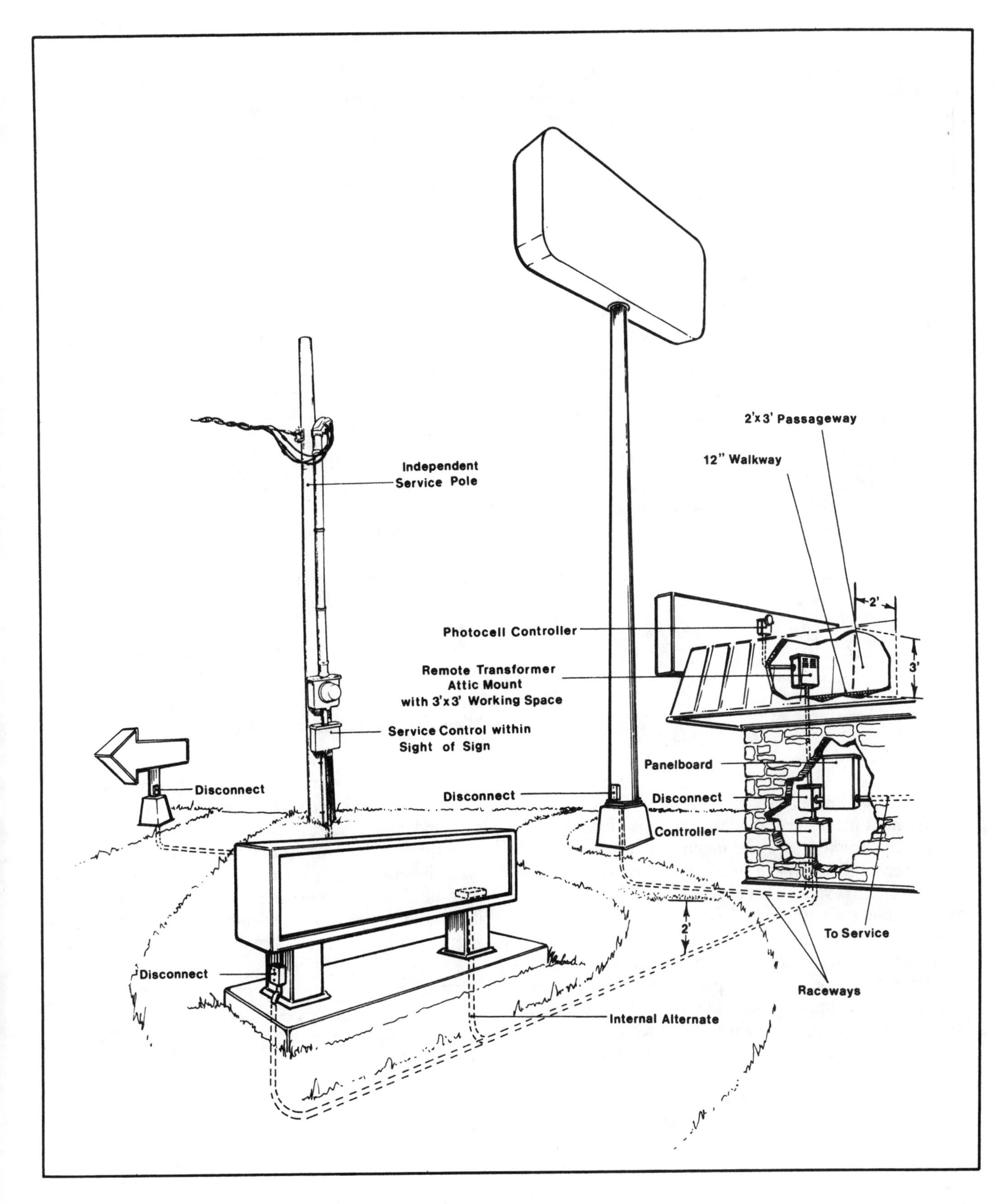

Fig. 8-2: NE Code regulations governing the installation of signs and outline lighting.

ly enclosed (unventilated), where passing through and for a minimum distance of 6 feet above the floor to provide adequate protection from physical damage."

Fig. 8-3: Ventilated busway is fully rated only when mounted in this position.

Totally enclosed busway complies with this NE Code requirement with no modification. In the case of ventilated busway, if the enclosure is not provided by a busway manufacturer, his busway may meet requirements of NE Code 364-6, but void the UL manifest, since UL cannot sanction modifications made to a product in the field. Ventilated busway requires expensive modification to satisfy both UL and NE Code requirement.

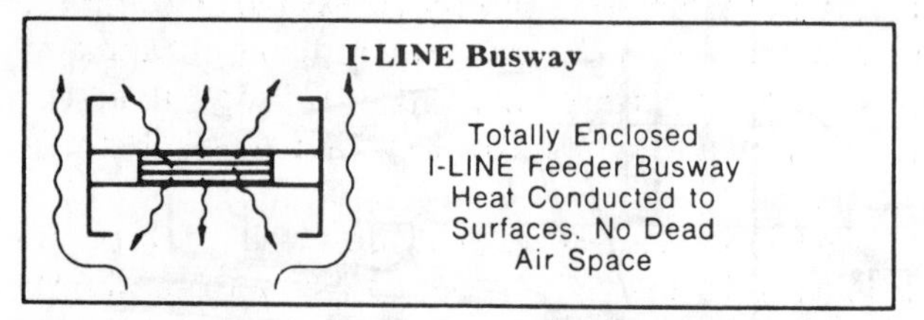

Fig. 8-4: Totally enclosed busway is fully rated in all mounting positions.

The safety precaution embodied in the NE Code requirement mentioned above is obvious. Totally enclosed construction affords much greater protection from mechanical damage to the bus bars and insulation. It also gives much better protection from dust and dirt accumulation in the housing than does ventilated duct.

Because totally enclosed busway construction does not require that bus bars be spaced apart for air flow between them, the physical size of the housing can, with proper design, be smaller, rating for rating.

The close spacing of bus bars in this type of busway gives it exceptionally low reactance. This is particularly true of feeder busway, where the spacing between bars is less than 1/16 inch. The low reactance thus achieved helps reduce voltage dips during the instant of a change in load, such as motor starting. Under such conditions, the high inrush current (up to 600% of load) is at a very low power factor and, consequently, the reactive component of voltage drop assumes greater importance.

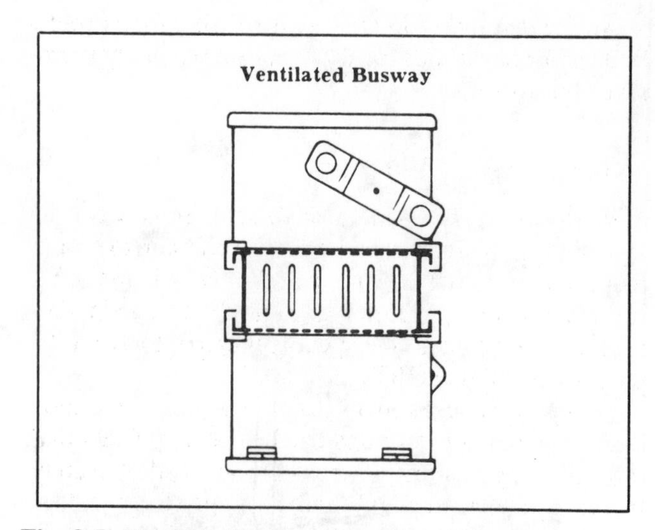

Fig. 8-5: Ventilated busway shown here in its preferred position. Full rating can be applied only in this mounting position.

If a busway can be designed to operate within satisfactory temperature rise limitations without relying on air convection currents through housing perforations, all the above advantages are gained. Ventilation is something one puts up with only when satisfactory operation cannot be achieved through more sophisticated design considerations.

Weatherproof Feeder Busway

Feeder busway is manufactured in weatherproof, as well as indoor construction. The weatherproof design incorporates gasketed covers for joint parts, vapor barriers and other features which make it possible to install weatherproof busway in exposed locations. Weatherproof busway can be connected to indoor feeder or to plug-in busway with the standard joint. Plug-in busway is manufactured only in an outdoor configuration and should not be used outdoors.

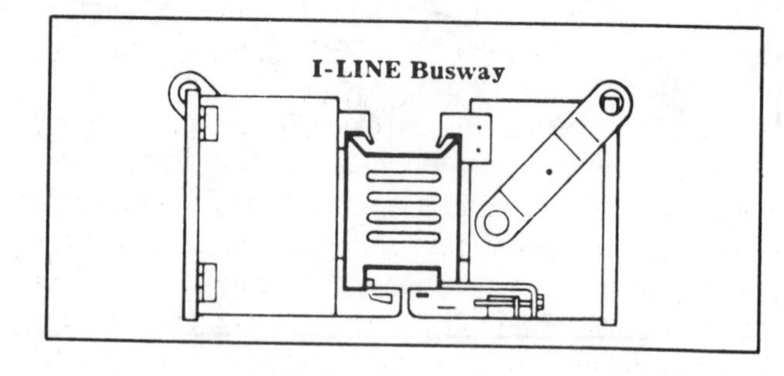

Fig. 8-6: Ventilated busway must be derated for mounting in this position.

No feeder or plug-in busway is suitable for extremely dusty or hazardous locations, or in extremely corrosive atmospheres.

Various Types of Wireway

Wireway is steel-enclosed trough (raceway) designed to carry electrical wires and cables and to protect them against possible damage. Wireway is manufactured for general purpose use, raintight use, and industrial use according to the specifications of the Joint Industry Conference (JIC).

Wireway can used in a variety of ways, and in most cases, is less expensive and can be installed faster than conduit. Applications where wireway is ideal include incoming cable runs and feeder circuits from switchboards to power and lighting panels; distribution of power in industrial plants; ganging equipment, such as motor control, safety switches and metering equipment; and vertical runs of cable in elevator shafts. When comparing wireway with conduit, wireway has distinct advantages. Among these are lighter weight, easier installation, reusability and the ease of making tap-offs.

Wireway is manufactured in four types for use in industrial, commercial and residential applications.

Typical wireway is a steel-enclosed wiring trough, wireway and auxiliary gutter with a removable hinged cover that can be used as either hinge-cover wireway or screw-cover trough. A complete set of fittings is available so that an entire wireway system can be installed, regardless of bends, offsets or other building contours which may be encountered.

Fittings have removable covers and sides to permit a complete "lay-in" installation and to provide access to wires throughout the entire length, without any alterations to the system. Most are Underwriters' Laboratories listed as wireways and auxiliary gutters.

Good electrical and mechanical continuity is assured through the direct connection of lengths by threaded screws at every connector. Some is manufactured in standard lengths of 1, 2, 3, 4, 5 and 10 feet in 2½″ x 2½″ through 12″ x 12″ sizes.

Raintight wireway is steel enclosed wiring trough designed to be used outdoors and in other areas where raintight construction is required. It has a removable cover with provisions for sealing and is UL listed for ganging meters, switches and other equipment suited for outdoor use. Raintight wireway is available in standard lengths of 1, 2, 3, 4 and 5 feet in 4″ x 4″, 6″ x 6″ and 8″ x 8″ sizes.

JIC lay-in wireway is a gasketed, metal wiring trough manufactured to meet specifications for Industrial Control Equipment and is used to protect electrical wiring from oil, water, dirt or dust. JIC lay-in wireway comes in standard lengths of 1, 2, 3, 4, 5, and 10 feet in 2½″ x 2½″, 4″ x 4″, 6″ x 6″, 8″ x 8″ and 12″ x 12″ sizes.

A complete line of elbows, tees and other fittings fully complement the line to meet all layouts. JIC wireway is manufactured with a minimum number of captive parts for quick assembly. Hex screws with slotted heads permit easy assembly requiring only the use of hand tools. Each length comes complete with a connector kit and hardware for joint connection. It is UL listed as wireways and auxiliary gutters.

Wireway can be used to advantage in most smaller distribution systems where multiple runs of conduit are required. It is superior to conduit for exposed work and where additions or alterations to the distribution system can be expected. Wireway can be installed without the expensive tools required in conduit systems. Normally, a screwdriver and wrench are all that are needed to install the lengths and fittings. Runs are assembled on the floor and hand-lifted into position. Wireway systems are adaptable and readily accessible at all times. Taps and splices can be made whenever an apparatus is to be added, moved or modified. It is simple to add or reroute a circuit anytime after the original installation is completed.

Conductors used in wireway do not have to be derated as is the case with conduit. The number of conductors used in a wireway run is limited to not more than 30 at any cross section, unless they are for signal currents or are controller conductors between a motor and its starter and used only for starting duty. Conductor cross-sectional area is limited to 20 percent of the wireway cross-sectional area. Wireway is considered by many electricians, designers and engineers to be the most economical wiring system for machine shops, laboratories, automated lines, lighting feeds, ganging gutter and many other electrical applications.

Wireway Connectors

Connectors form a closed section with a friction hinge (in the case of one manufacturer) on one side, to prevent the connector cover from falling closed. Besides connectors, there are also numerous combinations of hangers designed especially for wireway systems. Closing plates, elbows, telescopic fittings, tee fittings, junction boxes and crosses make up the majority of the related fittings. There are also transposition sections, gusset brackets (used in place of hangers), pull boxes and reducer fittings to make the transition from one size wireway to another. Some of these fittings are shown in Fig. 8-7. A practical application of wireway use is shown in Fig. 8-7.

Fig. 8-7: Typical application of wireway to feed various electrically-drive machines.

Underfloor Duct

Underfloor duct is a raceway system designed to be imbedded in the concrete floor of offices, classrooms, laboratories, manufacturing areas, supermarkets, etc., for the purpose of providing an enclosed raceway for wires and cables from their originating panel or closet to their point of use.

An underfloor raceway is composed of two types of ducts: feeder ducts and distribution ducts. Complementing these two types of ducts are junction boxes, support couplers and supports, horizontal and vertical elbows, power and telephone outlets and numerous cast and sheet metal fittings used for conduit adapters, change of direction of duct runs, "Y" take-offs, etc. A typical example appears in Fig. 8-8.

Underfloor duct is one of the many members of the family of "Conduit and Raceways" and as such, must conform to all of the requirements of the 1993 NE Code governing this family. In addition, it must conform specifically to the requirements of Article 354 of the NE Code.

Feeder Ducts: Feeder ducts are those ducts which provide the "feed" from the service terminal points

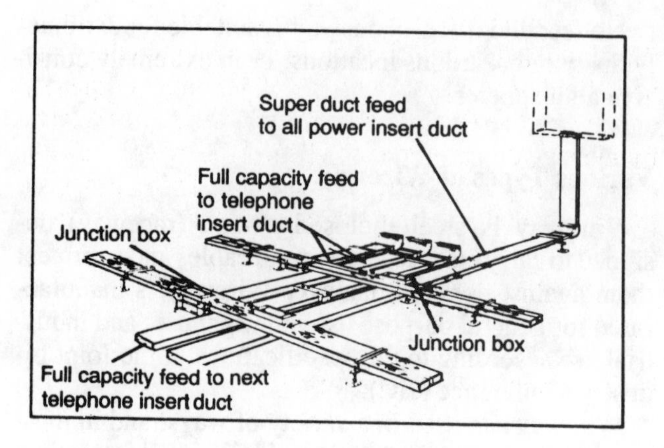

Fig. 8-8: Practical application of underfloor duct.

(lighting panelboards, telephone closet or cabinet, signal cabinet, etc.) to the duct that distributes the system to its point of use.

Feeder duct is referred to as "blank duct" because it has no "inserts" as does distribution duct. Feeder ducts are available in several sizes to meet most installation situations.

Distribution Ducts: Distribution ducts are those ducts that provide the "distribution" raceway from the point of juncture with the feed raceway to the point of use of the service. Distribution duct has a fabricated "insert" every 24 inches along its length for the usual installation and is also available in several sizes. Most are available in 10-foot lengths.

Junction boxes are used at the point of juncture of the feed raceways and the distribution raceways. Junction boxes, by the use of interior partitions, also maintain the separation of services where two or more non-compatible services form a junction. For instance, in a two-cut system, where one raceway is for telephone cables and one raceway is for branch circuit wiring, the junction box provides a separate raceway for each system, both through the box and at right angles to it.

There are a number of additional accessories of various uses that complete the cut system. These include vertical elbows, cabinet connectors, horizontal elbows, offsets, "Y" take-offs, conduit adapters, box opening plugs, duct end plugs, marker screws, sealing compound, and service fitting.

Induction and Dielectric Heating Equipment

The wiring for and connection of induction and dielectric heat generating equipment used in industrial and scientific applications (but not for medical or dental applications) are covered in Article 665.

The heating effect of such equipment is accomplished by placing the materials to be heated in the magnetic field of an electric voltage of very high frequency or between two electrodes connected to a source of high frequency voltage. Induction heating is used in heating metals and other conductive materials. Dielectric heating is used in the heating of materials that are poor conductors of electric current.

The equipment used consists either of motor-operated, high-frequency generators, or electric tube or solid-state oscillators. Such equipment is supplied by manufacturers or their representatives. Designers, electrical contractors, and electrical workers can benefit by contacting these manufacturers to obtain installation procedures, specifications, and the like.

The size of the supply circuit conductors, overcurrent protection, disconnecting means, type of grounding, and output circuits are covered in Article 665 of the NE Code.

Electronic Computer/Data-Processing Systems

The use of computers more than doubles each year. Consequently data-processing rooms or areas are becoming more common. Article 645 of the NE Code covers the installation of power supply wiring, grounding of equipment, and other such provisions that will insure a safe installation.

In many computer installations, the equipment operates continuously. For this reason, Section 645-5(a) of the NE Code requires that branch circuits supplying one or more computers have an ampacity not less than 125% of the total connected load. Therefore, if a computer system is rated at, say, 12 amperes, the branch circuit feeding this equipment should be rated at 12 x 1.25 = 15 amperes. The supply conductors are normally flexible cables or cords for easy moving or when adding new equipment. Article 645 lists the following as permissible:

- Computer/data processing cable and attachment plug cap.
- Flexible cord and an attachment plug cap.
- Cord-set assembly. When run on the surface of the floor, they shall be protected against physical damage.

Separate data processing units shall be permitted to be interconnected by means of cables and cable assemblies listed for the purpose. Where run on the surface of the floor, they shall be protected against physical damage.

Since there are so many interconnections for power, control, and communications, computer equipment is sometimes installed on raised floors with the cables running underneath the floor. The branch-circuit conductors in that case must in either rigid conduit, intermediate metal conduit, electrical metallic tubing, metal wireway, surface metal raceway with metal cover, flexible metal conduit, liquidtight flexible metal or nonmetallic conduit, Type MI cable, Type MC cable, or Type AC cable. Whichever wiring method is used, it must be installed in accordance with Section 300-11.

This underfloor area can also be used to circulate air for computer equipment cooling. Panels in the floor are removed under the equipment to allow air to enter the cabinets that house the equipment. In such an arrangement, a separate heating/ventilating/air conditioning (HVAC) system should be provided solely for the computer area. If this system services other areas, fire-/smoke dampers must be provided at the point of penetration of the room boundary. See Section 645-10.

A disconnecting means must be provided that will allow the operator to disconnect all computer equipment in the area. A disconnecting means must also be provided to disconnect the ventilation system serving the computer area.

Fig. 8-9 summarizes the installation rules for electronic computer data-processing areas.

Electric Welders

Electric welding equipment is normally treated as a piece of industrial power equipment for which are provided branch circuits adequate for the current and voltage of the equipment. Certain specific conditions, however, apply to circuits feeding ac transformers and dc rectifier arc welders, motor-generator arc welders, resistance welders, and the like; the requirements are found in Article 630 of the NE Code.

Electrical personnel should be aware that electrical arc welders require a relatively large amount of current at a relatively low voltage delivered to the welding electrode. The welding is accomplished with the melting and fusing of the welding rod or electrode to the material being welded. The reduction of the voltage and the increase in current are accomplished through the use of transformers or motor generators.

Resistance welding is accomplished by the melting and fusing of the metals due to the passage of a large amount of current from one electrode to another through the material being welded.

In arc welding, the current flows continuously as long as the welding rod is in contact with the metal being welded. In resistance welding, the flow of the current is interrupted intermittently either manually or automatically.

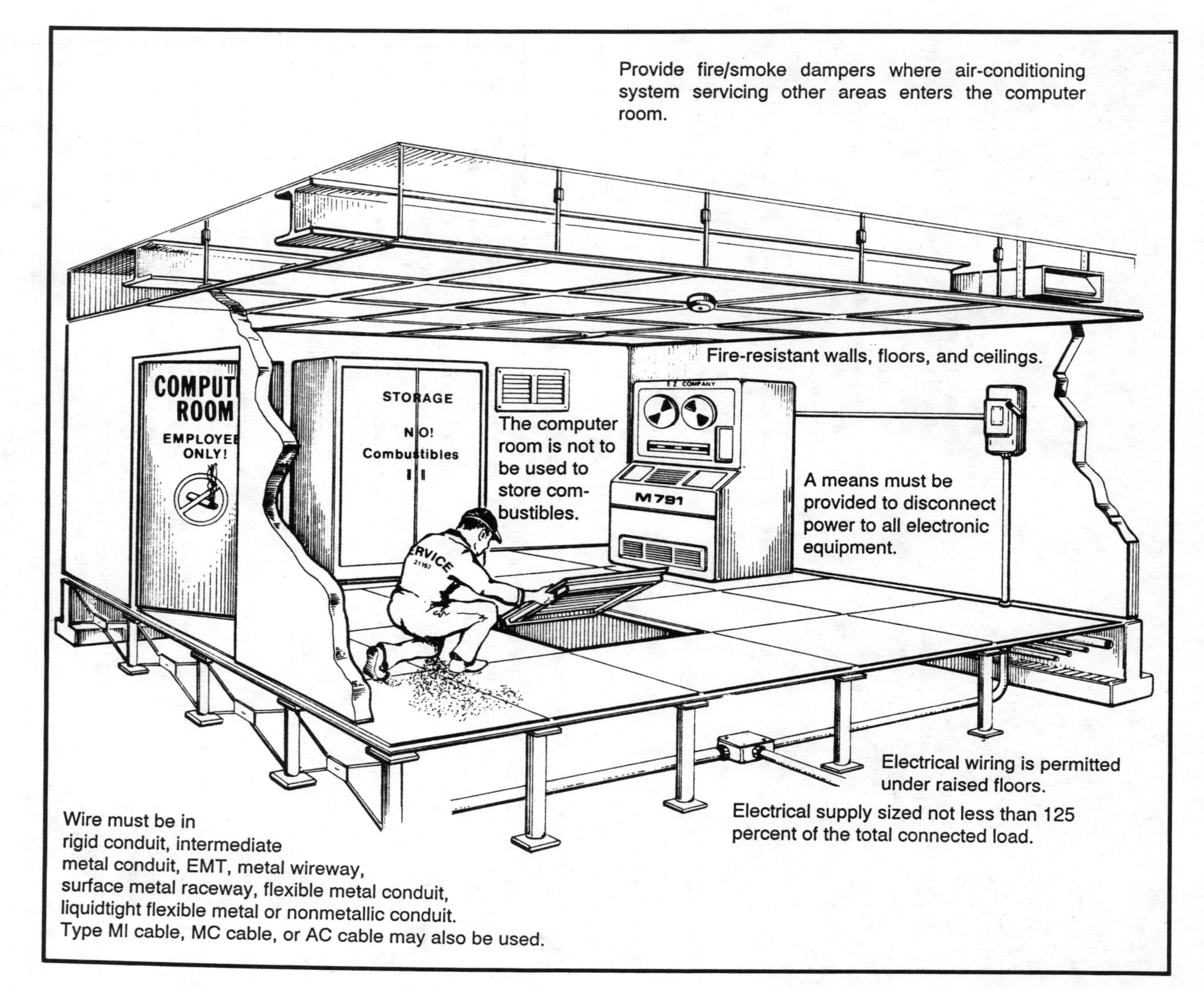

Provide fire/smoke dampers where air-conditioning system servicing other areas enters the computer room.
Fire-resistant walls, floors, and ceilings.
STORAGE
NO! Combustibles
The computer room is not to be used to store combustibles.
E Z COMPANY
M791
A means must be provided to disconnect power to all electronic equipment.
Electrical wiring is permitted under raised floors.
Electrical supply sized not less than 125 percent of the total connected load.
Wire must be in rigid conduit, intermediate metal conduit, EMT, metal wireway, surface metal raceway, flexible metal conduit, liquidtight flexible metal or nonmetallic conduit. Type MI cable, MC cable, or AC cable may also be used.

Fig. 8-9: NE Code guidelines for electrical installations in computer rooms.

Cranes and Hoists

Crane and hoist equipment is usually furnished and mechanically installed by crane manufacturing companies or their representatives. When working on such equipment, refer to Article 610 of the NE Code.

Crane and hoist wiring consists of the control and operating circuits on the equipment itself and the contact conductors or flexible conductors supplying electric power to the equipment. Electricians on the job normally install the feeder circuit at a point of connection to the contact conductor or flexible cables, including over-current protection and disconnecting means.

The motor-control equipment, control and operating circuits, and the bridge contact conductors are usually furnished and installed by the manufacturer; sometimes, it is the responsibility of the owner or electrical contractor.

Swimming Pools

The NE Code recognizes the potential danger of electric shock to persons in swimming pools, wading pools, and therapeutic pools or near decorative pools or fountains. This shock could occur from electric potential in the water itself or as a result of a person in the water or a wet area touching an enclosure that is not at ground potential. Accordingly, the NE Code provides rules for the safe installation of electrical equipment and wiring in or adjacent to swimming pools and similar locations. Article 680 of the NE Code covers the specific rules governing the installation and maintenance of swimming pools and similar installations.

The general requirements for the installation of outlets, overhead conductors and other equipment are summarized in Fig. 8-10.

Other installations falling under the category of special equipment are listed below along with the appropriate NE Code Article for further reference.

- Electrically driven or controlled irrigation machines, Article 675
- Electrolytic cells, Article 668
- Electroplating, Article 669
- Elevators, dumbwaiters, escalators and moving walks, Article 620
- Integrated electric systems, Article 685
- Manufactured wiring systems, Article 604
- Office furnishings, Article 605
- Organs, Article 650
- Solar photovoltaic systems, Article 690
- Sound-recording and similar equipment, Article 640
- X-ray equipment, Article 660

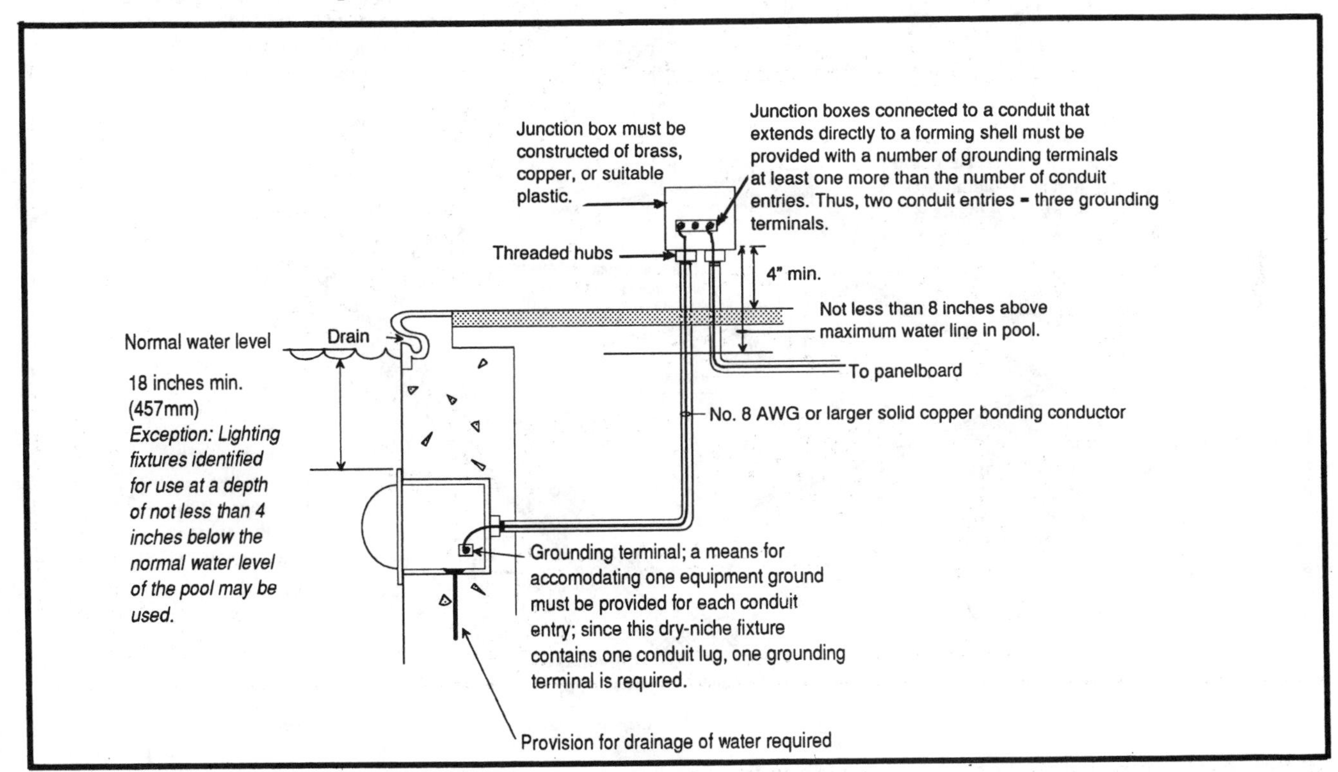

Fig. 8-10: NE Code regulations governing swimming pool electrical installations.

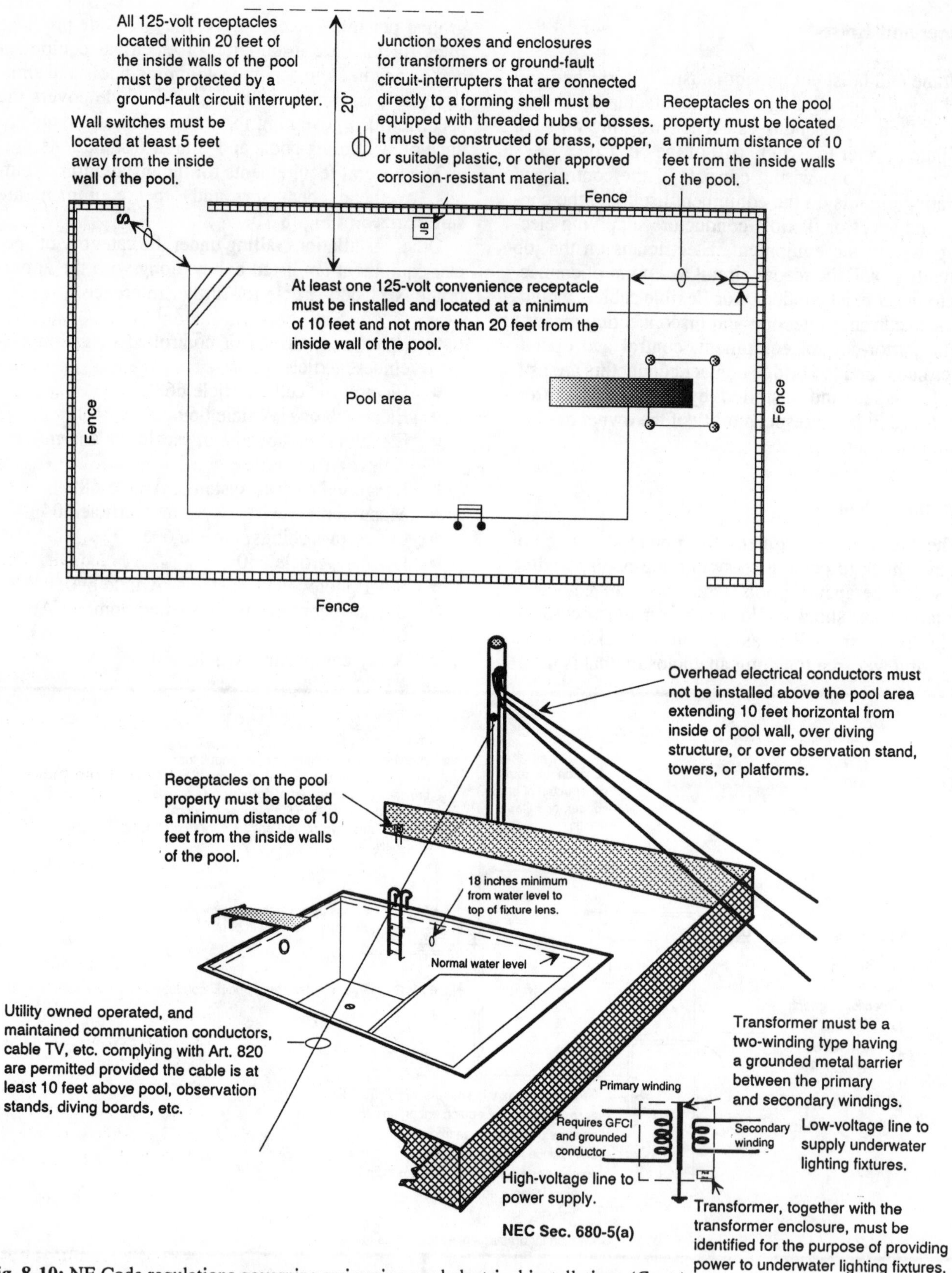

All 125-volt receptacles located within 20 feet of the inside walls of the pool must be protected by a ground-fault circuit interrupter.
20'
Junction boxes and enclosures for transformers or ground-fault circuit-interrupters that are connected directly to a forming shell must be equipped with threaded hubs or bosses. It must be constructed of brass, copper, or suitable plastic, or other approved corrosion-resistant material.
Receptacles on the pool property must be located a minimum distance of 10 feet from the inside walls of the pool.
Wall switches must be located at least 5 feet away from the inside wall of the pool.
Fence
S
JB
At least one 125-volt convenience receptacle must be installed and located at a minimum of 10 feet and not more than 20 feet from the inside wall of the pool.
Pool area
Fence
Fence
Fence
Overhead electrical conductors must not be installed above the pool area extending 10 feet horizontal from inside of pool wall, over diving structure, or over observation stand, towers, or platforms.
Receptacles on the pool property must be located a minimum distance of 10 feet from the inside walls of the pool.
18 inches minimum from water level to top of fixture lens.
Normal water level
Utility owned operated, and maintained communication conductors, cable TV, etc. complying with Art. 820 are permitted provided the cable is at least 10 feet above pool, observation stands, diving boards, etc.
Transformer must be a two-winding type having a grounded metal barrier between the primary and secondary windings.
Primary winding
Requires GFCI and grounded conductor
Secondary winding
Low-voltage line to supply underwater lighting fixtures.
High-voltage line to power supply.
Transformer, together with the transformer enclosure, must be identified for the purpose of providing power to underwater lighting fixtures.
NEC Sec. 680-5(a)

Fig. 8-10: NE Code regulations governing swimming pool electrical installations *(Cont.)*

Underground Systems

There are several methods used to install underground wiring, but the most common include direct-burial cables and the use of duct lines or duct banks. The method used depends on the type of wiring, soil conditions, allotted budget for the work, etc.

Direct-burial installations will range from small, single-conductor wires to multiconductor cables for power or communications or alarm systems. In any case, the conductors are installed in the ground either by placing them in an excavated trench, which is later backfilled, or by burying them directly by means of some form of cable plow, which opens a furrow, feeds the conductors into the furrow, and closes the furrow over the conductor.

Sometimes it becomes necessary to use lengths of conduit in conjunction with direct-burial installations, especially where the cables emerge on the surface of the ground or terminate at an outlet or junction box. Also, where the cables cross a roadway or concrete pavement, it is best to install a length of conduit under these areas in case the cable must be removed at a later date. By doing so, the road or concrete pad will not have to be disturbed.

Figure 8-11 shows a cross section of a trench with direct-burial cable installed. Note the sand base on which the conductors lie to protect them from sharp stones and such. A treated board is placed over the conductors in the trench to offer protection during any digging that might occur in the future. Also, a continuous warning ribbon is laid in the trench, some distance above the board, to warn future diggers that electrical conductors are present in the area.

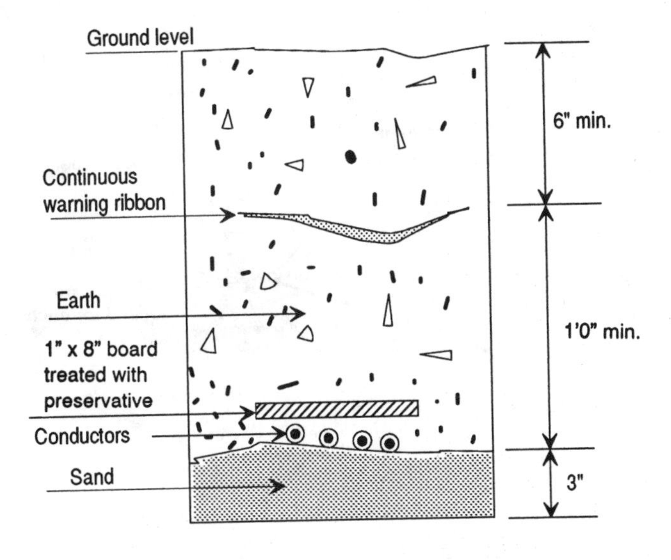

Fig. 8-11: Cross section of direct-burial cable.

Conductor Types

Type USE (Underground service-entrance cable): This type of cable is approved for underground use since it has a moisture-resistant covering.

Cabled single-conductor Type USE constructions may have a bare copper conductor cabled with the assembly. Type USE single, parallel, or cabled conductor assemblies may have a bare copper concentric conductor applied. These constructions do not require an outer overall covering.

Type USE cable may be used for underground services, feeders, and branch circuits.

Type UF (Underground feeder and branch-circuit cable): This type of cable is manufactured in sizes from No. 14 AWG copper through No. 4/0. In general, the overall covering of Type UF cable is flame-retardant, moisture-, fungus-, and corrosion-resistant, and is suitable for direct burial in the ground.

Type UF cable may be used for direct-burial, underground installations as feeders or branch circuits when provided with overcurrent protection of the rated ampacity as required by the NE Code.

Where single-conductor cables are installed, all cables of the feeder circuit, subfeeder circuit, or branch circuit—including the neutral and equipment grounding conductor (if any)—must be run together in the same trench or raceway.

Nonmetallic-armored cable: This type of cable is also used for underground installations. The interlocking armor consists of a single strip of interlocking tape that extends for the length of the cable. The surface of the cable is rounded, which allows it to deflect blows from picks and shovels much better than flat-bend armor. The cable must have an outer covering that will not corrode or rot. An asphalt-jute finish may be placed over the cable if it is to be subjected to particularly harsh corrosive environments.

Minimum Cover Requirements

The National Electrical Code specifies minimum cover requirements for direct-buried cable (Fig. 8-12). Furthermore, all underground installations must be grounded and bonded in accordance with NEC Article 250.

Where direct-buried cables emerge from the ground, they must be protected by enclosures or raceways extending from the minimum cover distance specified in the table in Fig. 8-12. However, in no cases will the protection be required to exceed 18 inches below the finished grade.

Practical Application

Methods of installing direct-burial underground cable vary according to the length of the installation, the size of

Minimum Cover Requirements for Direct-Burial Cable 0 to 600 Volts

Location of Underground Wiring	Direct-Burial Cables (Inches)	Residential Branch Circuits Rated 120 Volts or less with GFCI Protection and Maximum Over-current Protection of 20 Amperes (Inches)
All locations not specified below	24	12
In trench below 2-inch thick concrete or equivalent	18	6
Under a building	Not allowed; cables must be in raceway	Not allowed
Under minimum of 4-inch thick concrete exterior slab with no vehicular traffic and the slab extending not less than 6 inches beyond the underground installation	18	6
Under streets, highways, roads, alleys, parking lots	24	24
One-, and two-family dwelling driveways and parking areas, and used for no other purpose	18	12
In or under airport runways including adjacent areas where trepassing prohibited	18	--
In solid rock where covered by minimum of 2 inches concrete extending down to rock	Not allowed; must be encased in raceway at least 2 inches thick	Not allowed; must be encased in raceway at least 2 inches thick

Fig. 8-12: Minimum cover requirements for direct-burial cable, 0 to 600 volts.

cable being installed, and the soil conditions. For short runs, from, say, a residential basement to a garage located 20 or so feet away, the excavation is often done by hand. For longer runs, power equipment is almost always used.

In general, the trench is opened to the correct depth with an entrenching tractor or backhoe. All sharp rocks, roots, and similar items are then removed from the trench to prevent these objects from damaging the direct-burial cable. If soil conditions dictate, a 3-inch layer of clean sand is poured into the bottom of the trench to further protect the direct-burial cable.

Underground cable will almost always come from the manufacturer on either metal or wooden reels. In the case of relatively short runs of the smaller sizes of multiconductor cable, the reel containing the cable is set up at one end of the trench, using two reel jacks and a length of conduit through the center hole in the reel to allow the reel to rotate as the cable is "paid off." See Fig. 8-13.

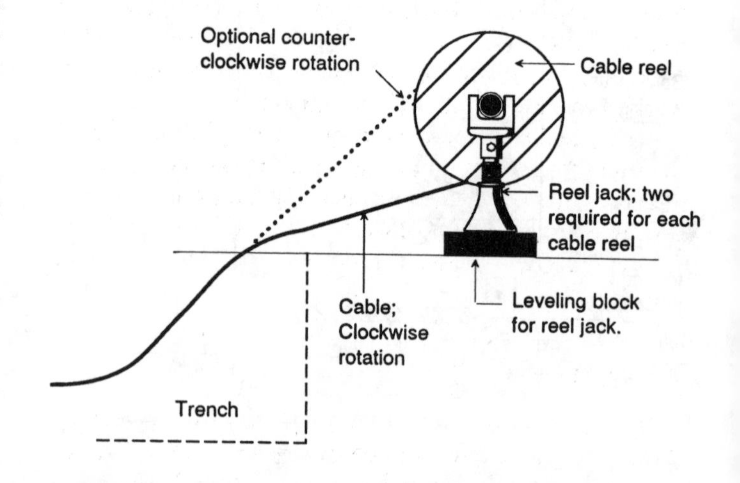

Fig. 8-13: Method of setting up cable reel for underground cable pull.

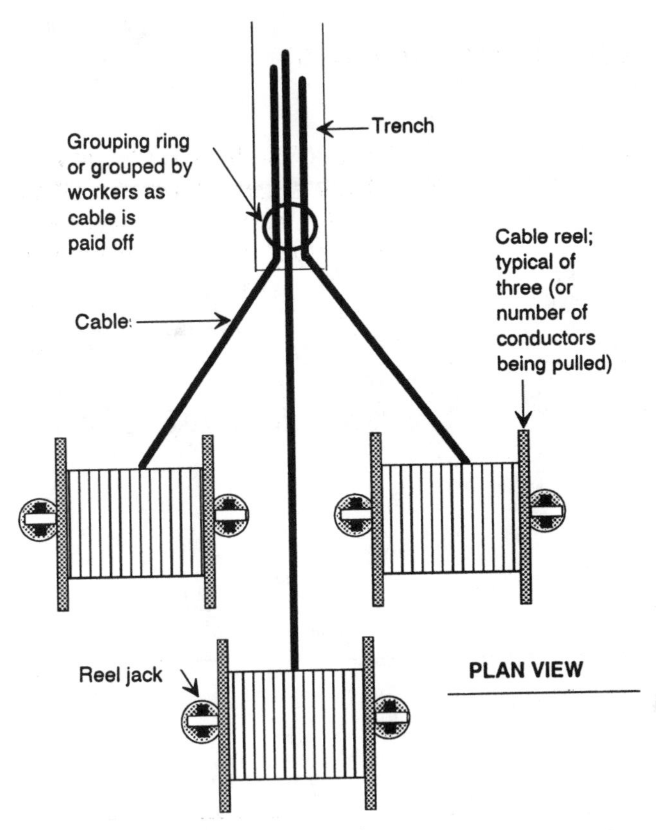

Fig. 8-14: Method of pulling three conductors at once for direct burial in an open trench.

Where several single conductors are installed, more than one reel is set up in a manner described above. Then cable is pulled from all the reels simultaneously as shown in Fig. 8-14.

For longer runs, or where the larger sizes of cable are installed, the weight of the cable dictates a different method. In this case, the end of the cable is secured at one end of the trench, while the reel is attached to a tractor or backhoe and arranged so that the reel will rotate freely. The tractor or backhoe then runs along the trench—either at the trench edge or straddling it—and the cable is paid out by workers and allowed to fall into the trench.

Once installed, another layer of sand may be placed over the cable for protection against sharp rocks; the trench is then backfilled. A treated wooden plank may also be used for cable protection; a yellow warning ribbon—designed for the purpose—is also a good idea; both are shown in Fig. 8-11.

Duct Systems

By definition, a *duct* is a single enclosed raceway through which conductors or cables are pulled. One or more ducts in a single trench is usually referred to as a *duct bank.* A duct system provides a safe passageway for power lines, communication cables, or both.

Depending upon the wiring system and the soil conditions, a duct bank may be placed in a trench and covered with earth or enclosed in concrete. Underground duct systems also include manholes, handholes, transformer vaults, and risers.

Manholes are set at various intervals in an underground duct system to facilitate pulling conductors or cables when first installed, and to allow for testing and maintenance later on. Access to manholes are provided through *throats* extending from the manhole compartment to the surface (ground level). At ground level, a manhole cover closes off the manhole area tightly.

In general, an underground cable run normally terminates at a manhole, where it is spliced to another length of cable. Manholes are sometimes constructed of brick and concrete. Most, however, are prefabricated, reinforced concrete, made in two parts—the base and the throat—for quicker installation. Their design provides room for workers to carry out all appropriate activities inside them, and they are also provided with a means for drainage. See Fig. 8-15.

There are three basic designs of manholes: two-way, three-way, and four-way. In a two-way manhole, ducts and cables enter and leave in two directions. A three-way manhole is similar to a two-way manhole, except that one additional duct/cable run leaves the manhole. Four duct/cable runs are installed in a four-way manhole. See Fig. 8-16. Also see Fig. 8-17 for specifications of a typical manhole.

Transformer vaults house power transformers, voltage regulators, network protectors, meters and circuit breakers. Other cables end at a substation or terminate as risers—connecting to overhead lines.

Types of Ducts

Ducts for use in underground electrical systems are made of fiber, vitrified tile, metal conduit, plastic or poured concrete. In some existing installations, the worker may find that asbestos/cement ducts have been used. In most areas, a contractor must be certified before removing or disturbing asbestos ductwork, and then extreme caution must be practiced at all times.

The inside diameter of ducts for specific installations is determined by the size of the cable that the ducts will house. Sizes from 2 to 6 inches are common.

Fiber duct: Fiber duct is made with wood pulp and various chemicals to provide a lightweight raceway that will resist rotting. It can be used enclosed in a concrete envelope with at least 3 inches of concrete on all sides. The extremely smooth interior walls of this type of duct facilitate cable pulling through them.

Vitrified clay duct: Vitrified clay tile is sometimes called *hollow brick.* Its main use is in underground systems for

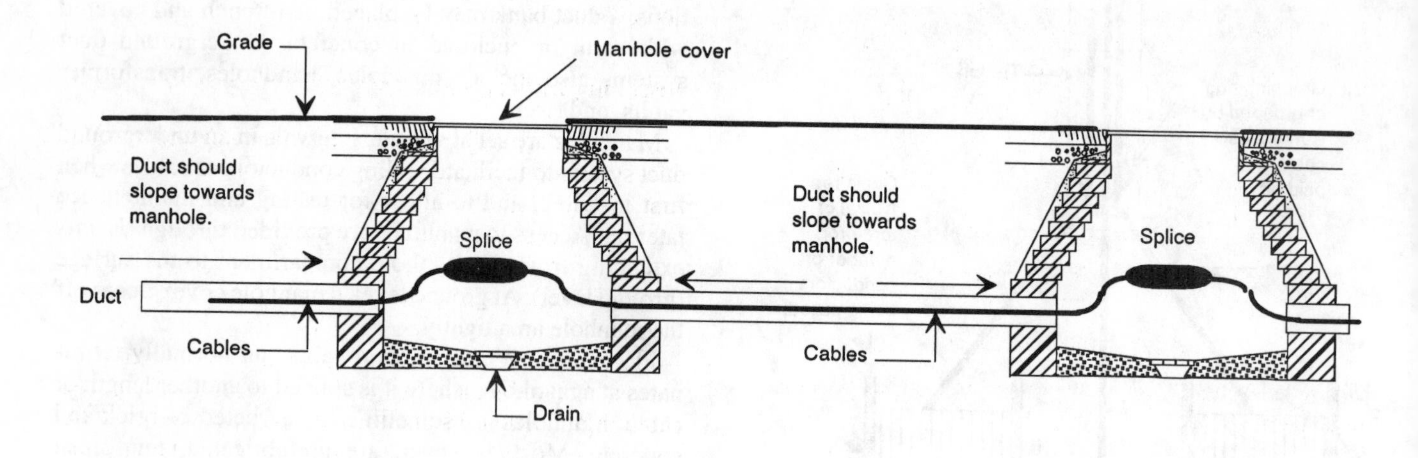

Fig. 8-15: Cross section of typical underground duct system run between manholes.

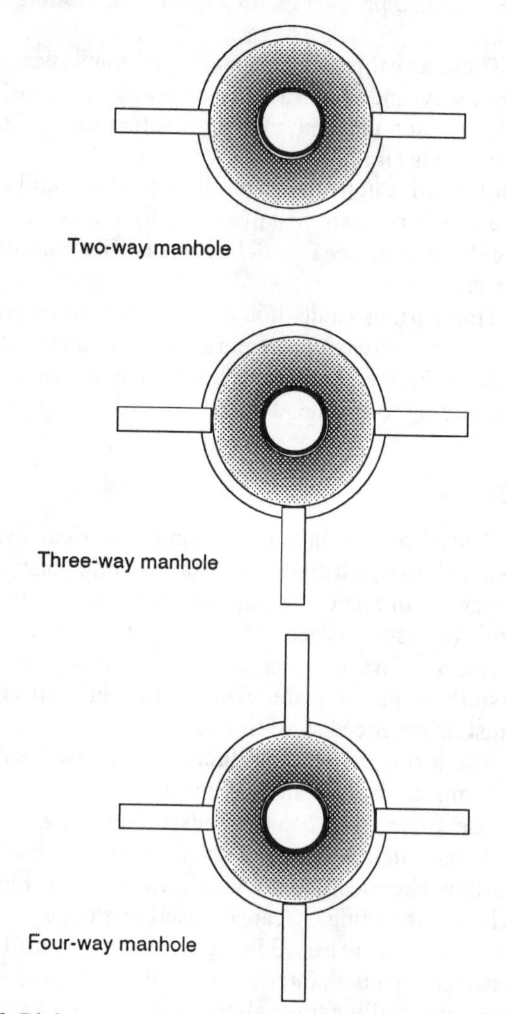

Fig. 8-16: Plain view of two-, three- and four-way manholes.

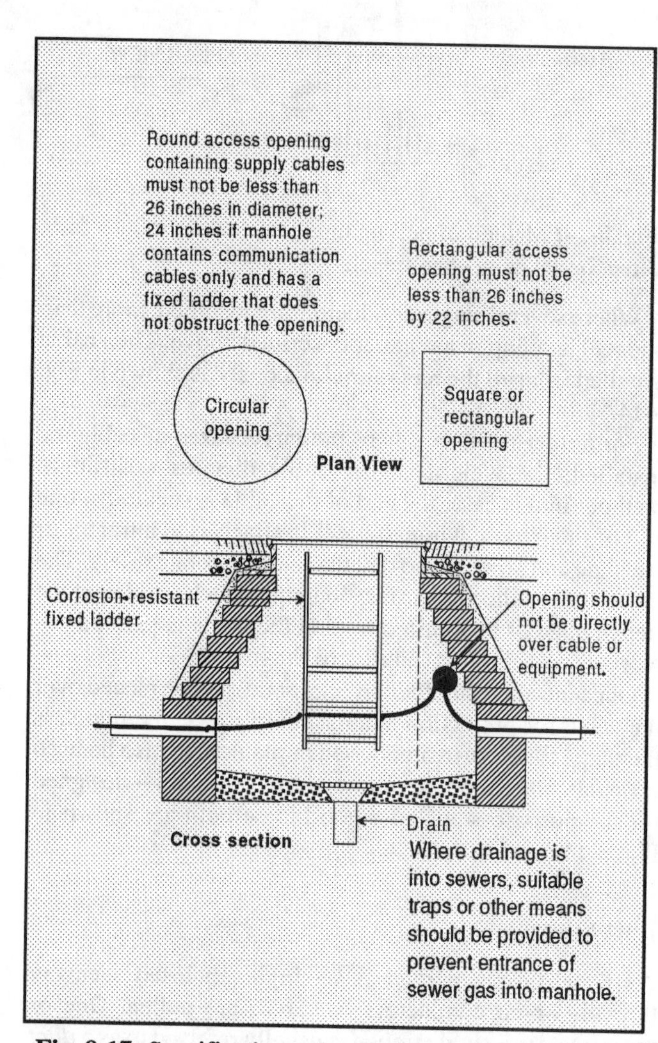

Fig. 8-17: Specifications of typical manhole.

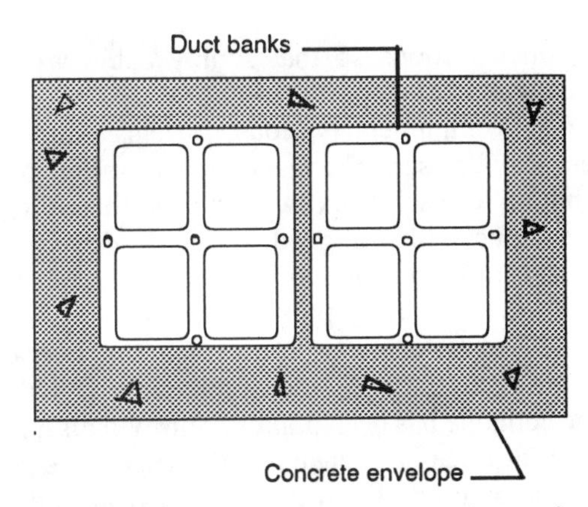

Fig. 8-18: Vitrified clay four-way square ducts—installed in tandem.

low-voltage and communication cables and is especially useful where the duct run must be routed around underground obstacles, because the individual pieces of duct are shorter than other types.

The four-way multiple duct (Fig. 8-18) is the type most often encountered. However, vitrified ducts are available in sizes up to 16 ducts in one bank. The square ducts are usually 3½ inches in diameter, while round ducts vary from 3½ to 4½ inches.

When vitrified clay ducts are installed, their joints should be staggered to prevent a flame or spark from a defective cable in one duct from damaging cable in an adjacent duct.

Metal conduit: Metal conduit, such as iron, rigid metal conduit, intermediate metal conduit, etc, is relatively more expensive to install than other kinds of underground ductwork. However, it provides better protection than most other types, especially against the hazards caused by future excavation.

Plastic conduit: Plastic conduit is made of polyvinylchloride (PVC), polyethylene (PE), or styrene. Since they are available in lengths up to 30 feet, fewer couplings are needed than with many types of duct systems. PVC conduit is currently very popular for underground electrical systems since it is light in weight, relatively inexpensive, and requires less labor to install.

Monolithic concrete ducts: This type of system is poured at the job site. Multiple duct lines can be formed using tubing cores or spacers. The cores may be removed after the concrete has set. Although relatively expensive, this system has the advantage of creating a very clean duct interior with no residue that can decay. It is also useful when curves or bends in duct systems are necessary.

Cable-in-duct: This is another popular duct type that offers a reduction in labor cost when installed. It is manufactured with cables already installed. Both the duct and the cable it contains is shipped on a reel to facilitate installing the entire system with ease. Once installed, the cables can be withdrawn or replaced in the future if it should become necessary.

Installing Underground Duct

The selection of high-voltage cables and their installation in underground ductwork is not part of the requirements of the NE Code. However, Appendix B of the NE Code provides installation criteria "for information purposes only." Furthermore, the National Electrical Safety Code (NESC) covers regulations governing high-voltage underground installations in detail. Section 33 of the NESC covers supply cable including detailed requirements of conductors, insulation, sheaths, jackets, and shielding. Section 34 also covers underground installations. Both of these Sections should be studied by anyone involved in underground duct systems—either design or actual installations.

In practical applications of underground duct systems, present needs and potential growth are both considered. Construction aspects such as trenching and pouring concrete are cost factors, so it is economically sound to provide for future growth as part of an original installation. Often the number of conduits laid is twice the number needed for present usage.

In general, there should be at least 1 to 3 inches of earth or concrete between adjacent conduits containing power cables. This insulation will ensure that the heat radiated by one line will not affect the surrounding ducts. Heat will cause insulation to deteriorate faster and, in general, the hotter the cable, the least amount of load it will carry. Consequently, duct banks should be designed so that heat from conductors is dissipated into either the surrounding concrete envelope or earth.

During the installation of a duct system, spacers should be used to hold the ducts in place while concrete is being poured. Figures 8-19 and 8-20 show how spacers are arranged prior to a concrete pour.

Duct Installation

Excavation is the first order of business; that is, a backhoe or other digging apparatus is employed to dig the trenches and ground openings for manholes. Manholes are then constructed at various intervals throughout the "run." In the case of two-piece, prefabricated manholes, only the bases are installed prior to installing the ductwork.

The bottom of the trenches must be flat and compacted prior to installing the ducts. This is to prevent the trench from settling and putting stress on the duct banks. In most cases, if a concrete envelope is to be used, the trench is first filled with 3 inches of concrete and finished at the appropriate grade. Once hardened, duct lines are then placed in the trench utilizing fiber or plastic spacers at various inter-

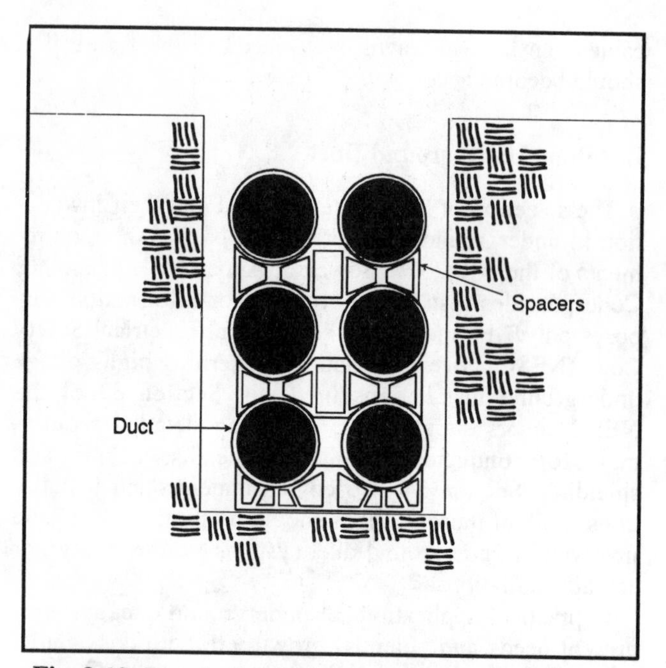

Fig. 8-19: Duct bank with spacer separation.

vals along the duct run. This process is repeated until the final row of duct is laid and embedded.

During the pouring process, and before the concrete sets, it is important to make absolutely certain that the ducts line up at the joints and couplings. All duct installations must join in a manner sufficient to prevent solid matter from entering the raceway. Furthermore, the joints must form a sufficiently continuous smooth interior surface between joining duct sections so that supply cable will not be damaged when pulled past the joint.

To ensure alignment, various dowels, mandrels, and scrapers may be used. The mandrel, for example, must be long enough to reach back two joints so that at leat three sections of duct will be aligned. A leather or rubber washer is attached to the mandrel which serves to clean out the conduit as the mandrel is pulled through.

A wire brush—slightly larger than the interior duct diameter—or a scraper should be pulled through the ducts after the concrete has been poured. This will eliminate any cement or dirt that has penetrated the duct at the joints.

One method used to prevent concrete from entering the joints during pouring is to wrap each joint with coarse muslin or some similar material prior to pouring the concrete envelope. The muslin is dampened to help it stick to the cut and then coated with cement. Although time-consuming, this procedure prevents concrete from entering into the raceway. It is also best to stagger the joints in a multiple duct run.

Soil conditions will dictate whether concrete encasement is required. If the soil is not firm, concrete encasement is mandatory. Concrete is also required with certain types of duct lines that are not able to withstand the pressure of an earth covering. If the soil is firm, and concrete encasement is still desired (or specified), the trench need only be wide enough for the ducts and concrete encaement. The concrete is then poured between the conduit and the earth wall. If the soil is not very firm and concrete is required, 3 additional inches should be allowed on each side of the duct banks to permit the use of concrete forms.

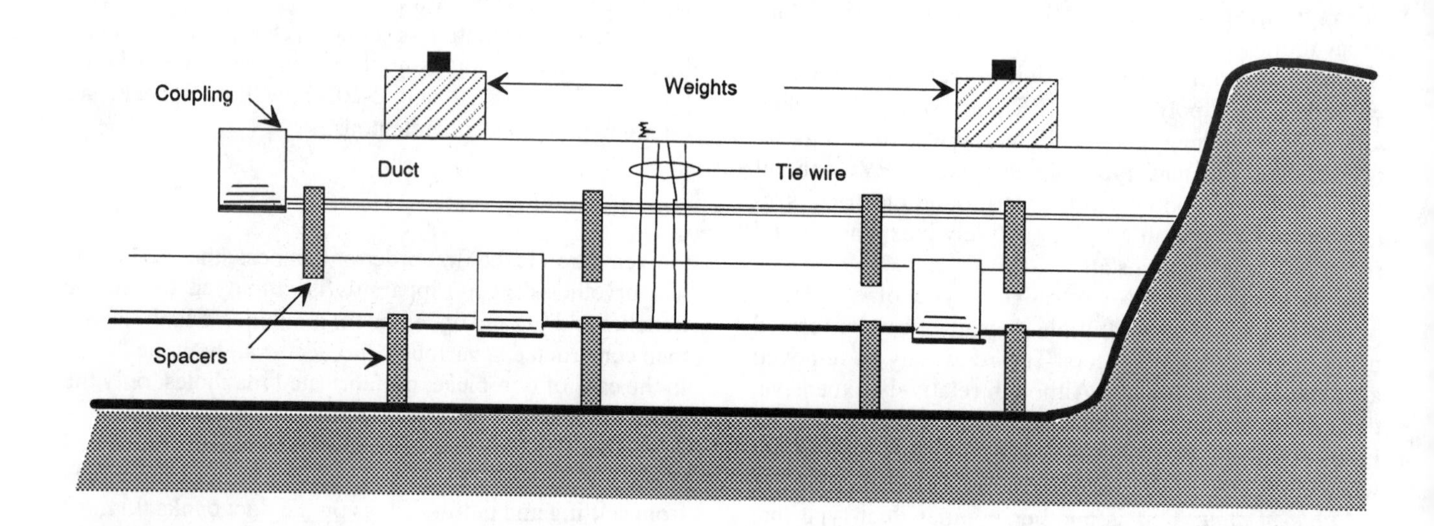

Fig. 8-20: Duct bank, prior to pouring concrete, using spacers, weight, and twine to keep duct in place during pouring operation.

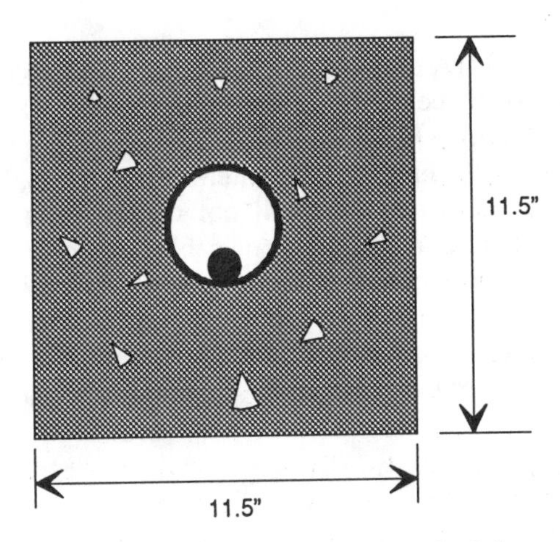

Fig. 8-21: Electrical duct bank for one electrical duct.

Ducts may be grouped in any of several different ways, but for power distribution, each duct should have at least one side exposed to the earth or the outside of the concrete envelope. This means that the pattern of ducts for power distribution should be restricted to either a two-conduit width or a two-conduit depth. This permits the heat generated by power transmission to dissipate into the surrounding earth. In other words, ducts for power distribution should not be completly surrounded by other ducts. When this type of situation exists, the inner ducts may be referred to as *dead ducts*. The heat that these ducts radiate is not dissipated as fast as from the ducts surrounding them.

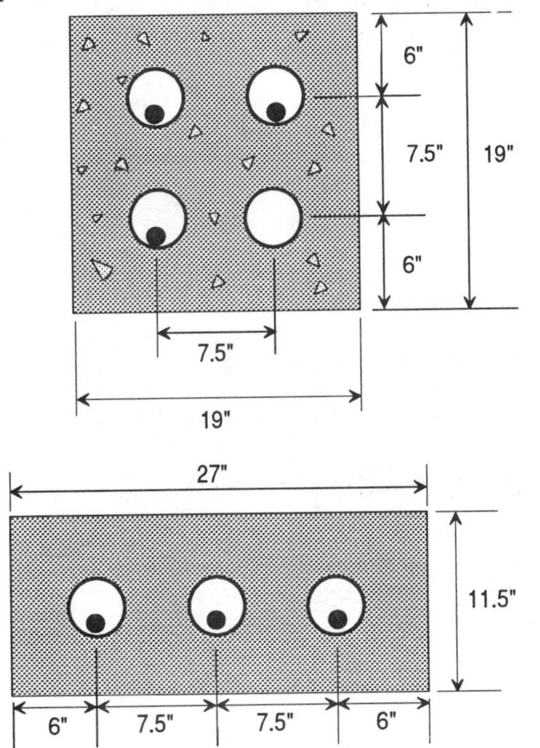

Fig. 8-22: Two arrangements for three electrical ducts.

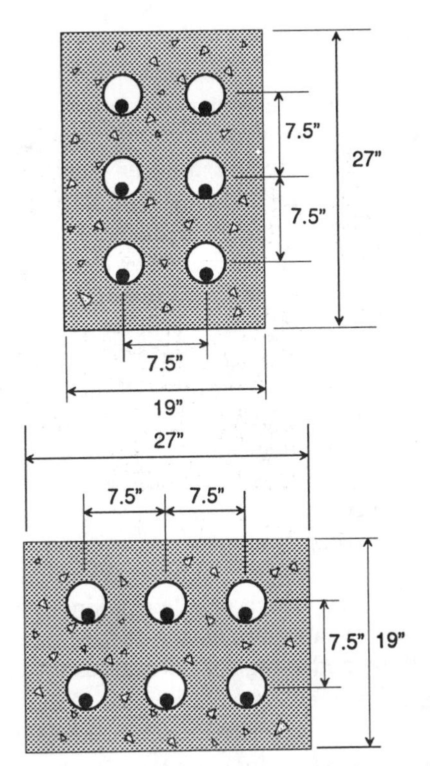

Fig. 8-23: Two arrangements for duct bank containing six electrical ducts.

While not suited for power cable, these dead ducts may be used for street lighting, control cable, or communication cable. The heat generated by these types of cables is relatively low, so the ducts can be arranged in any convenient configuration.

Figs. 8-21 through 8-24 show duct-bank configurations and dimensions given in Appendix B of the NE Code, and should serve as a guide.

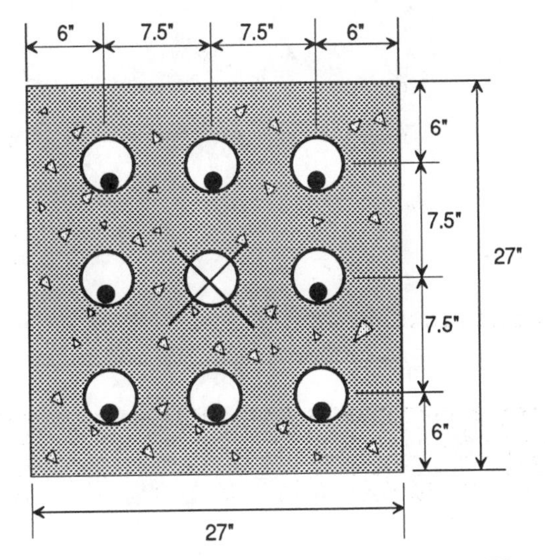

Fig. 8-24: Electrical duct bank with nine ducts. The center duct, however, should not be used for power transmission; only control or communication cables.

Pulling Cable

There are several preliminary operations prior to pulling cable through a duct bank. *Rodding* involves the use of many short wooden rods or dowels which are joined together on a long, flexible steel rod, stiff enough to be pushed or pulled through the duct. When the rod reaches the far end of the duct, steel "fish" wire is attached to the near end. Then the rod is pulled through, followed by the wire. This wire is then used to pull the cables through the ductwork.

Rigging of some kind must be set up to pull the cable through the duct. In general, a gripping device is attatched to the ends of the cable. This device consists of flexible steel mesh so that the harder the pull, the tighter the device grips the cable. Before the pulling operation is started, however, a wire lubrication known as "soap" in the trade is freely applied to the ends of the cable to reduce friction during the pull. A winch or other pulling mechanism is used to move the cable through the duct. More soap or wire lubricant is applied at regular intervals as the cable enters the ducts.

Part 341A of the National Electrical Safety Code covers control of bending, pulling, tensions and sidewall pressures during the installation of cable. This section also covers cleaning foreign material from ducts, selection of cable lubricants that will not damage any part of the installation, and restraint of cables in sloping or vertical runs to prevent them from moving downhill. It further specifies that power supply cables, control cables, and communication cables should not be installed in the same duct unless the same utility maintains or operates them.

Cable Splicing

High-voltage cable splicing is a specialized skill and is beyond the scope of this book to get into very much detail. In fact, most certified splicers do no other electrical work but splicing high-voltage cable. Factors that govern the splicing techniques include the type of cable, the type of insulating material, and whether the insulation has a conducting shield. Protection from the hazard of moisture during splicing is critical.

chapter 9

SIGNALING CIRCUITS

Electrical systems falling under the heading of "signaling" include such categories as security, fire alarm, and similar systems—employing a wide variety of techniques, often involving special types of equipment and materials designed for specific applications. Many of these systems operate on low-voltage circuits but are installed similarly to conventional electrical circuits for light and power. In all cases, however, when designing systems for use in buildings, the installations must conform to applicable NE Code requirements.

Several 1993 NE Code Sections will apply directly to signaling circuits. Some of these Sections follow:

- Alarm systems for health care facilities, Sections 517-32(c) and 517-42(c)
- Burglar Alarms, Sections 230-82 Ex. 5, and 230-94 Ex. 4.
- Fire Alarms, Sections 230-82 Ex. 5, and 230-94 Ex. 4
- Fire Protection, Article 760
- Remote Control Signaling, Article 725

All alarm systems have three functions in common:

1. Detection
2. Control
3. Annunciation (or alarm) signaling

Many systems incorporate switches or relays that operate because of entry, movement, pressure, infrared-beam interruption, and the like. The control senses operation of the detector with a relay and produces an output that may operate a bell, siren, silent alarm—such as a telephone dialer to law enforcement agencies—and similar devices. The controls frequently contain ON-OFF switches, test meters, time delays power supplies, standby batteries, and terminals for connecting the system together. The control output usually provides power on alarm to operate signaling devices or switch contacts for silent alarms.

One of the simplest and most common electric signal systems is the residential door-chime system. Such a system contains a low-voltage power source, one or more push buttons, wire, and a set of chimes.

The wiring diagram in Fig. 9-1 shows a typical two-note chime controlled at two locations. One button, at the main entrance, will sound the two notes when pushed, while the other button, at the rear door, will sound only one note when pushed.

Signal Circuit Components

Wire sizes for the majority of low-voltage systems range from No. 22 to No. 18 AWG. However, where larger-than-normal currents are required or when the distance between the outlets is longs, it may be necessary to use wire sizes larger than specified to prevent excessive voltage drop. Voltage-drop calculations should be made to determine the correct wire size for a given appliation—even on low-voltage circuits.

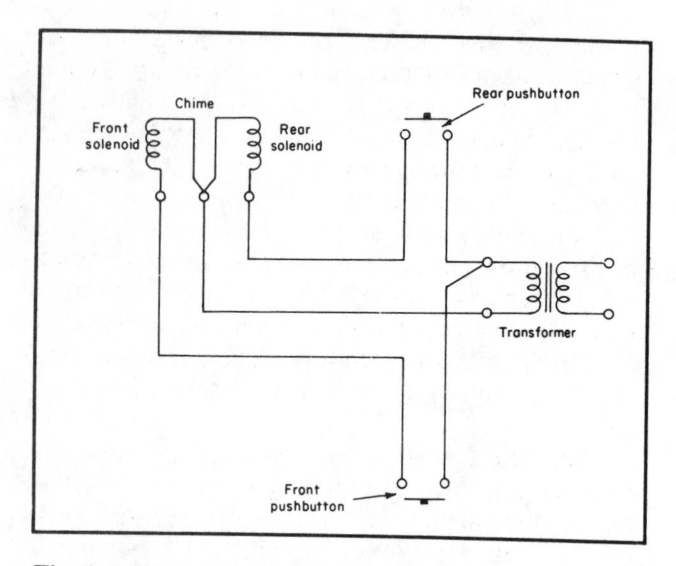

Fig. 9-1: Wiring diagram of a two-note chime.

The wiring of any alarm system is installed like any other type of low-voltage system; that is, locating the outlets, furnishing a power supply, and finally interconnecting the components with the proper size and type of wire.

Smoke and Fire Detectors

Fire signatures: Any product of a fire that changes the ambient conditions is called a fire signature and is potentially useful for detection. The principle fire signature used in residential smoke detectors is aerosol. Aerosols are particles suspended in air. The process of combustion releases into the atmosphere large numbers of such solid and liquid particles that may range in size from 10 microns (a micron is one thousandth of a millimeter) down to 0.001 microns. Aerosols resulting from a fire represent two different fire signatures. Those particles less than 0.3 microns do not scatter light efficiently and are classified as invisible. Those larger thanh 0.3 microns scatter light and are classified as visible. The invisible aerosol signature is usually referred to as the "products of combustion" and the visible aerosol signature as "smoke." Invisible aerosol is the earliest appearing fire signature.

Types of Fire Detection Devices

Thermal Detectors: Thermal detectors are devices that operate on high heat—typically 135°F. These units consist of a bi-metallic element which bends to complete a circuit under high heat conditions. Since these units do not detect smoke or products of combustion, they are not recommended for living areas of a residence. They do have value for use in attics, unheated garages and furnace rooms.

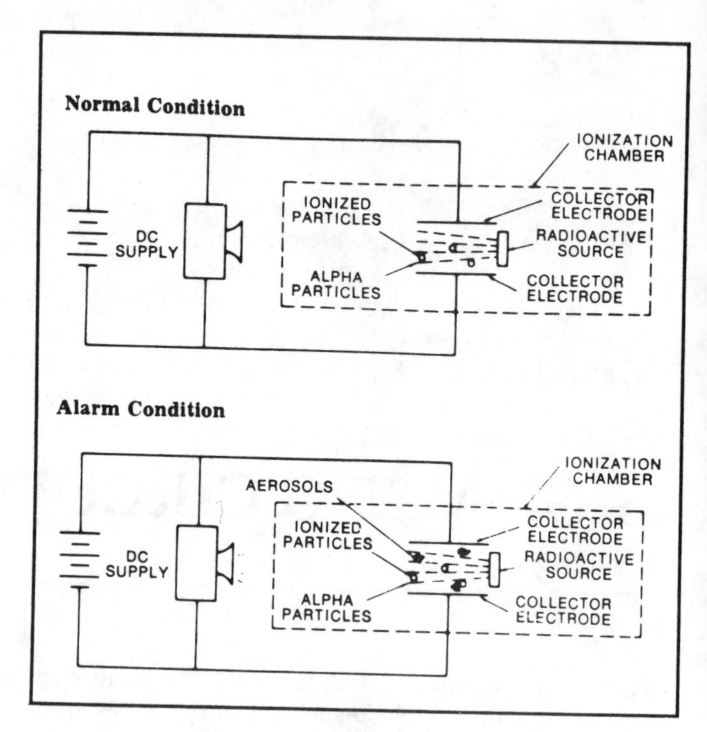

Fig. 9-2: Diagrams of ionization detectors. The top diagram is under normal conditions. The bottom diagram shows aerosols, such as products of combustion or smoke, entering the sensor. In this condition, the alarm is activated.

Flame Detectors: Flame detectors detect actual flames by sensing the ultra-violet emissions. These devices would not be used in residential applications.

Gas Detectors: These units respond to certain gases (propane, carbon monoxide, LP, butane, gasoline vapors, etc.) that would not be detected by a smoke and fire detector. While these detectors do have some uses, they should not be a substitute for a smoke and fire detector. They will not respond to aerosols produced by the majority of residential fires.

Ionization Detectors—Basic Operation: Inside the ionization chamber, the radioactive source emits radiation, mainly alpha particles, which bombard the air and ionize the air particles, which, in turn, are attracted by the voltage on the collector electrodes. This action results in a minute current flow. If aerosols, such as products of combustion or smoke, enter the chamber, the ionized air particles attach themselves to the aerosols and the resultant particles, being of larger mass than ionized air, move more slowly, and thus, per unit of time, fewer reach the electrodes. A decrease in current flow, therefore, takes place within the chamber whenever aerosols enter. The decrease in current flow is electronically converted into an alarm signal output.

An ionization type of detector responds best to invisible aerosols where the particles from burning materials are in the range of 1.0 microns in size down to 0.01 microns. A tremendous amount of these particles are produced by a flaming fire as opposed to a smoldering fire which produces large and small particles, but, because of low heat, the low thermal lift tends to allow particles to agglomerate into larger particles if the detector is some distance from the fire.

High air flows will affect the operation of this type unit by reducing the ion concentration in the detector chamber. In fact, with a high enough air flow, the unit will respond and alarm even though a fire does not exist. For this reason, locations near windows, direct air flows from vents and comparable areas should be avoided. Devices installed in these environments create a fail-safe condition.

This unit is virtually trouble free, because it uses a minute quantity of radioactive material (such as Americium 241) in its operations. It is, however, recommended that in dusty atmospheres the unit be periodically vacuumed. The detector will become more sensitive as large quantities of dust accumulate. Failure to clean the unit results in a "fail-safe" condition because the unit may alarm unnecessarily.

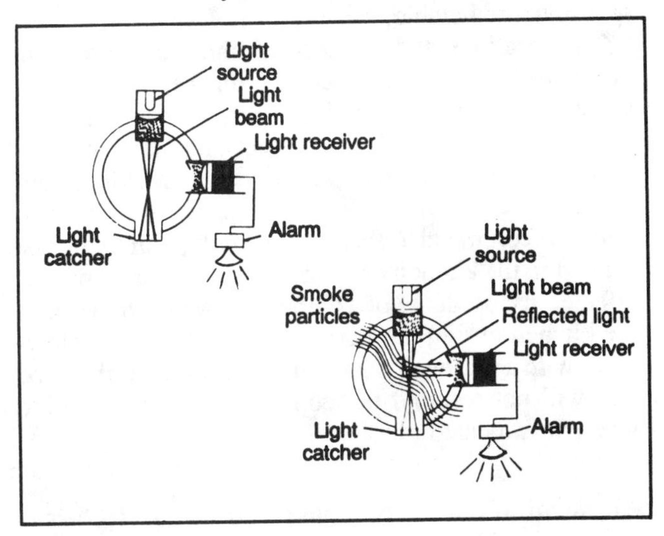

Fig. 9-3: Basic operating principles of photoelectric detectors.

Photoelectric Detectors—Basic Operation: A beam from the light source is projected across a chamber into a light catcher. The chamber is designed to permit access of smoke, but not access of external light. A photo resistive cell or light sensitive device is located in a recessed area perpendicular to the light beam. When smoke enters the chamber, smoke particles will scatter or reflect a small portion of the light beam to the light receiving device, which, in turn, will provide a signal for amplification to the alarm. This description of operation is the basic operating principle of photoelectric detectors. Some variations in design are used.

Some photoelectric detectors are adversely affected by dirt films. Any accumulated dirt, dust, film or foreign matter collecting on either or both lenses of the light source or the photocell will cause an opaque effect and the detector will then become less and less sensitive. It, therefore, will require more smoke in order to respond.

While latest photoelectric models utilize solid state light emitting and receiving devices which have a longer life than previous light devices, the problem of failure of the light source still exists. UL requires an audible alarm if light failure occurs.

Photoelectric units respond best to fires producing visible aerosols where the particles range from 10 microns down to 0.3 micron. These particles would be produced by a smoldering fire where very little heat is produced.

Ionization and Photoelectric Devices—Basic Operation: The diagram in Fig. 9-4 can be used to illustrate both types of devices—the difference is the use of either an ionization sensor or a photoelectric sensor in the reference chamber and detector portions of the circuit. Under normal conditions, the voltage across the reference chamber and the detection chamber is the same. However, when fire occurs, the detection chamber then functions as described in the previous explanations. Thus, when there is sufficient voltage difference between the two chambers, the alarm is activated through the switching circuit.

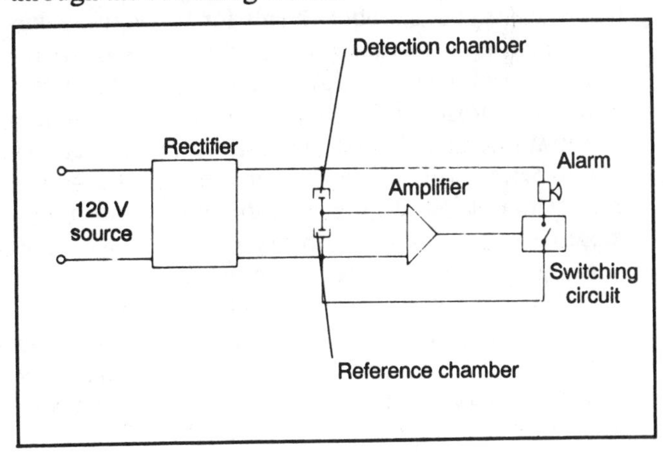

Fig. 9-4: Diagram of ionization and photoelectric devices.

Not all ionization or photoelectric devices have a reference chamber. The purpose of the reference chamber is to compensate automatically for changes in relative humidity, temperature or atmosphere pressure. Some devices use either a fixed or a variable resistor, instead of the reference chamber. Units which use the dual chamber (reference chamber and detection chamber) are much more constant in their response to a fire.

In comparison of photoelectric and ionization devices, claims have been made that one is superior to the other. While tests performed by organizations such as the National Bureau of Standards and others have shown that ionization units react faster to flaming fires and photoelectric units react faster to low heat smoldering fires, their conclusions state that either type could "provide more than adequate lifesaving potential under real residential fire conditions when properly installed."

Models of Residential Detectors

Direct Wired Single Station: Direct wired single station units are 120 volt ac powered with two wires for connection to the electrical system. They usually include a power indicator lamp and a push-to-test-button, but normally do not have other extra features. They are designed to be used where only one detector is desired or where units are not to be interconnected.

Direct Wired Multiple Station: These devices are 120 volt ac powered and designed to be interconnected so that when one unit senses a fire and alarms, all units will alarm. Two types of interconnect methods are used. One method utilizes a three wire (hot, neutral and interconnect) interconnection at 120 volts ac. This system must meet code requirements for 120 volts ac wiring and all units in the system must be installed on the same 120 volt ac circuit. The second method incorporates a transformer in the unit to allow interconnection at a low voltage. This will allow the interconnect wiring to be low voltage cable (signaling cable). Article 725 of the NE Code specifies this voltage to be not more than 30 volts. Also, this type of interconnect allows the units in the system to be installed on different 120 volt ac circuits if desired. Only two wires are required for interconnection. While the 120 volt ac interconnect units are somewhat less expensive than the low voltage interconnect units, the overall installation costs with the low voltage units will be less.

Direct Wired Multiple Station with Auxiliary Contacts: These models are 120 volt ac powered interconnect units that also include a set of auxiliary contacts for connection to remote devices. The auxiliary contacts can be used to operate remote horns, lights, or shut off exhaust or ventilating fans. Such units have an additional pair of leads for the auxiliary contacts.

Battery Powered Single Station: Most battery powered units are single station devices powered by a 9 volt battery. UL requires that batteries last at least one year under normal use and that such units produce an audible signal indication of a low battery condition for seven days. Use of batteries other than ones designated by the manufacturer can cause the device to fail to operate, or operate improperly. This is due to the circuitry of the unit being designed to match the performance of specified batteries.

UL Listing: Residential smoke and fire detectors are tested for compliance with UL Standard No. 217. All devices submitted to UL since January 1976 were tested in accordance with this standard. Periodically, changes in this standard are made and manufacturers are required to resubmit their units in order to maintain their listing.

Installation: A specific mounting position for smoke and fire detectors is necessary to permit the proper entrance of aerosols to activate the detector. Most units are designed for wall or ceiling mounting and recommended positions are indicated in the instruction manual supplied with the unit. Mounting in a position not specified may make the unit less sensitive or the response unpredictable.

The locations and quantity of units required in an installation can vary, depending on the authority involved in the installation. However, almost all require one detector outside sleeping areas and many authorities are requiring interconnect units in multi-floor residences.

It is recommended that ionization type units not be located in the kitchen or in any area where they will be affected by products of combustion which may leave the kitchen during the normal cooking process. However, with a reasonable amount of ventilation, this type unit will not respond to people smoking in the room where it is located.

Authorities Requiring Smoke and Fire Detectors

The following are some of the authorities requiring smoke and fire detectors.

- BOCA—Building Officials and Code Administrators
- UBC—Uniform Building Code
- FHA—Federal Housing Administration
- NFPA—National Fire Protection Association

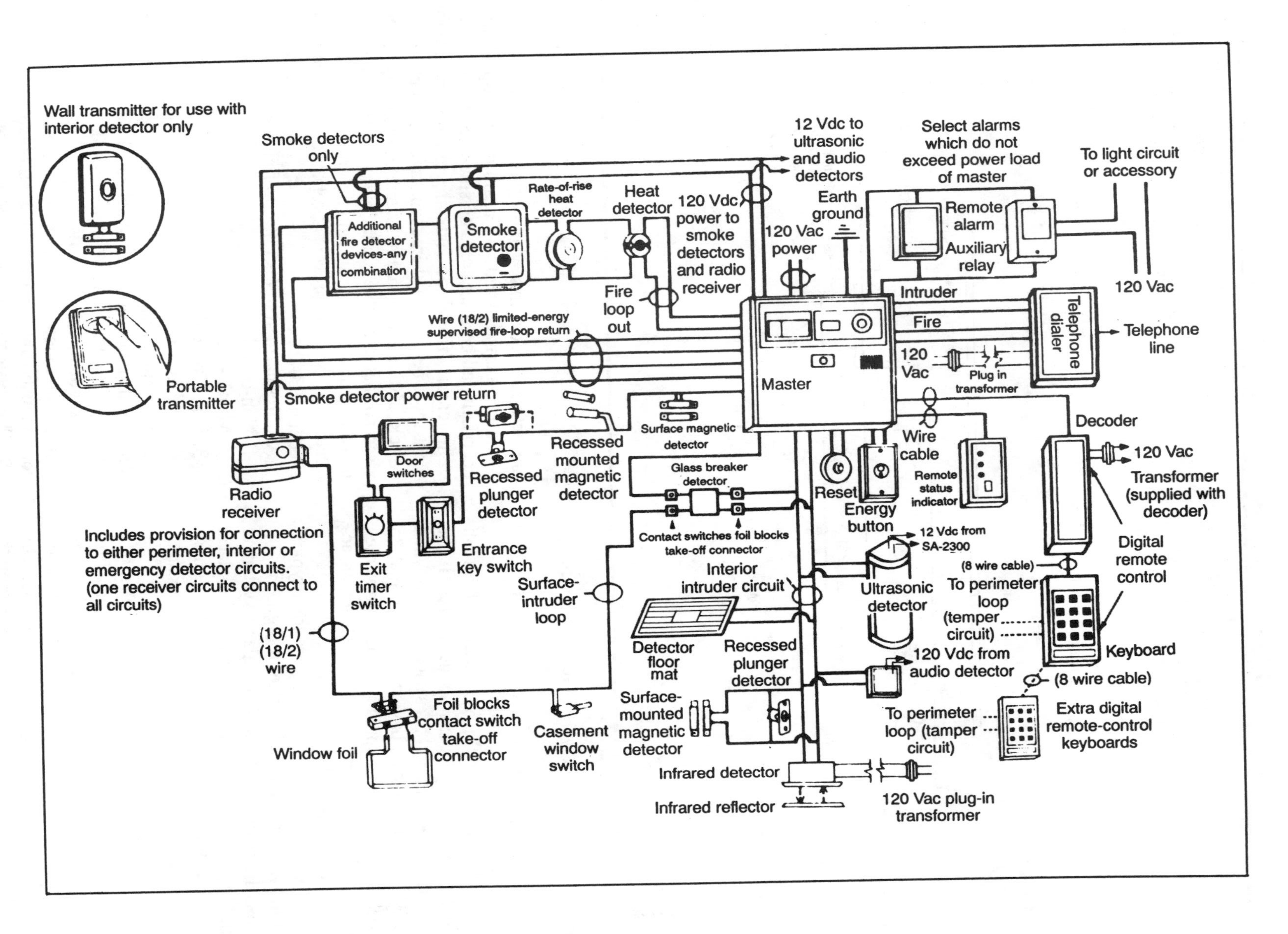

Fig. 9-5: Summary of components for one type of security/fire-alarm system.

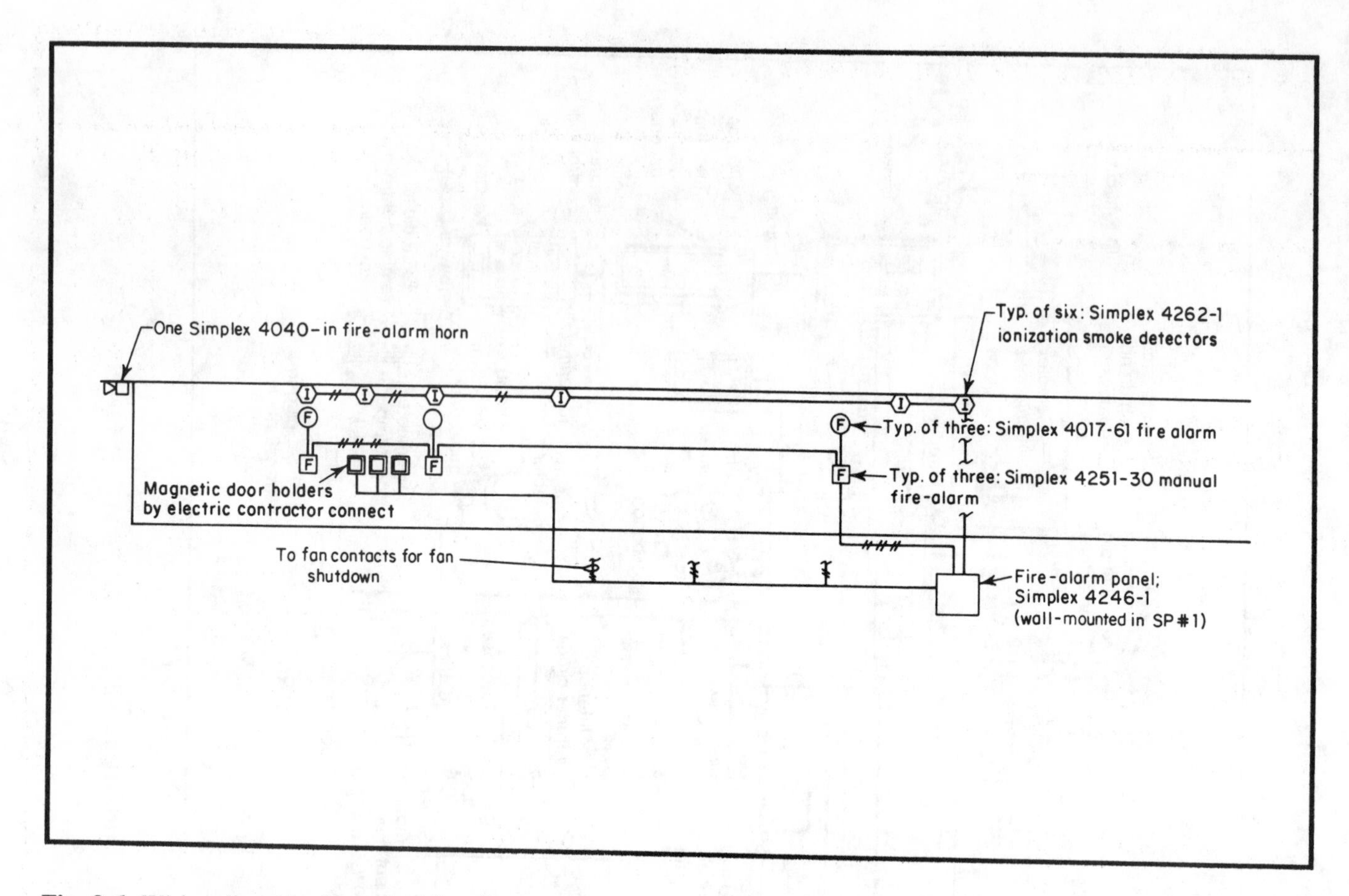

Fig. 9-6: Wiring diagram of a typical fire-alarm system.

In addition, many state and local codes now also have requirements for smoke and fire detectors.

According to a study by NFPA, residential fires kill more than half of all the people who die in fires. It has also been established that fatal fires usually occur when the family is asleep. Studies have also shown that 85 percent of all cases of death could have been prevented if smoke and fire detectors had been installed in accordance with recommended standards.

A summary of the various components for a typical security/fire-alarm system is depicted in Fig. 9-5.

A wiring diagram for a typical fire-alarm system is shown in Fig. 9-6. This diagram shows the fire-alarm panel smoke detectors (SD), striking stations, gongs (bells), magnetic door-release switches, and the interconnecting wiring.

Basically, if any smoke detector senses smoke, or if any manual striking station is operated, all bells within the building will ring, indicating a fire. At the same time, the magnetic door switches will release the smoke doors to help block smoke and/or drafts. This system is also connected to a water-flow switch on the sprinkler system, so that if any of the sprinkler valves are activated—causing water to flow in the system —the fire-alarm system will again go into operation, energizing all bells and closing smoke doors.

In general, fire-alarm signaling circuits and equipment must be grounded in accordance with NE Code Article 250, and all circuits must be electrically supervised so that a trouble signal will indicate the occurrence of a single-open or a single-ground fault on any circuit that would prevent proper alarm operation (Section 760-22).

All fire-alarm systems fall into four basic types:

1. Non-coded
2. Master-coded
3. Selective-coded
4. Dual-coded

Each of these four types has several functional features, so that electrical personnel can design a specific system to a client's needs within local and state fire codes, statutes, and regulations.

In a non-coded system, an alarm signal is sounded continuously until manually or automatically turned off. This system is used mainly in remote areas or for unoccupied buildings.

■ GUIDE CHART FOR FIRE-ALARM SYSTEMS

Type of system	General alarm annunciation			Presignal annunciation		
	None	"A" type	"D" type	None	"A" type	"D" type
Noncoded system	N-1	NA-1	ND-1	NM-1	NAM-1	NDM-1
	N-2	NA-2	ND-2	NM-2	NAM-2	NDM-2
Common-coded systems	C-1	CA-1	CD-1	CM-2	CAM-2	CDM-1
	C-2	CA-2	CD-2	CM-2	CAM-2	CDM-2
Selective-coded systems:						
Series noninterfering	S-1	-	-	SM-1	-	-
successive	S-2	-	-	SM-2	-	-
Positive noninterfering, P-1	-	PD-2	PM-1	-	PDM-1	-
successive	P-2	-	PD-2	PM-2	-	PDM-2
Zone-coded	-	ZA-1	ZD-1	-	ZAM-1	ZDM-1
	-	ZA-2	ZD-2	-	ZAM-2	ZDM-2
Dual-coded system:						
Noncode + series noninterfering	NS-1	-	-	-	-	-
successive	NS-2	-	-	-	-	-
Noncode + positive	NP-1	-	NPD-1	-	-	-
non-interfering, successive	NP-2	-	NPD-2	-	-	-
Nocoded + zone-coded	-	NZA-1	NCD-1	-	-	-
	-	NZA-2	NZD-2	-	-	-
Common-coded + series	CS-1	-	-	-	-	-
noninterfering, successive	CS-2	-	-	-	-	-
Common-coded + positive	CP-1	-	CPD-1	-	-	-
noninterfering, successive	CP-2	-	CPD-2	-	-	-
Common-coded + zone-coded	-	CZA-1	CZD-1	-	-	-
	-	CZA-2	CZD-2	-	-	-

Fig. 9-7: Practical application of various fire-alarm systems.

In a master-coded system, a common-coded alarm signal is sounded for not less than three rounds. The same code is sounded regardless of the alarm-initiating device activated.

In a selective-coded system, a unique coded alarm is sounded for each firebox or fire zone on the protected premises.

In a dual-coded system, a unique coded alarm is sounded for each firebox or fire zone to notify the owner's personnel of the location of the fire, while non-coded or common-coded alarm signals are sounded separately to notify other occupants to evacuate the building.

The most difficult aspect of applying the fire-alarm systems is the selection of the basic system along with the functional features required by different occupancies. The chart in Fig. 9-7 simplifies this selection process, for the chart reads easily yet covers 64 standard systems.

The system numbers used in the table consist of up to three alphabetical characters followed by the suffix 1 or 2. The alphabetical characters denote:

N. A no-code system.

M. A multialarm or presignal system where the initial operation of any manual or automatic alarm-initiating device sounds an alarm on certain primary alarm signals. Authorized personnel with special keys may initiate a general evacuation alarm from any manual firebox. On automatic fire-alarm systems, auxiliarized city fireboxes (if used) are tripped on the first alarm, and general-evacuation alarms are all non-coded.

C. A common-code or master-code system.

S. A selective-code system using coded fireboxes in which all fireboxes are in one common wired loop, and the coded firebox nearest the control unit has precedence on the loop. This is commonly called series non-interfering, and successive.

P. A selective-code system using coded fireboxes in which all fireboxes are on one common wired loop and arranged so that only one code at a time can sound even if two or more fireboxes are activated simultaneously. This prevents confusion resulting from incomplete, lost, or mutilated codes. This is commonly referred to as positive, non-interfering, and successive.

Z. A selective-code system using non-coded fireboxes or automatic devices that are point-wired to the control unit and with the coding done with the control unit. All codes sound in a positive, non-interfering, and successive manner as described for the P systems. All Z systems are furnished with annunciators in the control unit; that is, they are self-contained and ready for immediate use.

A. An annunciator feature on some systems in which each alarm zone illuminates a backlighted window. In addition, each alarm-zone window has an adjacent trouble window light in the event of an "open" condition on its associated alarm-zone wiring. Two lamps are used in the alarm window and one lamp in the trouble window. A flashing alarm-lamp feature is available with the A-line annunciators.

D. An annunciator feature on some systems in which each alarm zone illuminates a backlighted window. Two lamps are used in parallel so that the window can be read even if one lamp is burned out.

The suffixes 1 and 2 have to do with the alarm-signal circuits:

1. Alarm-signal operating voltages derived from the primary power source of 120 V ac.

2. Alarm-signal voltages of 24 V dc derived from a transformer and rectifier isolated from the 120 V ac.

Bank Alarm Systems

Figure 9-8 shows various outlets for a bank security system including camera junction boxes, smoke detectors, sound receivers, alarm buttons, etc. They are shown on the floor plan only to indicate the approximate location of each. Details of wiring are not shown in this drawing, but the riser diagram in Fig. 9-9 indicates clearly how each outlet is to be installed.

Communication Circuits

Central station communication circuits are installed overhead on system poles or jointly with power distribution systems on jointly owned poles or underground. Protective devices are required in communication circuits entering buildings to prevent the higher-voltage power current from passing into the building in case of accidental contact of the two systems (Section 800-2). When installed in raceways, communication circuits must be installed in separate raceways from the power circuit, and except when introduced for the sole purpose of a power supply, power circuits must not enter the same outlet boxes used for communication circuits unless they are separated by a partition (Section 800-3).

Proper clearances from the power circuits—as well as clearance over streets, roads, and driveways—must be maintained when communication conductors are installed overhead.

Communication systems conductors vary in size, type of insulation, and combination of conductors, depending upon the system requirements. They are installed as single conductors, but most often as

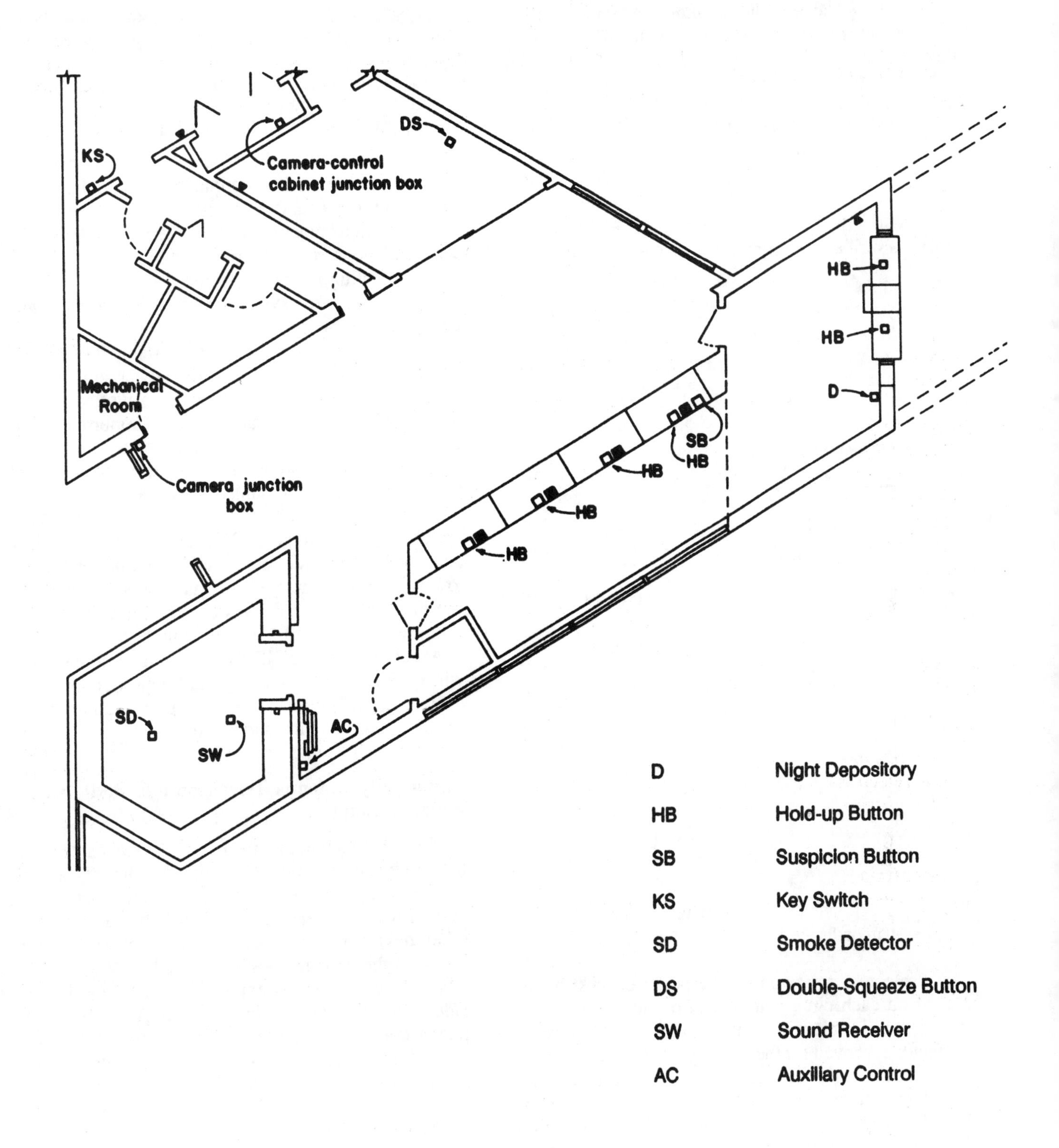

Fig. 9-8: Floor plan of a bank showing security system outlets.

nonmetallic sheathed cables, often containing a large number of color-coded paired conductors for easy identification.

Conductors are connected either directly to the equipment or to terminal blocks. Nonmetallic or lead-sheathed cables are spliced when necessary to make long continuous runs and are usually terminated in special terminal boxes to prevent moisture from entering the insulation. Lead-sheathed cables are installed overhead, suspended from steel messenger cables as well as underground in raceway systems.

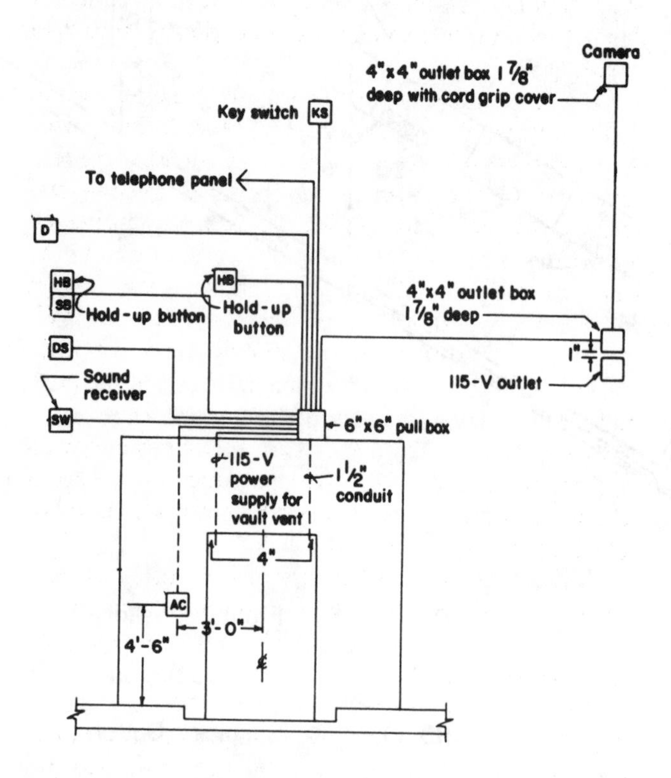

Fig. 9-9: Riser diagram showing details of the security system outlets in Fig. 9-8.

A protector, approved for the purpose, must be installed on each circuit run partly or entirely in aerial wire or aerial cable not confined within a block. Also, a protector approved for the purpose must be provided on each circuit, aerial or underground, so located within the block containing the building served as to be exposed to accidental contact with light or power conductors operating at over 300 V to ground.

The protector should be located in, on, or immediately adjacent to the structure or building served and as close as practical to the point at which the exposed conductors enter or attach. The protector should not be located in the vicinity of easily ignitable material or in any area classified as hazardous as defined in NE Code Article 500.

Sound Recording and Similar Equipment

Buildings such as theaters, schools, hospitals, hotels, apartment houses, commercial establishments, office buildings, and industrial plants will normally contain various types of sound systems. These systems will include sound-recording, public-address, intercommunicating telephone, nurses' call, and radio receiver and speaker systems and other audible signal equipment and wiring.

Most of the circuits on sound systems will be low-voltage, with the exception of the power supply. Conductors of varying sizes and types of insulation are used and either installed in raceways or supported by approved cable straps. Special outlet boxes or cabinets are usually provided with the equipment, although some items may be mounted in or on standard outlet boxes.

Some systems require direct current, which is usually obtained by electronic rectifying equipment. The low-voltage alternating current is obtained through the use of relatively small transformers connected on the primary side to a 120-V power source.

Article 640 of the NE Code covers equipment and wiring for sound recording and reproduction, centralized distribution of sound, public-address and speech-input systems, and electronic organs.

Community Antenna Television and Radio Distribution Systems

The NE Code permits the use of coaxial cable to deliver low-energy power to equipment directly associated with an antenna (radio frequency) distribution system if the current supply is from a transformer or other device having energy-limiting characteristics, provided the voltage does not exceed 60 V.

Where the system is exposed to lightning or to accidental contact with lightning-arrestor conductors or power conductors operating at a voltage of 300 V to ground, the outer conductive shield of the coaxial cable must be grounded at the building premises as close to the point of cable entry as practicable. Where the outer conductive shield of a coaxial cable is grounded, no other protective devices will be required.

Where practicable, coaxial conductors on poles should be located below the light or power conductors and must not be attached to a crossarm that carries light or power conductors. Lead-in cables from a pole or other support, including the point of initial attachment to a building or structure, must be kept away from electric light or power circuits so as to avoid the possibility of accidental contact.

Conductors installed over buildings, between buildings, and on buildings must be installed in the same way as service entrance cables or other outside overhead wiring.

Remote Control System for Lighting Circuits

Remote-Control Relay Switching: Relays were developed several years ago to meet the demands for more convenient and reliable control of industrial, commercial, and residential equipment. The relays permitted more flexible control for residential lighting, and low voltage and low current in the control circuit provided for added safety.

The basic circuit of a remote-control wiring system is as shown in Fig. 9-10.

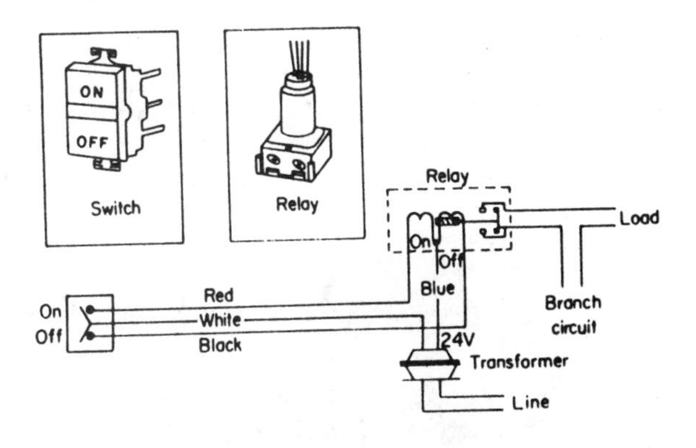

Fig. 9-10: Basic circuit and components of a remote-control wiring system.

The split-coil type of relay permits positive control for on and off. The relay can be located near the lighting load, or installed in centrally located distribution panel boxes, depending on the application.

Because no power flows through the control circuits and low voltage is used for all switch and relay wiring, it is possible to place the controls at a great distance from the source or load, thus offering many advantages through this modern system of wiring. The average installation is usually most costly, however.

Power Source: A single-main power source is used to provide the correct voltage for operation of all relays in a remote-control system. The power supply usually consists of a pulse-type transformer, a selenium rectifier, and an electrolytic capacitor. Input voltage is 120 or 277 V while the momentary output voltage occurring when a switch is pressed is 24 V. This power supply is usually mounted on one of the relay cabinets, but it may be installed at other locations, such as at the service entrance, if required for a particular application.

Relay: The relay normally used in residential remote-control systems is a mechanical latching type with a single coil. Momentary impulses from the power supply alternately open and close the contacts in the load circuit. Each relay may control several outlets if desired, but a single switch may not control more than one relay.

Switches: One basic type of switch, the single-pole, double-throw, momentary-contact push button, does the work of single-pole, double-pole, three-way, and four-way switches used in conventional wiring—and with less wiring. Pressing the ON position of the switch energizes the related relay and closes the circuit to the load; pressing the OFF position of the switch again energizes the relay and opens the circuit to the load. One type of remote-control switch does not have "on-off" designations, but with each touch of the push button, the relay alternately opens and closes.

Master-Selector Switch: Where many circuits must be controlled from one convenient location, master-selector switches perform the necessary functions of selecting only those circuits wanted.

A "dial-type" master-selector switch for 12 circuits permits individual control of the circuits; alternatively, by pressing and then sweeping all circuits, master control is accomplished.

Motor-Master Control Units: Motor-driven master-control units are used to turn on or off up to 25 individual circuits; when these units are cascaded, the pressing of a single master switch can control any number of circuits, depending on the number of Motor-Master units that are ganged together. Ganging is accomplished by having the last position of the first unit wired to start the third unit, and so on. These units can be activated by photoelectric cell relays, for automated lighting control.

Combining Remote Controls with Dimmers: Sometimes it is desirable to dim certain lighting fixtures which are controlled by a remote-control system. This is accomplished by wiring a dimmer between the relay and the fixture, as shown in the wiring diagram in Fig. 9-11. The dimmer adjusts the intensity of the light, and the remote controls turn it on or off. Several lights may be controlled at one time.

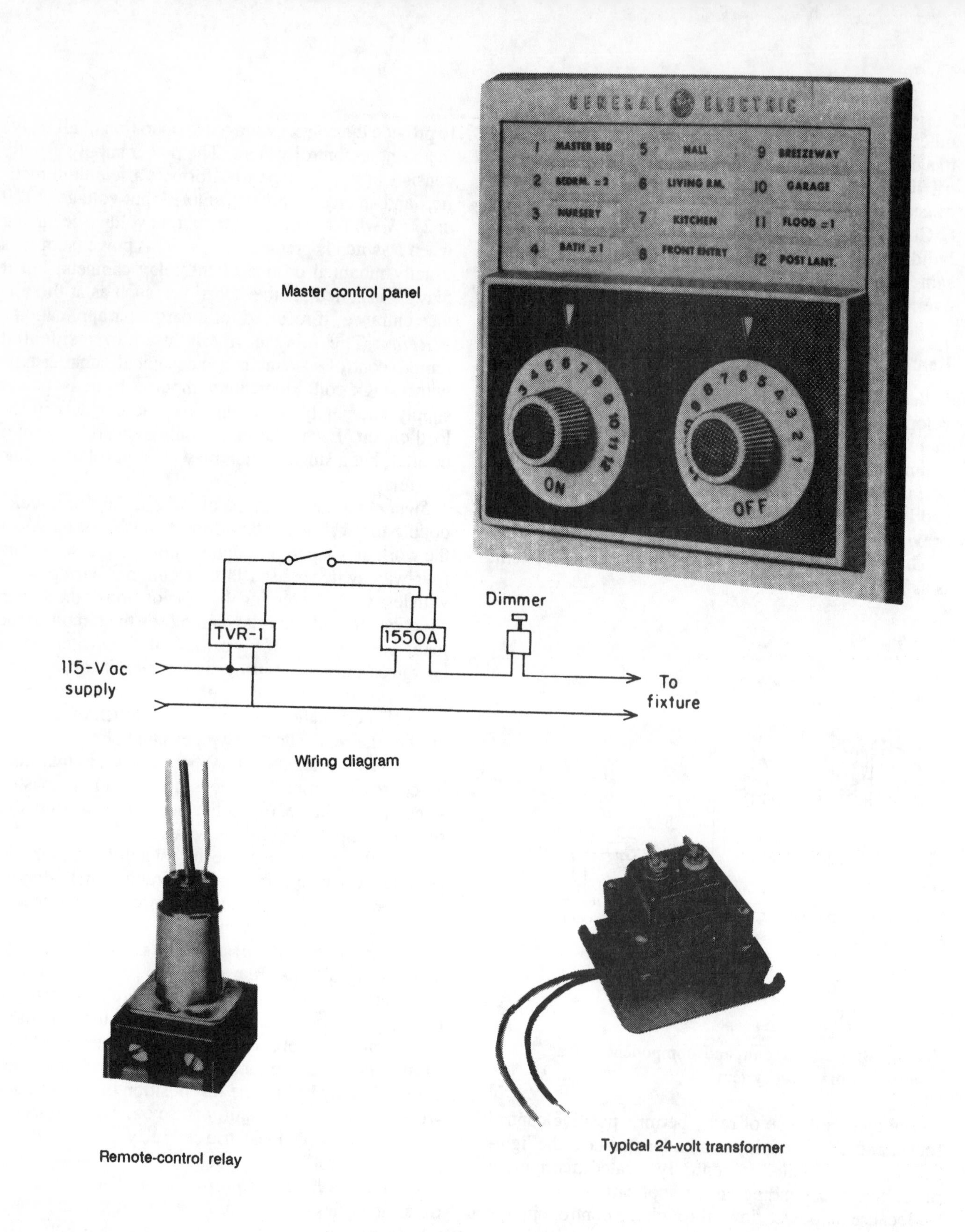

Fig. 9-11: Wiring diagram and circuit components required for dimming lights with a remote-control system.

■ POWER LIMITATIONS FOR AITERNATING-CURRENT SIGNALING CIRCUITS						
	Inherently Limited Power Source (Overcurrent protection not required)			Not Inherently Limited Power Source (Overcurrent protection required)		
Circuit Voltage	0 thru 20	Over 20 and thru 30	Over 30 and thru 100	0 thru 20	Over 20 and thru 100	Over 100 and thru 150
Power Limitations	-	-	-	250	250	N.A.
Current Limitations	8.0	8.0	150/V max	1000/V max	1000/V max	1.0
Maximum Overcurrent Protection (Amps)	—	-	-	5.0	100/V max	1.0
VA (volt-amps)	5.0 x V max	100	100	5.0 x V max	100	100
Current (Amps)	5.0	100/V max	100/V max	5.0	100/V max	100/V max

■ POWER LIMITATIONS FOR DIRECT-CURRENT SIGNALING CIRCUITS						
	Inherently Limited Power Source (Overcurrent protection not required)			Not Inherently Limited Power Source (Overcurrent protection required)		
Circuit Voltage	0 thru 20	Over 20 and thru 30	Over 30 and thru 100	0 thru 20	Over 20 and thru 100	Over 100 and thru 150
Power Limitations	-	-	-	250	250	N.A.
Current Limitations	8.0	8.0	150/V	1000/V	1000/V	1.0
Maximum Overcurrent Protection (Amps)	-	-	-	5.0	100/V	1.0
VA	5.0 x V	100/V	100/V	5.0 x V	100	100
Current	5.0	100/V	100/V	5.0	100/V	100/V

Fig. 9-12: NE Code rules governing signaling circuits.

chapter 10

TRANSFORMERS AND CAPACITORS

The electric power produced by alternators in a generating station is transmitted to locations where it is utilized and distributed to users. Many different types of transformers play an important role in the distribution of electricity. Power transformers are located at generating stations to step up the voltage for more economical transmission. Substations with additional power transformers and distribution transformers are installed along the transmission line. Finally, distribution transformers are used to step down the voltage to a level suitable for utilization.

Transformers are also used quite extensively in all types of security/fire-alarm systems and heating/air-conditioning controls, to raise and lower ac voltages. It is important for anyone working with electricity to become familiar with transformer operation; that is, how they work, how they are connected into circuits, their practical applications and precautions to take while using them.

The NE Code rules for the installation and protection of transformers operating at voltages up to 600 volts are covered in Article 450 of the 1993 NE Code. The NE Code rules listed in this Article cover all transformers except special transformers used in specific applications; these are covered in other Sections of the NE Code. For example, rules for arc welders are found in Article 630; transformers for electric discharge lighting are covered in Article 410; rules for transformers serving signs and outline lighting may be found in Section 600-32 and look at Section 680-5 for rules governing underwater lighting. Transformer connections at services are covered in Section 230-82.

Transformer Basics

A very basic transformer consists of two coils or windings formed on a single magnetic core. Such an arrangement will allow transforming a large alternating current at low voltage into a small alternating current at high voltage, or vice versa. Transformers, therefore, enable changing or converting power from one voltage to another. For example, generators that produce moderately large alternating currents at moderately high voltages utilize transformers to convert the power to very high voltage and proportionately small current in transmission lines, permitting the use of smaller cable and providing less power loss.

When alternating current (ac) flows through a coil, an alternating magnetic field is generated around the coil. This alternating magnetic field expands outward from the center of the coil and collapses into the coil as the ac through the coil varies from zero to a maximum and back to zero again, as discussed earlier in this book. Since the alternating magnetic field must cut through the turns of the coil, a self-inducing voltage occurs in the coil which opposes the change in current flow.

If the alternating magnetic field generated by one coil cuts through the turns of a second coil, voltage will be generated in this second coil just as voltage is induced in a coil which is cut by its own magnetic field. The induced voltage in the second coil is called the *voltage of mutual induction*, and the action of generating this voltage is called *transformer action*. In transformer action, electrical energy is transferred from one coil (called the primary) to another (the secondary) by means of a varying magnetic field.

The basic parts of a transformer are shown in Fig. 10-1. Two windings are shown on a rectangular core made of iron. The source of alternating current and voltage is connected to the primary winding of the transformer. The secondary winding is connected to the circuit in which there is to be a higher voltage and consequently, a smaller current—although it could be just the opposite; that is, a higher current and smaller voltage.

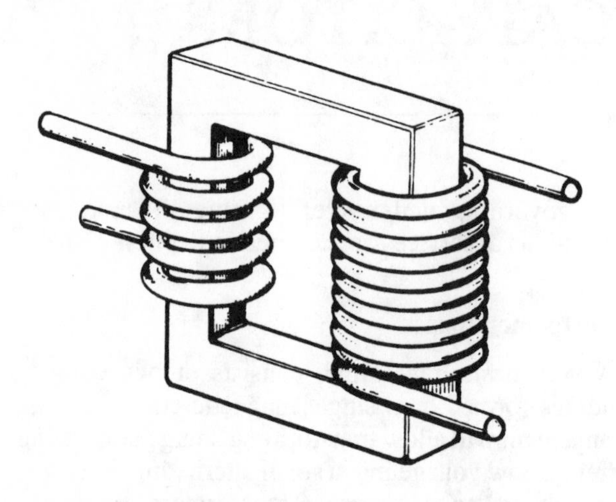

Fig. 10-1: The basic components of a transformer consist of two separate coils of wire wrapped around an iron core.

Assuming that all the primary magnetic lines of force cut through all the turns of the secondary, the amount of induced voltage in the secondary will vary with the ratio of the number of turns in the secondary to the number of turns in the primary.

If there are more turns on the secondary than on the primary winding, the secondary voltage will be higher than that in the primary and by the same proportion as the number of turns in the winding. The secondary current, in turn, will be proportionately smaller than the primary current. With fewer turns on the secondary than on the primary, the secondary voltage will be proportionately lower than that in the primary, and the secondary current will be that much larger. Since alternating current continually increases and decreases in value, every change in the primary winding of the transformer produces a similar change of flux in the core. Every change of flux in the core and every corresponding movement of magnetic field around the core produce a similarly changing voltage in the secondary winding, causing an alternating current to flow in the circuit that is connected to the secondary.

For example, if there are, say, 1000 turns in the secondary and only 100 turns in the primary, the voltage induced in the secondary will be 10 times the voltage applied to the primary. See Fig. 10-2.

$$1000/100 = 10$$

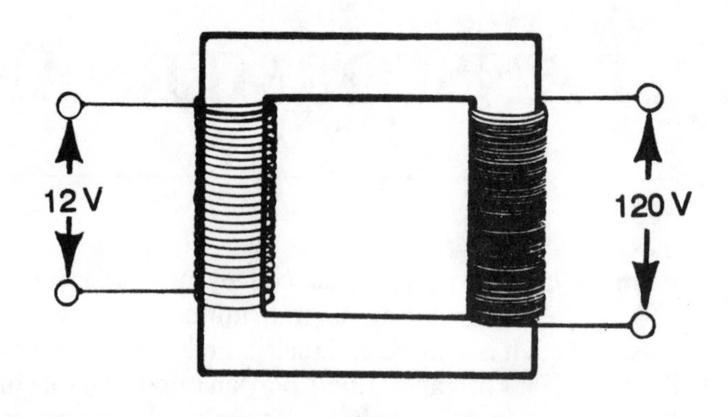

Fig. 10-2: With 100 turns in the primary and 1000 turns in the secondary, the voltage in this transformer is stepped up from 12 volts to 120 volts.

Since there are more turns in the secondary than in the primary, the transformer is called a "step-up transformer." If the reverse were true (100 turns in the secondary and 1000 turns in the primary) the voltage induced in the secondary will be 1/10 of the voltage applied to the primary. See Fig. 10-3.

$$10/100 = 0.10$$

Since there are now less turns in the secondary than in the primary, the transformer is called a "step-down transformer."

A transformer does not generate electrical power. It simply transfers electric power from one coil to another by magnetic induction. Transformers are rated in kilovolt-amperes.

Theoretical study of conditions in a transformer includes the use of phasor diagrams that represent graphically voltages and currents in transformer windings. Calculations of impedance can be simplified by using equivalent circuits. Reactance voltage drop is governed by the leakage flux, and the voltage regulation depends on the power factor of the load. To determine transformer efficiency at various loads, it is necessary to first calculate the core loss, hysteresis loss, eddy-current loss, and load loss.

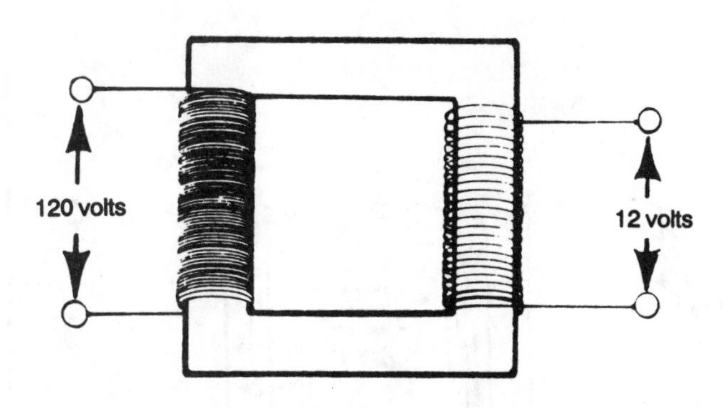

Fig. 10-3: Since there are 1000 turns in the primary and only 100 turns in the secondary, the voltage is stepped down from 120 volts to 12 volts.

However, for all practical purposes in electrical applications for building construction, a transformer's efficiency may be considered to be 100%. Therefore, for our purposes, a transformer may be defined as a device that transfers power from its primary circuit to the secondary circuit without any significant loss.

Since power (W) equals voltage (E) times current (I), if E_pI_p represents the primary power and E_sI_s represents the secondary power, then $E_pI_p = E_sI_s$ (the subscript p = primary, and the subscript s = secondary). See Fig. 10-4. If the primary and secondary voltages are equal, the primary and secondary currents must also be equal. Let's assume that E_p is twice as large as E_s. For E_pI_p to equal E_sI_s, I_p must be one-half of I_s as shown in Fig. 10-5. Therefore, a transformer that steps voltage down always steps current up. Conversely, a transformer that steps voltage up always steps current down. However, transformers are classified as step-up or step-down only in relation to their effect on voltage. The wattage or volt-ampere ratings also remain the same.

Transformer Construction

Transformers designed to operate on low frequencies have their coils, called "windings," wound on iron cores. Since iron offers little resistance to magnetic lines, nearly all the magnetic field of the primary flows through the iron core and cuts the secondary.

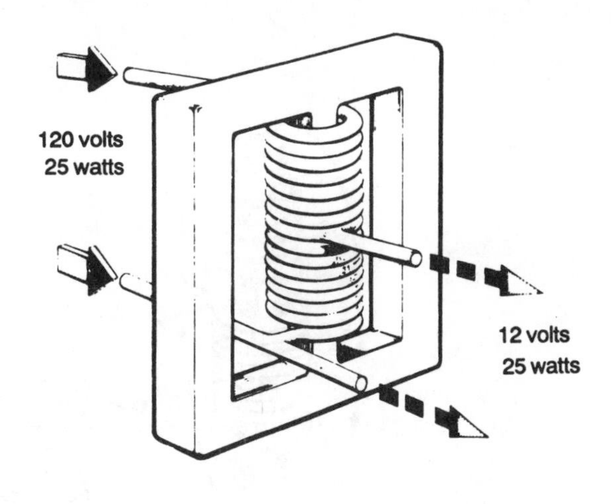

Fig. 10-4: Primary power equals secondary power in a transformer.

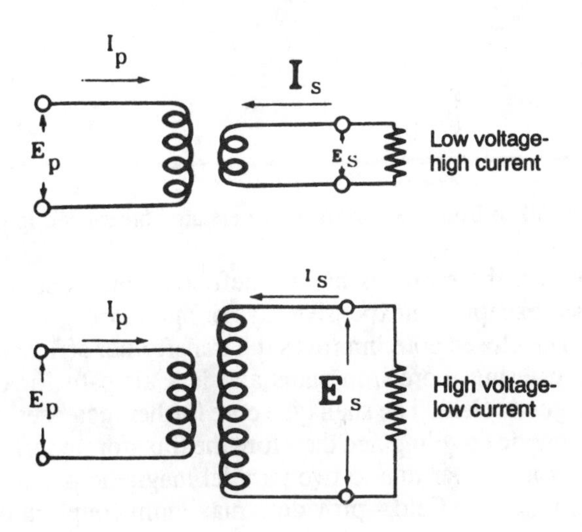

Fig. 10-5: A transformer that steps down the voltage always steps up the current. Conversely, a transformer that steps voltage up (step-up transformer) always steps the current down.

Iron cores of transformers are constructed in three basic types—the open core, the closed core and the shell type. See Fig. 10-6. The open core is the least expensive to manufacture as the primary and secondary are wound on one cylindrical core. The magnetic path, as shown in Fig. 10-6, is partially through the core and partially through the surrounding air. The air path opposes the magnetic field, so that the magnetic interaction or *linkage* is weakened. The open core trans-

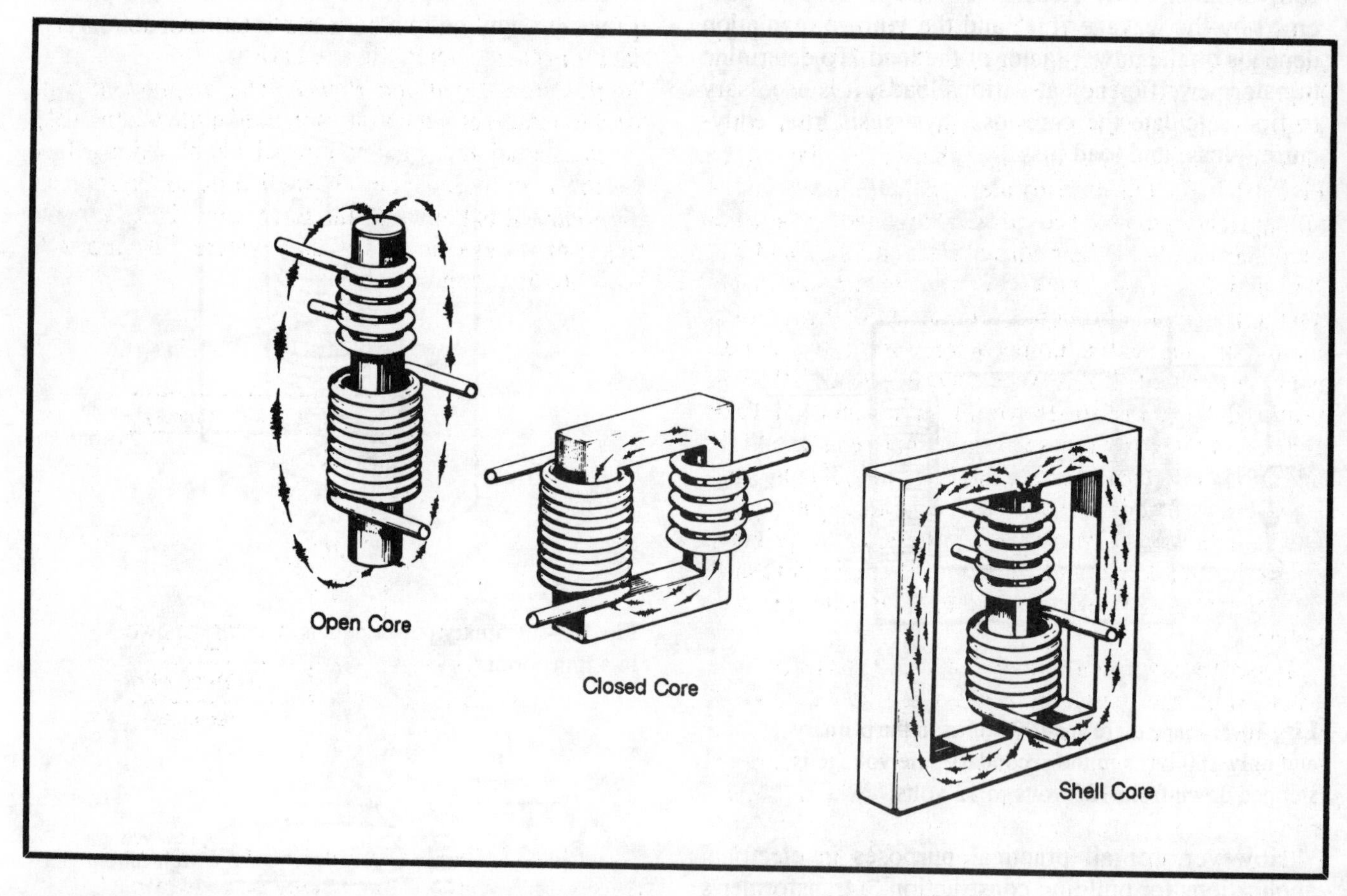

Fig. 10-6: Iron cores of transformers are constructed in three different types.

former, therefore, is highly inefficient and is seldom used except in inexpensive toys or appliances.

The closed core improves the transformer efficiency by offering more iron paths and less air path for the magnetic field. The shell type core further increases the magnetic coupling and therefore the transformer efficiency is greater due to two parallel magnetic paths for the magnetic field—providing maximum coupling between the primary and secondary.

Transformer Characteristics

In a well-designed transformer, there is very little magnetic leakage. The effect of the leakage is to cause a decrease of secondary voltage when the transformer is loaded. When a current flows through the secondary in phase with the secondary voltage, a corresponding current flows through the primary in addition to the magnetizing current. The magnetizing effects of the two currents are equal and opposite.

In a perfect transformer, that is, one having no eddy-current losses, no resistance in its windings, and no magnetic leakage, the magnetizing effects of the primary load current and the secondary current neutralize each other, leaving only the constant primary magnetizing current effective in setting up the constant flux. If supplied with a constant primary pressure, such a transformer would maintain constant secondary pressure at all loads. Obviously, the perfect transformer has yet to be built; the closest is one with very small eddy-current loss where the drop in pressure in the secondary windings is not more than 1 to 3 percent, depending on the size of the transformer.

Transformer Taps

If the exact rated voltage could be delivered at every transformer location, transformer taps would be unnecessary. However, this is not possible, so taps are provided on the secondary windings to provide a means of either increasing or decreasing the secondary voltage.

Generally, if a load is very close to a substation or power plant, the voltage will consistently be above normal. Near the end of the line the voltage may be below normal.

In large transformers, it would naturally be very inconvenient to move the thick, well-insulated primary leads to different tap positions when changes in source-voltage levels make this necessary. Therefore, taps are used, such as shown in the wiring diagram in Fig. 10-7. In this transformer, the permanent high-voltage leads would be connected to H_1 and H_2, and the secondary leads, in their normal fashion, to X_1 and X_2, X_3, and X_4. Note, however, the tap arrangements available at taps 2 through 7. Until a pair of these taps is interconnected with a jumper wire, the primary circuit is not completed. If this were, say, a typical 7200-volt primary, the transformer would have a normal 1620 turns. Assume 810 of these turns are between H_1 and H_6 and another 810 between H_3 and H_2. Then, if taps 6 and 3 were connected together with a flexible jumper on which lugs have already been installed, the primary circuit is completed, and we have a normal ratio transformer that could deliver 120/240 volts from the secondary.

Between taps 6 and either 5 or 7, 40 turns of wire exist. Similarly, between taps 3 and either 2 or 4, 40 turns are present. Changing the jumper from 3 to 6 to 3 to 7 removes 40 turns from the left half of the primary. The same condition would apply on the right half of the winding if the jumper were between taps 6 and 2. Either connection would boost secondary voltage by 2½ percent. Had taps 2 and 7 been connected, 80 turns would have been omitted and a 5 percent boost would result. Placing the jumper between taps 6 and 4 or 3 and 5 would reduce the output voltage by 5 percent.

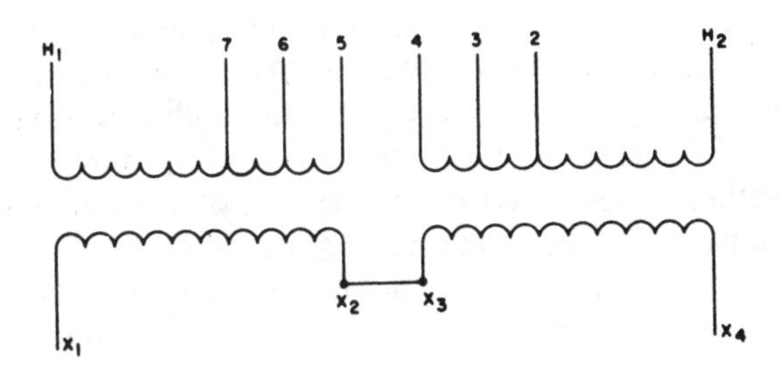

Fig. 10-7: Diagram showing taps on a typical transformer to adjust the voltage level on the secondary.

Transformer Connections

Transformer connections are many, and space does not permit the description of all of them here. However, an understanding of a few will give the basic requirements and make it possible to use manufacturer's data for others should the need arise.

Single-Phase Light and Power: The diagram in Fig. 10-8 is a transformer connection used quite extensively for residential and small commercial applications. It is the most common single-phase distribution system in use today. It is known as the three-wire, 240/120-volt single-phase system and is used where 120 and 240 volts are used simultaneously.

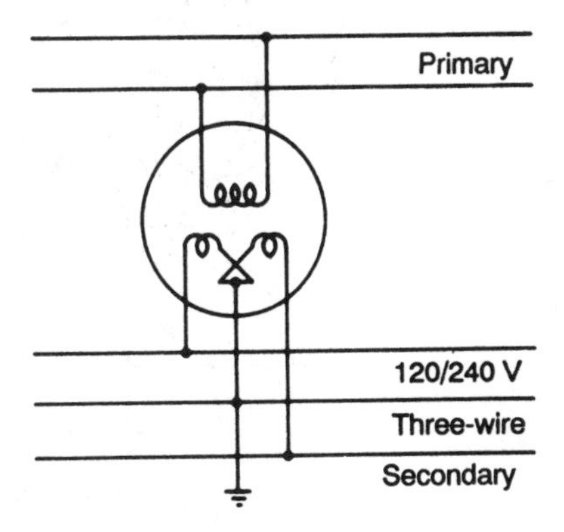

Fig. 10-8: 120/240-volt, three-wire, single-phase transformer connection that is used on residential and small commerical applications.

Y-Y for Light and Power: The primaries of the transformer connection in Fig. 10-9 are connected in Y. When the primary system is 2400/4160Y volts, a 4160-volt transformer is required when the system is connected in delta-Y. However, with a Y-Y system, a 2400-volt transformer can be used, offering a saving in transformer cost. It is necessary that a primary neutral be available when this connection is used, and the neutrals of the primary system and the transformer bank are tied together as shown in the diagram. If the three-phase load is unbalanced, part of the load current flows in the primary neutral. For these reasons, it is essential that the neutrals be tied together as shown. If this tie were omitted, the line to neutral voltages on the secondary would be very unstable. That is, if the load on one phase were heavier than on the other two, the voltage on this phase would drop excessively and the voltage on the other two phases would rise. Also, large third-harmonic voltages would appear between lines and neutral, both in the transformers and in the secondary system, in addition to the 60-hertz component of voltage. This means that for a given value of rms voltage, the peak voltage would be much higher than

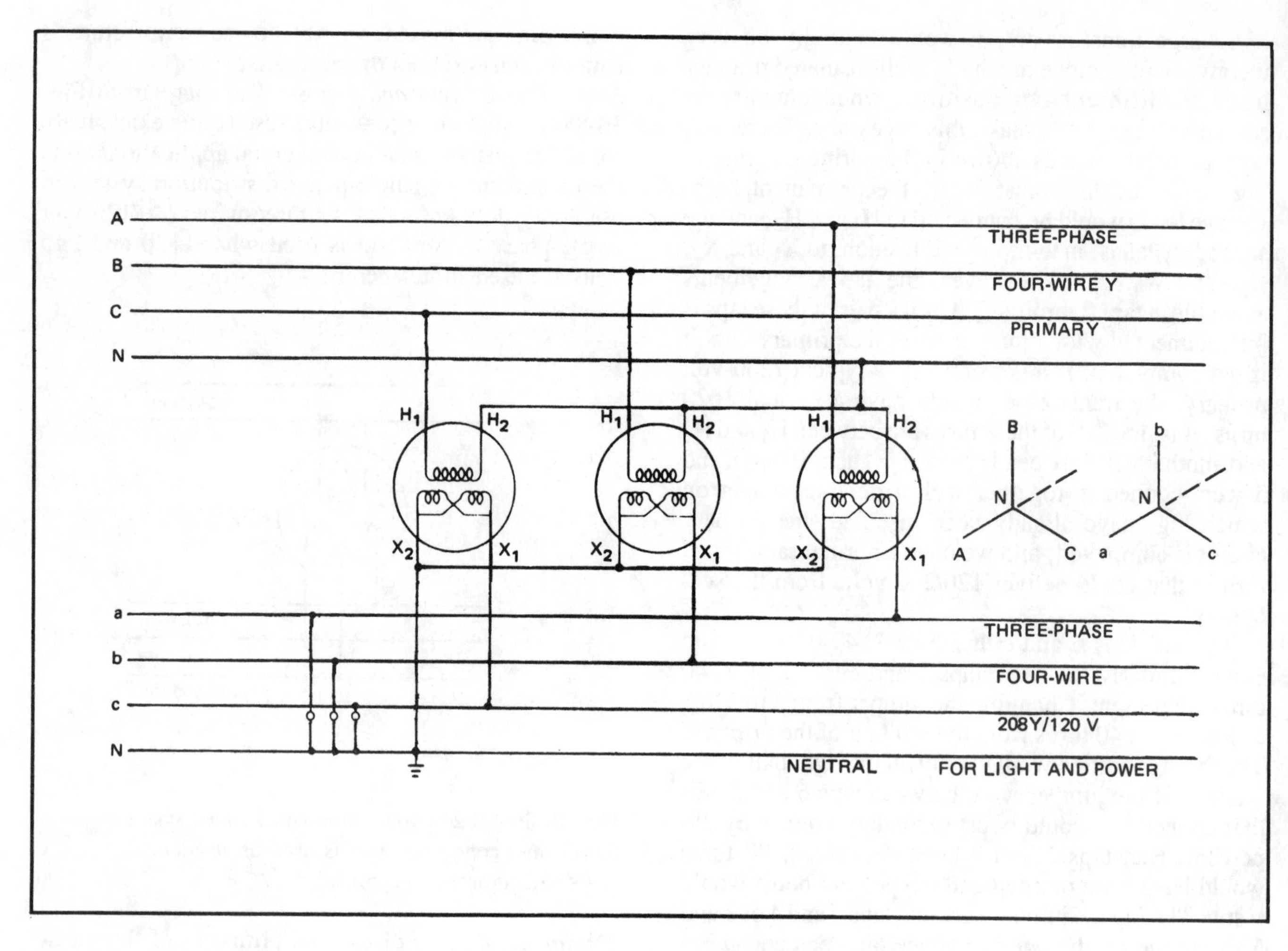

Fig. 10-9: Three-phase, 4-wire Y-Y connected transformer for use on light and power.

for a pure 60-hertz voltage. This overstresses the insulation both in the transformers and in all apparatus connected to the secondaries.

Parallel Operation of Transformers

Transformers will operate satisfactorily in parallel on a single-phase, three-wire system if the terminals with the same relative polarity are connected together. However, the practice is not very economical because the individual cost and losses of the smaller transformers are greater than one larger unit giving the same output. Therefore, paralleling of smaller transformers is usually done only in an emergency. In large transformers, however, it is often practical to operate units in parallel as a regular practice.

Two three-phase transformers may also be connected in parallel provided they have the same winding arrangement, are connected with the same polarity, and have the same phase rotation.

Autotransformers

An autotransformer is a transformer whose primary and secondary circuits have part of a winding in common and therefore the two circuits are not isolated from each other. See Fig. 10-10. The application of an autotransformer is a good choice for some users where a 480Y/277- or 208Y/120-volt, three-phase, four-wire distribution system is utilized. Some of the advantages are as follows:

- Lower purchase price
- Lower operating cost due to lower losses
- Smaller size; easier to install
- Better voltage regulation
- Lower sound levels

An autotransformer, however, cannot be used on a 480- or 240-volt, three-phase, three-wire delta system. A grounded neutral phase conductor must be available

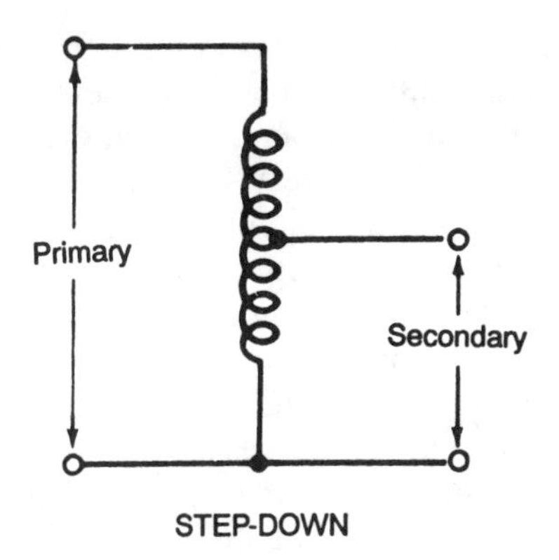

Fig. 10-10: Schematic diagram of an auto-transformer.

in accordance with NE Code Section 210-9, which states:

210-9: Circuits Derived from Autotransformers.
Branch circuits shall not be derived from auto-transformers unless the circuit supplied has a grounded conductor that is electrically connected to a grounded conductor of the system supplying the autotransformer.
Exception No. 1: An autotransformer shall be permitted to extend or add a branch circuit for an equipment load without the connection to a similar grounded conductor when transforming from a nominal 208 volts to a nominal 240-volt supply or similarly from 240 volts to 208 volts.
Exception No. 2: In industrial occupancies, where conditions of maintenance and supervision ensure that only qualified persons will service the installation, auto-transformers shall be permitted to supply nominal 600-volt loads from nominal 480-volt systems, and 480-volt loads from nominal 600-volt systems, without the connection to a similar grounded conductor.

Transformers must normally be accessible for inspection except for dry-type transformers under certain specified conditions. Certain types of transformers with a high voltage or kVA rating are required to be enclosed in transformer rooms or vaults when installed indoors. The construction of these vaults is covered in NE Code Sections 450-41 through 450-48.

In general, the NE Code specifies that the walls and roofs of vaults must be constructed of materials that have adequate structural strength for the conditions with a minimum fire resistance of 3 hours. However, where transformers are protected with an automatic sprinkler syster, water spray, carbon dioxide, or halon, the fire resistance construction may be lowered to only 1 hour. The floors of vaults in contact with the earth must be of concrete and not less than 4 inches thick. If the vault is built with a vacant space or other floors (stories) below it, the floor must have adequate structural strength for the load imposed thereon and a minimum fire resistance of 3 hours. Again, if the fire extinguishing facilities are provided, as outlined above, the fire resistance construction need only be 1 hour. The NE Code does not permit the use of studs and wall board construction for transformer vaults.

The overcurrent protection for transformers is based on their rated current, not on the load to be served. The primary circuit may be protected by a device rated or set at not more than 125% of their rated primary current of the transformer for transformers with a rated primary current of 9 amperes or more.

Instead of individual protection on the primary side, the transformer may be protected only on the secondary side if all the following conditions are met.

- The overcurrent device on the secondary side is rated or set at not more than 125% of the rated secondary current.
- The primary feeder overcurrent device is rated or set at not more than 250% of the rated primary current.

For example, if a 12 kVA transformer has a primary current rating of

12,000 VA/480 volts = 25 amperes

and a secondary current rated at

12,000 VA/120 volts = 100 amperes

the individual primary protection must be set at

1.25 x 25 amperes = 31.25 amperes

In this case, a standard 30-ampere cartridge fuse rated at 600 volts could be used, as could a circuit breaker approved for use on 480 volts. However, if certain conditions are met, individual primary protection for the transformer is not necessary in this case if the feeder overcurrent protective device is rated at not more than

2.5 x 25 amperes = 62.5 amperes

and the protection on the secondary side is set at not more than

1.25 x 100 amperes = 125 amperes

A standard 125 ampere circuit breaker could be used.

Please note that the example cited above is for the transformer only; not the secondary conductors. The secondary conductors must be provided with overcurrent protection as outlined in Chapter 4 of this book and also as specified in other Sections of the NE Code.

Control Transformers

Control transformers are available in numerous types, but most control transformers are dry-type step-down units with the secondary control circuit isolated from the primary line circuit to assure maximum safety. See Fig. 10-11. These transformers and other components are usually mounted within an enclosed control box or control panel, which has a push-button station or stations independently grounded as recommended by the NE Code. Industrial control transformers are especially designed to accommodate the momentary current inrush caused when electromagnetic components re energized, without sacrificing secondary voltage stability beyond practical limits.

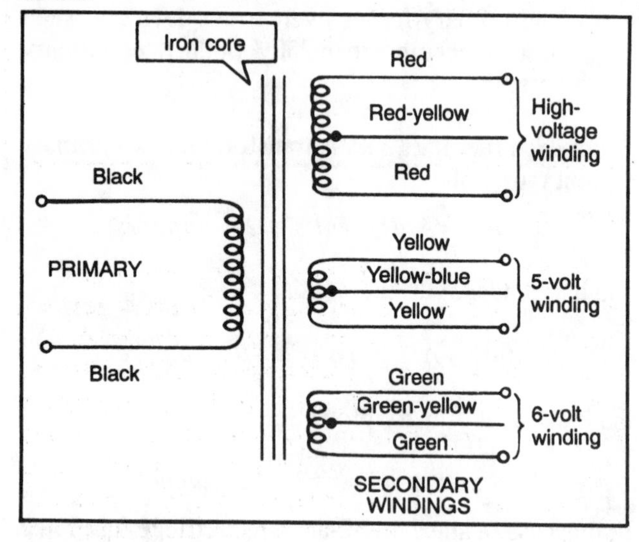

Fig. 10-11: Diagram of a typical control transformer.

Other types of control transformers, sometimes referred to as control and signal transformers, normally do not have the required industrial control transformer regulation characteristics. Rather, they are constant-potential, self-air-cooled transformers used for the purpose of supplying the proper reduced voltage for control circuits of electrically operated switches or other equipment and, of course, for signal circuits. Some are of the open type with no protective casing over the winding, while others are enclosed with a metal casing over the winding.

In seeking control transformers for any application, the loads must be calculated and completely analyzed before the proper transformer selection can be made. This analysis involves every electrically energized component in the control circuit. To select an appropriate control transformer, first determine the voltage and frequency of the supply circuit. Then determine the total inrush volt-amperes (watts) of the control circuit. In doing so, do not neglect the current requirements of indicating lights and timing devices that do not have inrush volt-amperes, but are energized at the same time as the other components in the circuit. Their total volt-amperes should be added to the total inrush volt-amperes.

Potential Transformers

In general, a potential transformer is used to supply voltage to instruments such as voltmeters, frequency meters, power-factor meters, and watt-hour meters. The voltage is proportional to the primary voltage, but it is small enough to be safe for the test instrument. The secondary of a potential transformer may be designed for several different voltages, but most are designed for 120 volts. The potential transformer is primarily a distribution transformer especially designed for good voltage regulation so that the secondary voltage under all conditions will be as nearly as possible a definite percentage of the primary voltage.

Current Transformers

A current transformer (Fig. 10-12) is used to supply current to an instrument connected to its secondary, the current being proportional to the primary current, but small enough to be safe for the instrument. The secondary of a current transformer is usually designed for a rated current of 5 amperes.

A current transformer operates in the same way as any other transformer in that the same relation exists between the primary and the secondary current and voltage. A current transformer is connected in series with the power lines to which it is applied so that line current flows in its primary winding. The secondary of the current transformer is connected to current devices such as ammeters, wattmeters, watt-hour meters, power-factor meters, some forms of relays, and the trip coils of some types of circuit breakers.

When no instruments or other devices are connected to the secondary of the current transformer, a short-circuit device or connection is placed across the secondary to prevent the secondary circuit from being opened while the primary winding is carrying current.

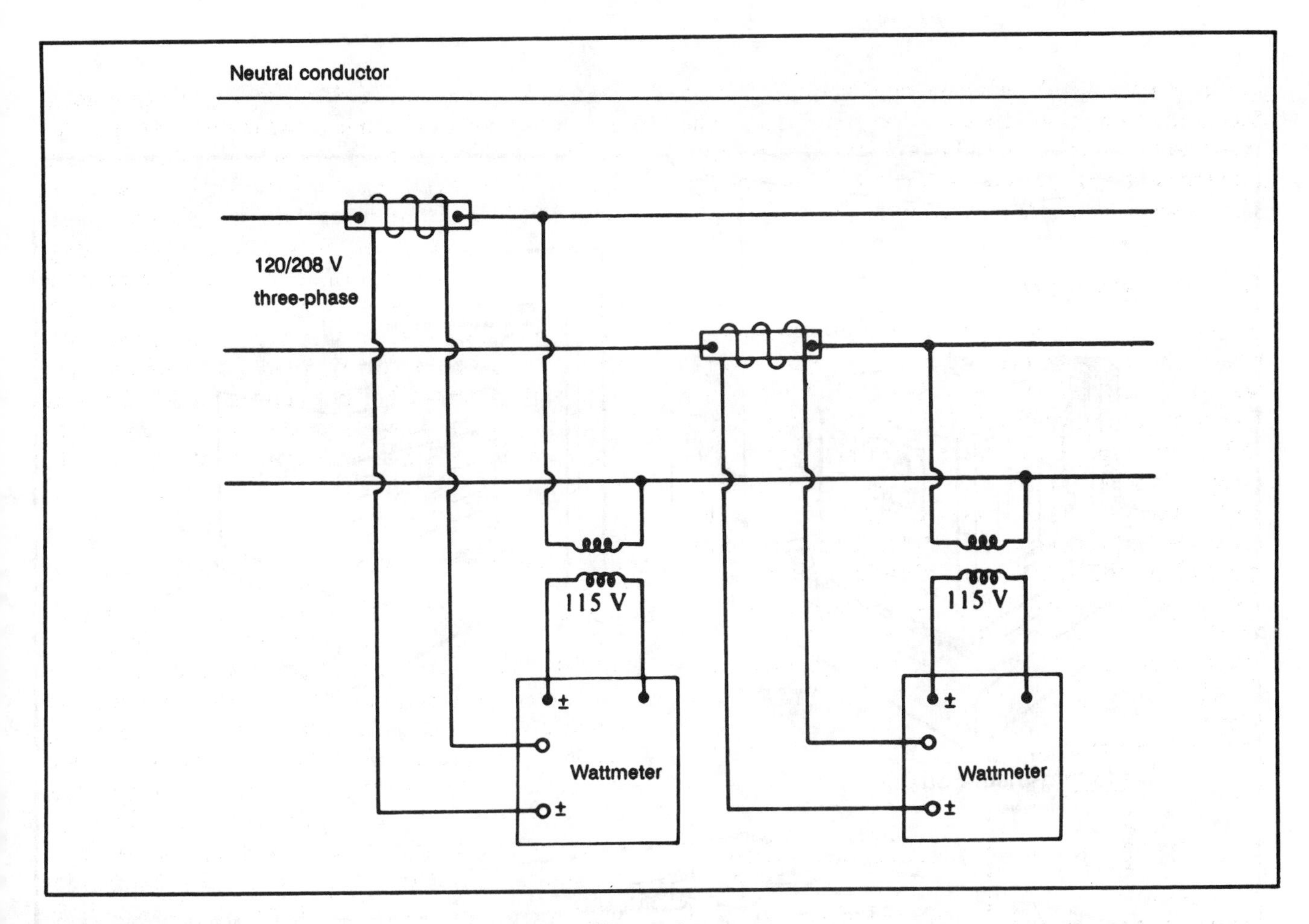

Fig. 10-12: Current transformers used in conjunction with watt-hour meter.

There will be no secondary ampere turns to balance the primary ampere turns, so the total primary current becomes exciting current and magnetizes the core to a high flux density. This produces a high voltage across both primary and secondary windings and endangers the life of anyone coming in contact with the meters or leads.

Transformer Grounding

Grounding is necessary to remove static electricity and also as a precautionary measure in case the transformer windings accidently come in contact with the core or enclosure. All should be grounded and bonded to meet NE Code requirements and also local codes, where applicable.

The tank or housing of every power transformer should be grounded to eliminate the possibility of obtaining static shocks from it or being injured by accidental grounding of the winding to the case. A grounding lug is provided on the base of most transformers for the purpose of grounding the case and fittings.

The NE Code specifically states the requirements of grounding and should be followed in every respect. Furthermore, certain advisory rules recommended by manufacturers provide additional protection beyond that of the NE Code. In general, the NE Code requires that exposed noncurrent-carrying metal parts of transfomer installations, including fences, guards, etc., must be grounded where required under the conditions and in the manner specified for electrical equipment and other exposed metal parts in Article 250.

A summary of NE Code regulations pertaining to transformer installations is illustrated in Fig. 10-13 and 10-14.

Capacitors

Article 460 of the NE Code also states specific rules for the installation and protection of capacitors other than surge capacitors or capacitors that are part of another apparatus. The chief use of capacitors is to improve the power factor of an electrical installation or an individual piece of electrically-operated equipment. This efficiency lowers the cost of power.

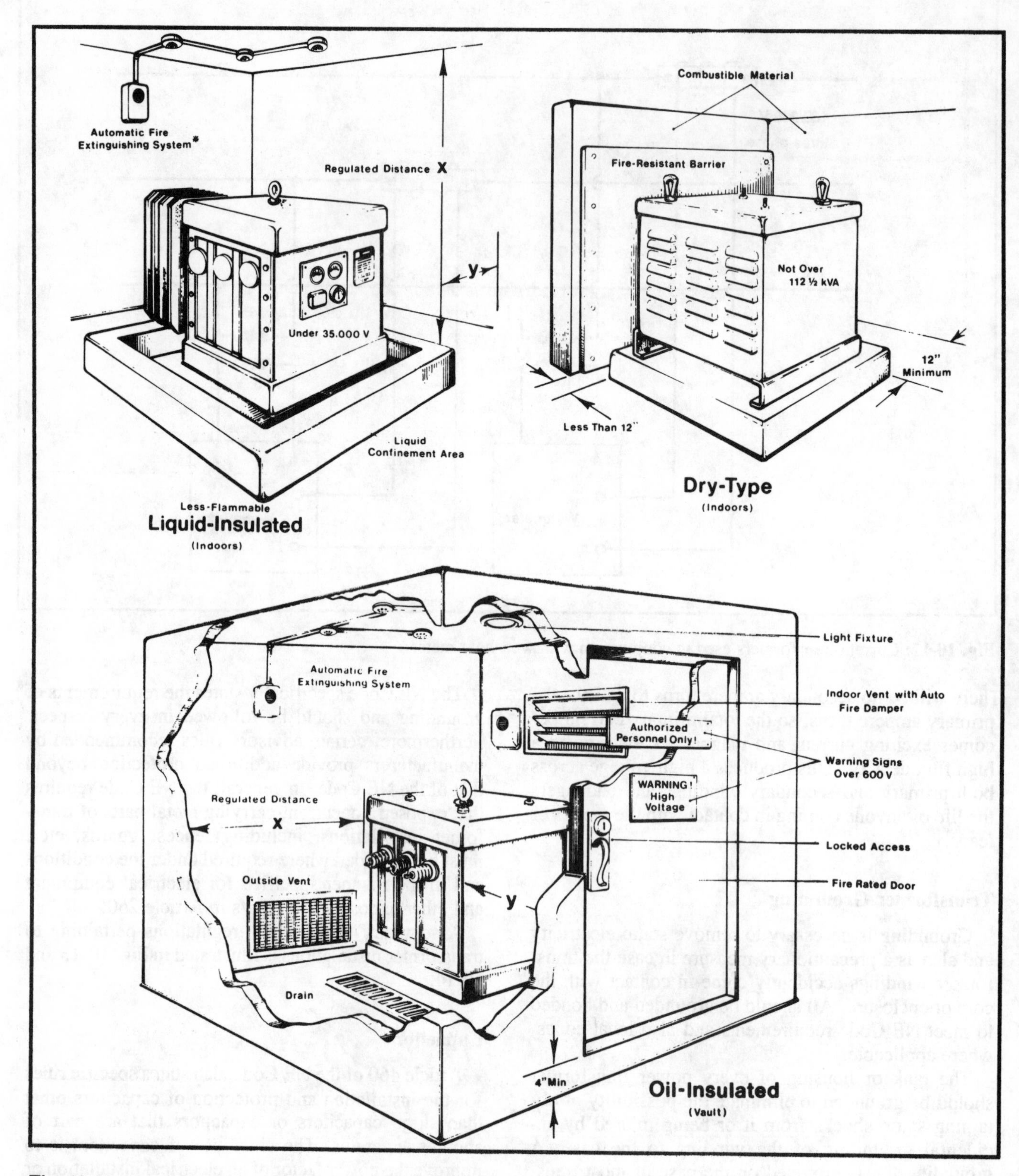

Fig. 10-13: NE Code regulations governing transformer installations.

■ INSTALLATION RULES FOR TRANSFORMERS

Application	NE Code Regulation	NE Code Section
Location	Transformers must be readily accessible to qualified personnel for maintenance and replacement. Dry-type transformers may be located in the open. Dry-type transformers not exceeding 600 volts and 50 kVA are permitted in fire-resistant hollow spaces of buildings under conditions as specified in the NE Code.	450-13
	Liquid-filled transformers must be installed as specified in the NE Code and usually in vaults when installed indoors.	Article 450, part B
Overcurrent protection	The primary protection must be rated or set as follows: 9 amperes or more, 125% Less than 9 amperes, 167% Less than 2 amperes, 300% If the primary current (line side) is 9 amperes or more, the next higher standard size overcurrent protective device greater than 125% of the primary current may be used. For example, if the primary current is 15 amperes, 125% of 15 amperes equals 18.75 amperes. The next standard size circuit breaker is 20 amperes. Therefore, this size (20 amperes) may be used. Conductors on the secondary side of a single-phase transformer with a two-wire secondary may be protected by the primary overcurrent device under certain NE Code conditions.	Article 450, part C
Over 600 volts	Special NE Code rules apply to transformers operating at over 600 volts.	450-3(a)

Fig. 10-14: NE Code regulations governing transformer installations

The NE Code rules for capacitors operating under 600 volts are summarized in Figs. 10-15 and 10-16.

Since capacitors may store an electrical charge and hold a voltage that is present even when a capacitor is disconnected from a circuit, capacitors must be enclosed, guarded, or located so that persons cannot accidentally contact the terminals. In most installations, capacitors are installed out of reach or are placed in an enclosure accessible only to qualified persons. The stored charge of a capacitor must be drained by a discharge circuit either permanently connected to the capacitor or automatically connected when the line voltage of the capacitor circuit is removed. The windings of a motor or a circuit consisting of resistors and reactors will serve to drain the capacitor charge.

Capacitor circuit conductors must have an ampacity of not less than 135% of the rated curent of the capacitor. This current is determined from the kVA rating of the capacitor as for any load. A 100 kVA (100,000 watts) three-phase capacitor operating at 480 volts has a rated current of

$$100{,}000 \text{ kVA}/1.73 \times 480 \text{ volts} = 120.3 \text{ amperes}$$

The minimum conductor ampacity is then

$$1.35 \times 120.3 \text{ amperes} = 162.4 \text{ amperes}$$

When a capacitor is switched into a circuit, a large inrush current results to charge the capacitor to the circuit voltage. Therefore, an overcurrent protective device for the capacitor must be rated or set high enough to allow the capacitor to charge. Although the exact setting is not specified in the NE Code, typical settings vary between 150% and 250% of the rated capacitor current.

In addition to overcurrent protection, a capacitor must have a disconnecting means rated at not less than 135% of the rated current of the capacitor unless the capacitor is connected to the load side of the motor-running overcurrent device. In this case, the motor disconnecting means would serve to disconnect the capacitor and the motor.

A capacitor connected to a motor circuit serves to increase the power factor and reduce the total kVA required by the motor-capacitor circuit. The power factor is defined as the true power in kilowatts divided by the total kVA or

$$pf = kW/kVA$$

where the power factor is a number between .0 and 1.0. A power factor less than one represents a lagging current for motors and inductive devices. The capacitor introduces a leading current that reduces the total kVA and raises the power factor to a value closer to unity. If the inductive load of the motor is completely balanced by the capacitor, a maximum power factor of unity results and all of the input energy serves to perform useful work.

The capacitor circuit conductors for a power factor correction capacitor must have an ampacity of not less than 135% of the rated current of the capacitor. In addition, the ampacity must not be less than one-third the ampacity of the motor circuit conductors.

The connection of a capacitor reduces current in the feeder up to the point of connection. If the capacitor is connected on the load side of the motor-running overcurrent device, the current through this device is reduced and its rating must be based on the actual current, not on the full-load current of the motor.

Troubleshooting

Since transformers are an essential part of the equipment used in electrical systems, electricians and others involved in this type work should know how to test and locate troubles that develop in transformers—especially in the smaller power supply or control transformers.

Open Circuit: Should one of the windings in a transformer develop a break or "open" condition, no current can flow and therefore, the transformer will not deliver any output. The symptoms of an open circuited transformer is that the circuits which derive power from the transformer are de-energized or "dead." Use an AC voltmeter to check across the transformer output terminals as shown in Fig. 10-17. A reading of zero volts indicates an open circuit.

Then take a voltage reading across the input terminals. If a voltage reading is present, then the conclusion is that one of the windings in the transformer is open. However, if no voltage reading is on the input terminals either, then the conclusion is that the open is elsewhere on the line side of the circuit; perhaps a disconnect switch is open.

However, if voltage is present on the line or primary side and none on the secondary or load side, open the switch to de-energize the circuit, and place a warning tag on this switch so that it is not inadvertently closed again while someone is working on the circuit. Disconnect all of the transformer primary and secondary leads, check each winding in the transformer for continuity (a continuous circuit), as indicated by a resistance reading taken with an ohmmeter as shown in Fig. 10-18. A megger may also be used for this purpose.

■ INSTALLATION RULES FOR CAPACITORS

Application	NE Code Regulation	NE Code Section
Enclosing and guarding	Capacitors must be enclosed, located, or guarded so that persons cannot come into accidental contact or bring conducting materials into accidental contact with exposed energized parts, terminals, or buses associated with them. However, no additional guarding is required for enclosures accessible only to authorized and qualified persons.	460-2(b)
Stored charge	Capacitors must be provided with a means of draining the stored charge. The discharge circuit must be either permanently connected to the terminals of the capacitor or capacitor bank, or provided with automatic means of connecting it to the terminals of the capacitor bank on removal of voltage from the line. Manual means of switching or connecting the discharge circuit shall not be used.	460-6
Capacitors on circuits over 600 volts	Special NE Code regulations apply to capacitors operating at over 600 volts.	Article 460
Conductor ampacity	The ampacity of capacitor circuit conductors must not be less than 135% of the rated current of the capacitor.	460-8(a)
Capacitors on motor circuits	The ampacity of conductors that connect a capacitor to the terminals of a motor or to motor circuit conductors shall not be less than one third the ampacity of the motor circuit conductors and in no case less than 135% of the rated current of the capacitor.	460-8(a)
Overcurrent protection	Overcurrent protection is required in each ungrounded conductor unless the capacitor is connected on the load side of a motor-running overcurrent device. The setting must be as low as practicable.	460-8(b)

Fig. 10-15: NE Code guidelines for capacitor installations.

■ INSTALLATION RULES FOR CAPACITORS		
Application	**NE Code Regulation**	**NE Code Section**
Enclosing and guarding	Capacitors must be enclosed or guarded unless accessible only to qualified persons.	460-2(b)
Stored charge	Capacitors must be provided with a means of draining the stored charge.	460-6
Capacitors on circuits over 600 volts	Special NE Code regulations apply to capacitors operating at over 600 volts.	Article 460
Conductor ampacity	The ampacity of capacitor circuit conductors must not be less than 135% of the rated current of the capacitor in any case.	460-8(a)
Capacitors on motor circuits	When connected to a motor, the ampacity of conductors connecting the capacitor must not be less than ⅓ the ampacity of the motor circuit conductors.	460-8(a)
Overcurrent protection	Overcurrent protection is required in each ungrounded conductor unless the capacitor is connected on the load side of a motor-running overcurrent device. The setting must be as low as practicable	460-8(b)
Disconnecting means	A disconnecting means is required for a capacitor unless it is connected to the load side of a motor-running overcurrent device. The rating must not be less than 135% of the rated current of the capacitor.	460-8(c)
Improved power factor	The total kilovar rating of capacitors connected to the load side of a motor controller must not exceed the value required to raise the no-load power factor to 1.	460-7
Overcurrent protection for improved power factor	If the power factor is improved, the motor-running overcurrent device must be selected based on the reduced current drawn; not the full-load current of the motor.	460-9
Grounding	Capacitor cases must be grounded except when the system is designed to operate at other than ground potential.	460-10

Fig. 10-16: NE Code guidelines for capacitor installations.

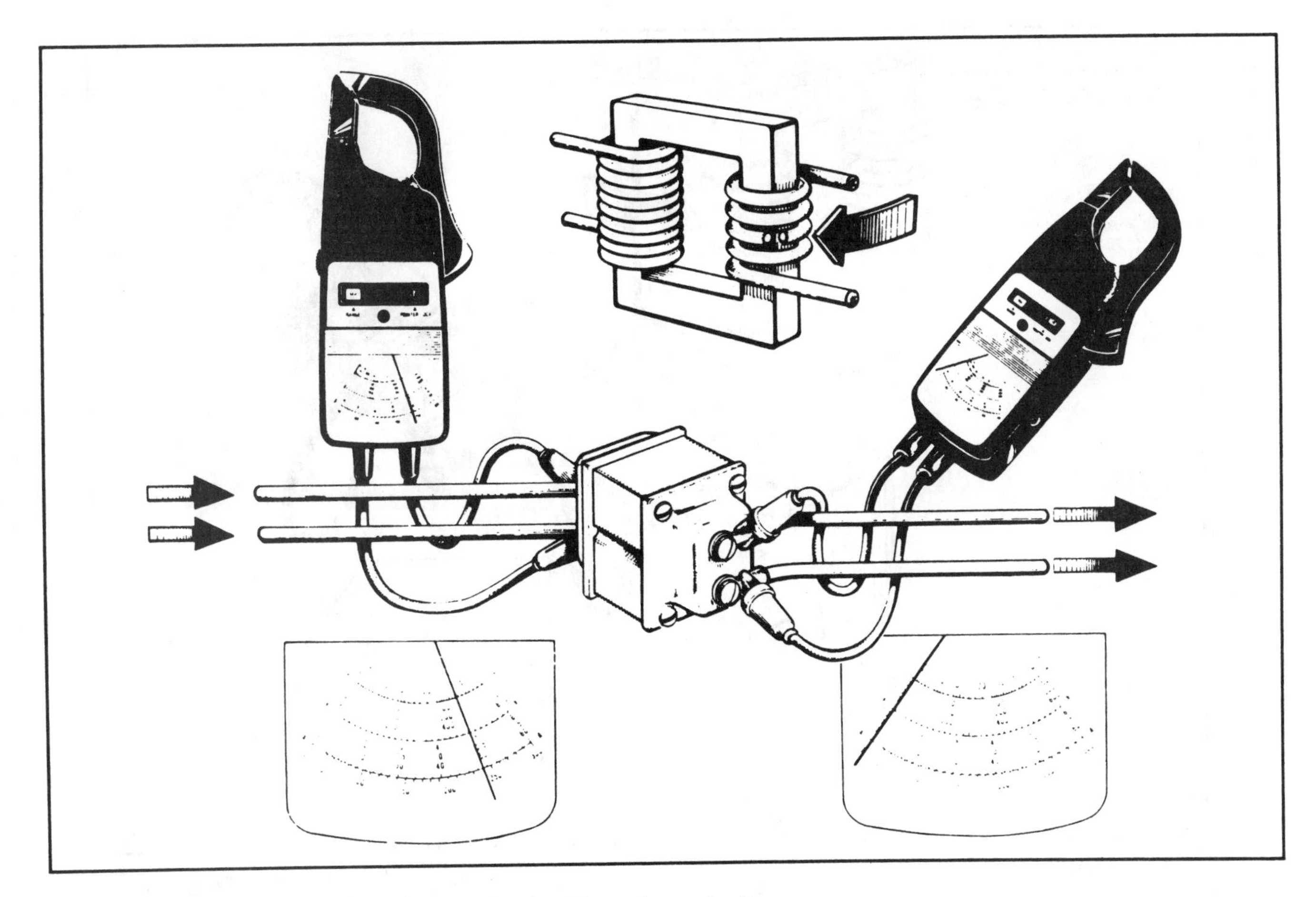

Fig. 10-17: Detecting and finding an open circuit with a voltage check.

Continuity is indicated by a relatively low resistance reading on control transformers, while an open winding will be indicated by an infinite resistance reading on the ohmmeter. In most cases, such small transformers will have to be replaced, unless of course the break is accessible and can be repaired.

Ground Fault: Sometimes a few turns in the secondary winding of a transformer will acquire a partial short, which in turn will cause a voltage drop across the secondary. The symptom of this condition is usually overheating of the transformer caused by large circulating currents flowing in the shorted windings.

The easiest way to check this condition is with a voltmeter set at the proper voltage scale. Take a reading on the line or primary side of the transformer first to make certain normal voltage is present. Then take a reading on the secondary side. If the transformer has a partial short or ground fault, the voltage reading should be lower than normal. See Fig. 10-19.

Replace the faulty transformer with a new one and again take a reading on the secondary. If the voltage reading is now normal and the circuit operates satisfactorily, leave the replacement transformer in the circuit, and either discard or repair the original transformer.

A highly sensitive ohmmeter may also be used to test this condition when the system is de-energized and the leads are disconnected; a lower reading on the ohmmeter than normal indicates this condition. However, the reading will usually be so slight that the average ohmmeter is not sensitive enough to detect the difference. Therefore, the recommended way is to use the voltmeter test.

Complete Short: Occasionally a transformer winding will become completely shorted. In most cases, this will activate the overload protective device and de-energize the circuit, but in other instances, the transformer may continue trying to operate with excessive overheating—due to the very large circulating current. This heat will often melt the wax or insulation inside the transformer, which is easily detected by the odor. Also, there will be no voltage output across the shorted winding and the circuit across the winding will be dead.

The short may be in the external secondary circuit or it may be in the transformer's winding. To determine

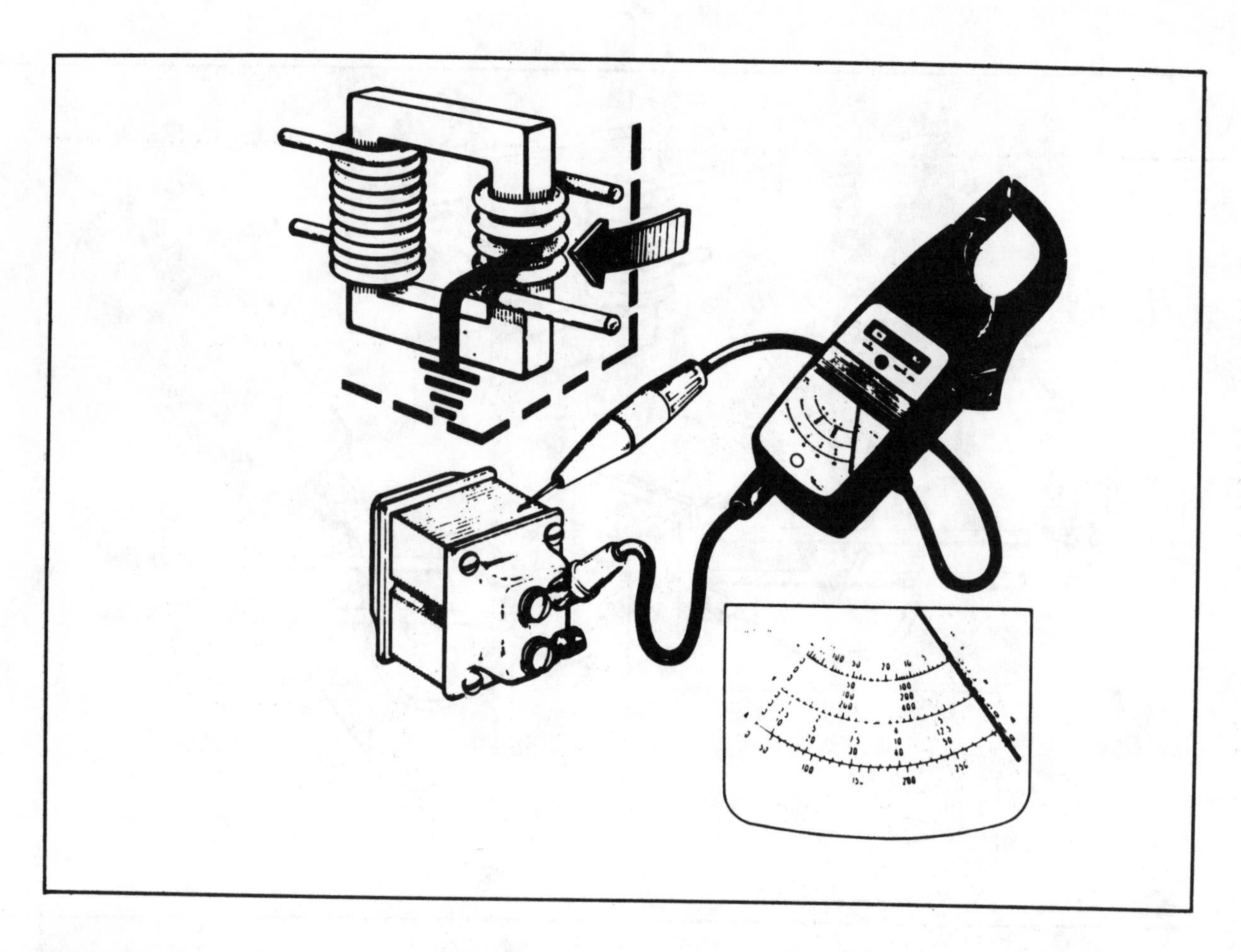

Fig. 10-18: Detecting and finding an open circuit with a continuity check.

its location, disconnect the external secondary circuit from the winding and take a reading with a voltmeter as shown in Fig. 10-20. If the voltage is normal with the external circuit disconnected, then the problem lies within the external circuit. However, if the voltage reading is still zero across the secondary leads, the transformer is shorted and will have to be replaced.

Grounded Windings: Insulation breakdown is quite common in older transformers—especially those that have been overloaded. At some point, the insulation breaks or deteriorates and the wire becomes exposed. The exposed wire often comes into contact with the transformer housing and grounds the winding.

If a winding develops a ground, and a point in the external circuit connected to this winding is also grounded, part of the winding will be shorted out. The symptoms will be overheating, which is usually detected by feel or smell, and a low voltage reading as indicated on a voltmeter reading as shown in Fig. 10-21. In most cases, transformers with this condition will have to be replaced.

A megohmmeter is the best test instrument to check for this condition. Disconnect the leads from both the primary and secondary windings. Tests can then be performed on either winding by connecting the megger negative test lead to an associated ground and the positive test lead to the winding to be measured as shown in Fig. 10-22.

Insulation resistance should then be measured between the windings themselves. This is accomplished by connecting one test lead to the primary and the second test lead to the secondary. All such tests should be recorded on a record card under proper identifying labels.

Summary

Transformer problems include:

- One of the windings can develop an open circuit.
- Part or all of one winding can become shorted.
- A ground can develop.

In troubleshooting a transformer, a voltmeter and ohmmeter are used to locate the trouble. If a transformer is defective, it usually must be replaced.

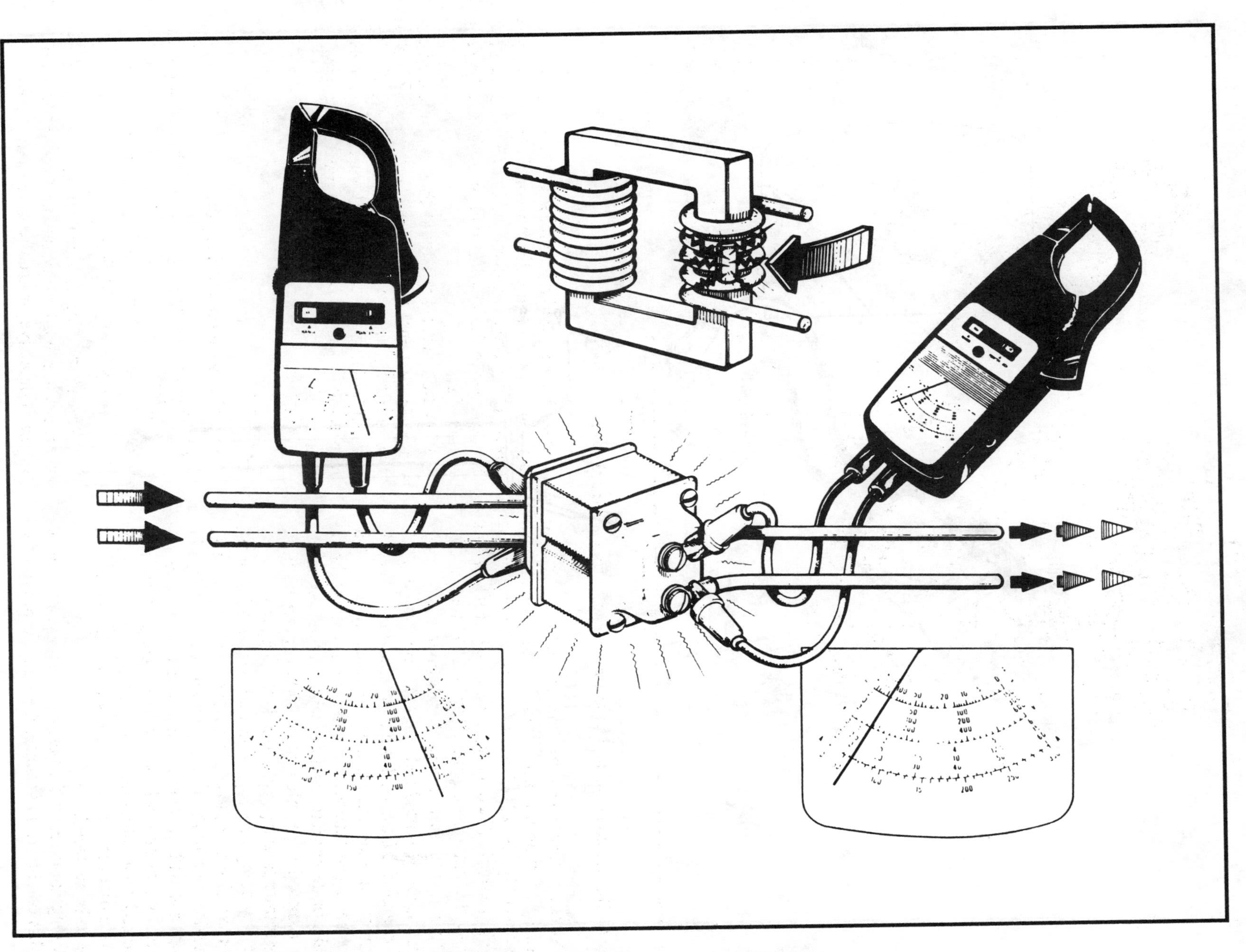

Fig. 10-19: A voltage reading lower than normal usually indicates a partial short or ground fault.

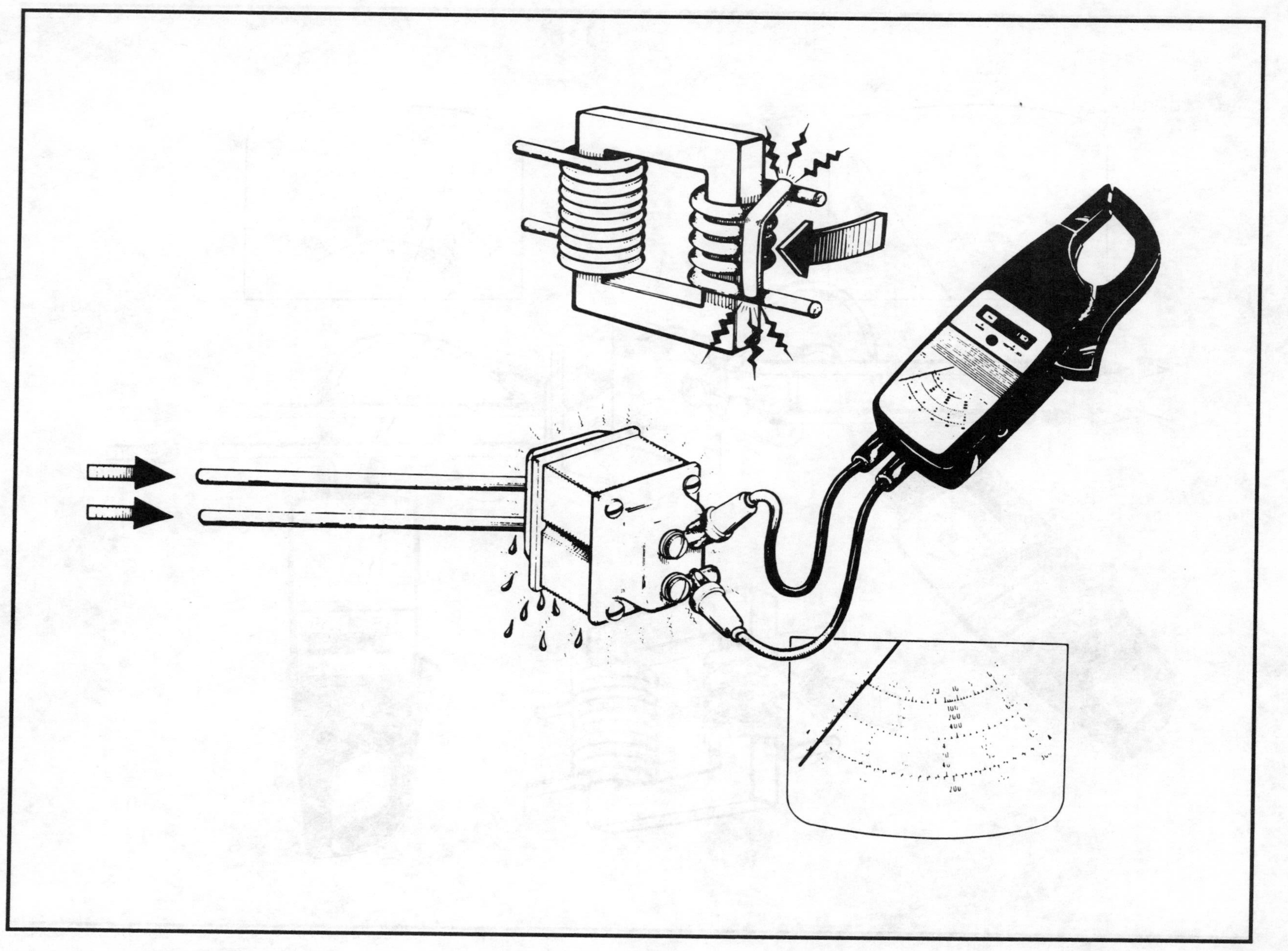

Fig. 10-20: If the voltage is normal with the external circuit disconnected, then the problem lies within the external circuit.

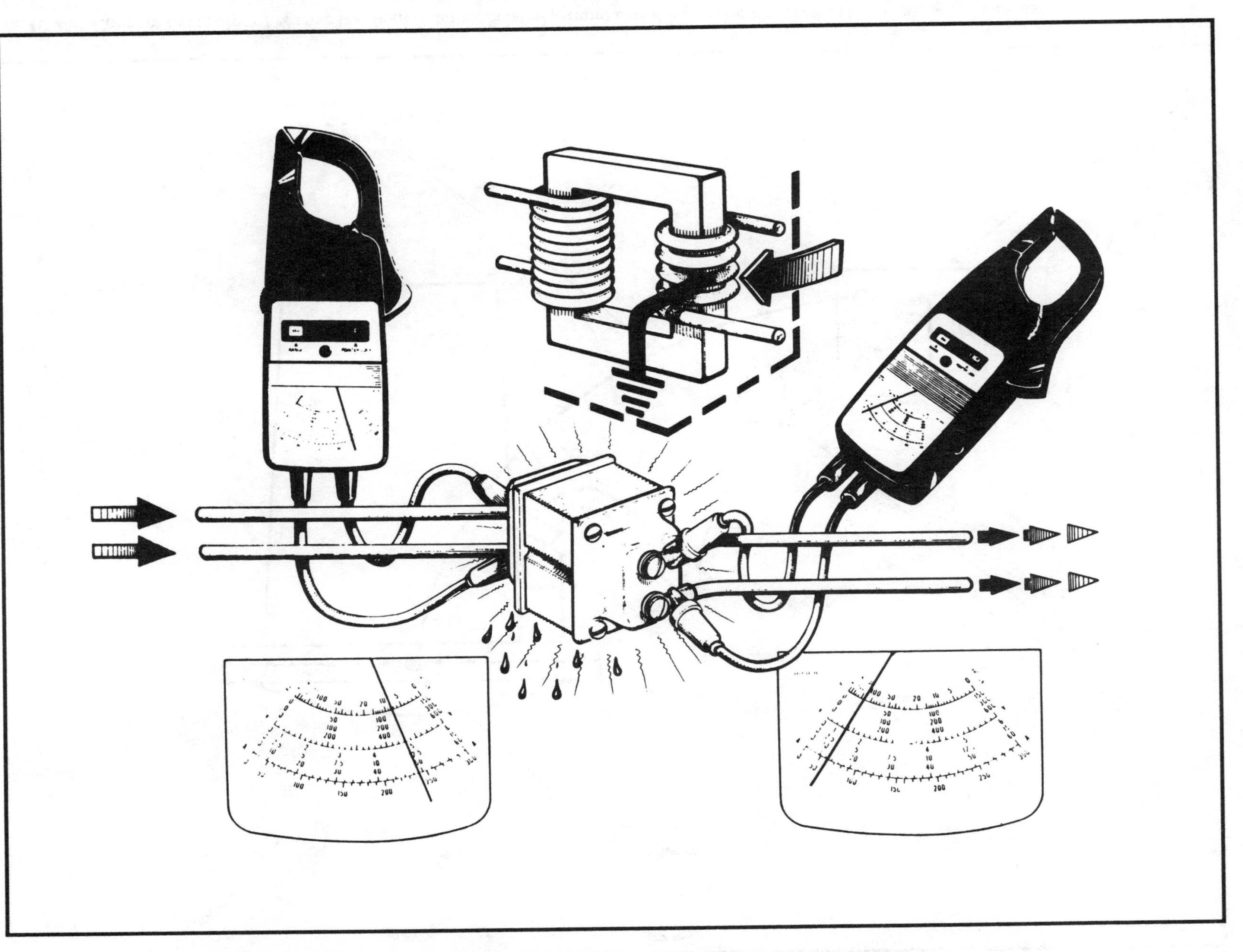

Fig. 10-21: When a transformer overheats and a low-voltage reading is indicated, it will have to be replaced.

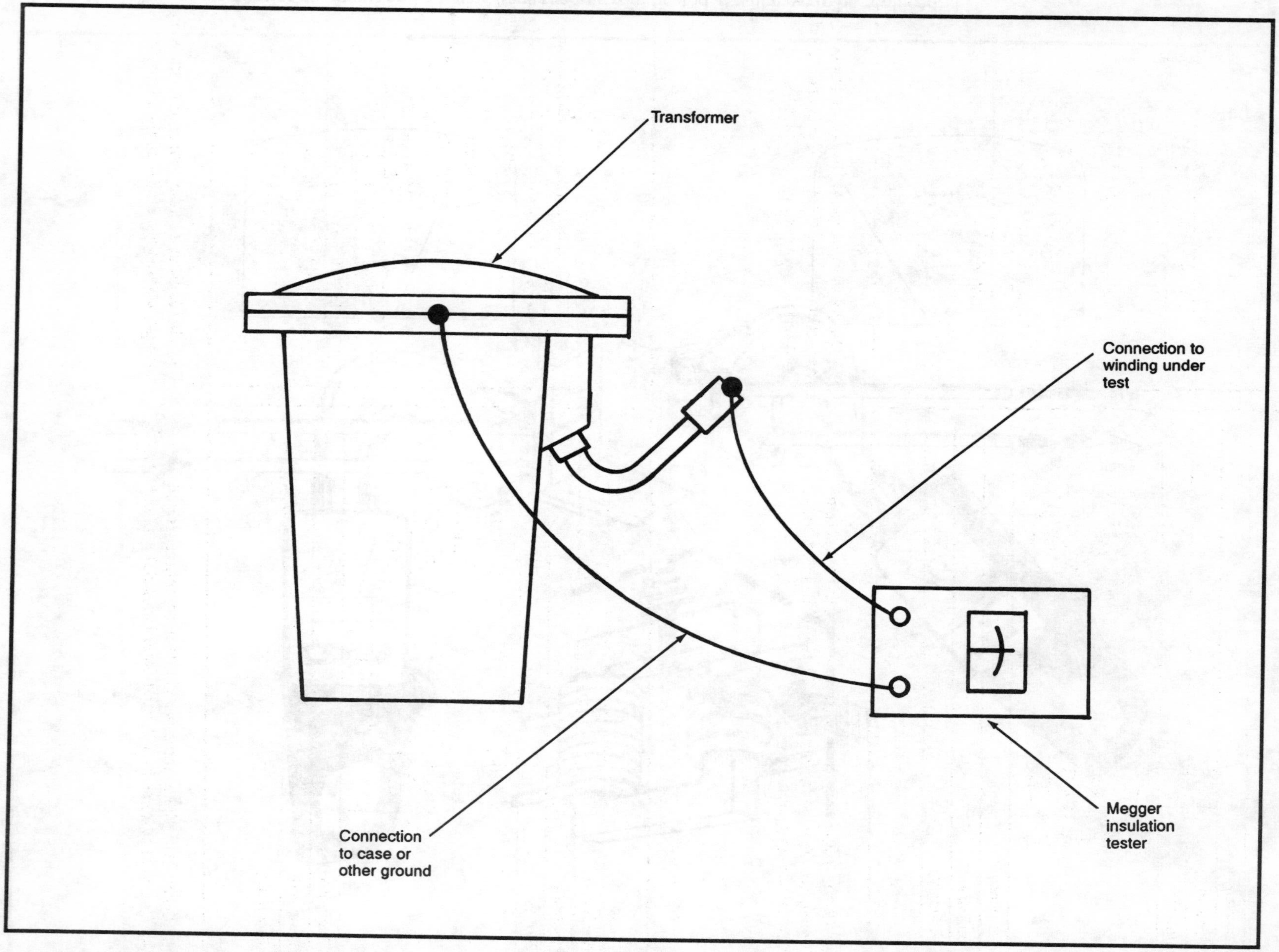

Transformer

Connection to winding under test

Megger insulation tester

Connection to case or other ground

Fig. 10-22: Connect the megger negative test lead to an associated ground and the positive test lead to the winding to be measured.

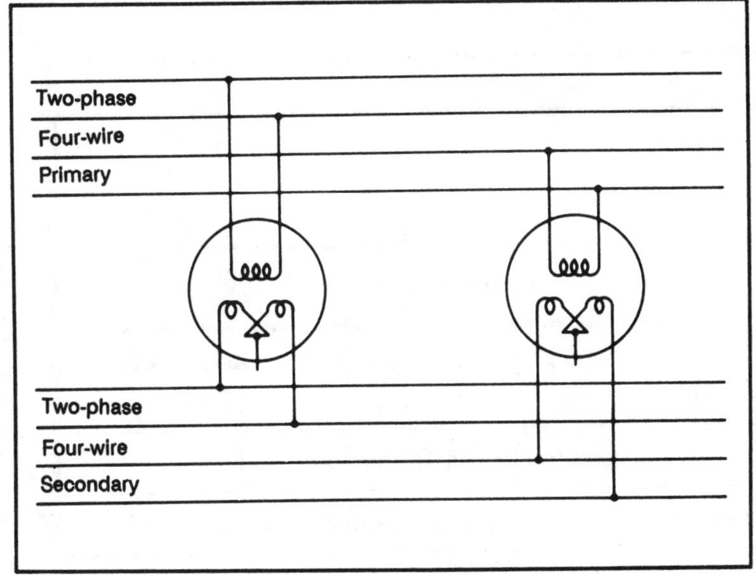

Fig. 10-23: Two-phase, four-wire transformer connection. Obsolete for new work, but sometimes found in existing, industrial wiring systems.

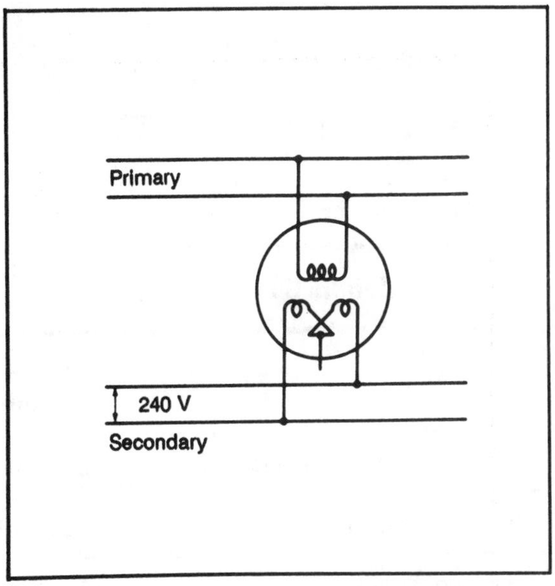

Primary
240 V
Secondary

Fig. 10-24: Single-phase, two-wire transformer connection giving 240 V on the secondary side. Used mostly in industrial applications.

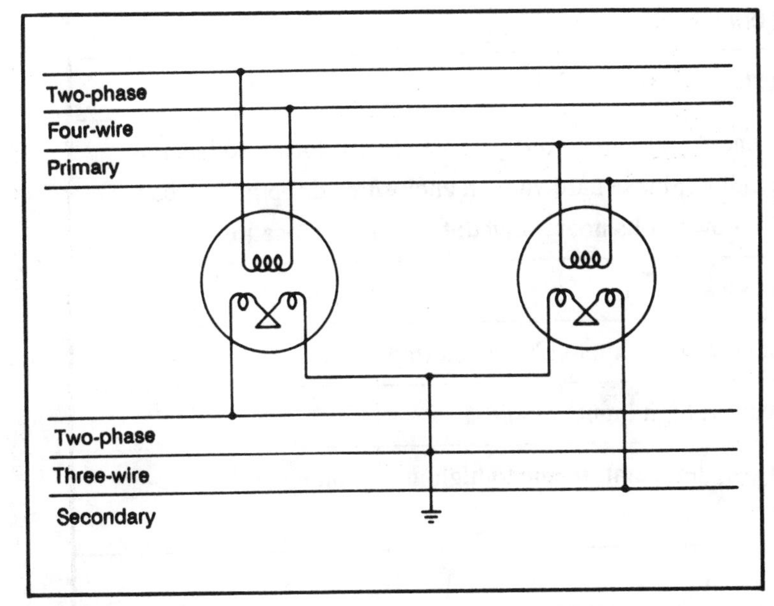

Fig. 10-25: Two-phase, three-wire system, sometimes found on older electrical systems in mills and other industrial applications.

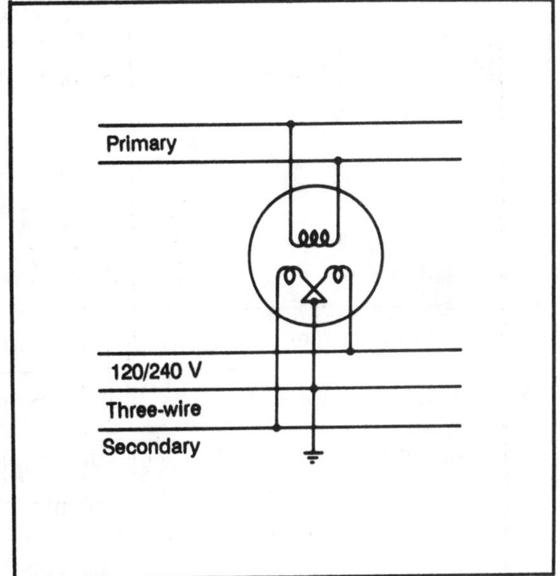

Fig. 10-26: Single phase connection that is the most popular for residential applications.

■ TROUBLESHOOTING CHART

Dry-Type Transformers

Malfunction	Probable cause
Overheating	Continuous overload; wrong external connections; poor ventilation; high surrounding air temperature.
Reduced to zero voltage	Short turns; loose connections to transformer terminal board
Excess secondary voltage	Input voltage high; dirt accumulations on primary terminal board
High conductor loss	Overload; terminal boards not on identical tap positions
Coil distortion	Coils short-circuited
Insulation failure	Continuous overloads; dirt accumulations on coils; mechanical damage in handling; lightning surge
Breakers or fuses opening	Short circuit; overload
Excessive cable heating	Improper bolted connection
High voltage to ground	Usually a static charge condition
Vibration and noise	Low frequency; high input voltage; core clamps loosened in shipment or handling; loose hardware on enclosure; shipping braces and/or hold-down bolts not removed; transformer location
Overheating	High input voltage
High exciting current	Low frequency; high input voltage; shorted turns
High core loss	Low frequency; high input voltage
Insulation failure	Very high core temperature due to high input voltage or low frequency
Smoke	Insulation failure
Burned insulation	Lightning surge; switching or line disturbance; broken bushings
Overheating	Clogged air ducts or inadequate ventilation

appendix I

GLOSSARY

AA (Aluminum Association): A manufacturers' association which promotes the use of aluminum.

AAC: All aluminum conductor.

AASC: Aluminum alloy stranded conductors.

Abrasion: The process of rubbing, grinding, or wearing away by friction.

Abrasion resistance: Ability to resist surface wear.

ac (alternating current): 1) A periodic current, the average of which is zero over a period; normally the current reverses after given time intervals and has alternately positive and negative values. 2) The type of electrical current actually produced in a rotating generator (alternator).

Accelerated life tests: Subjecting a product to operating conditions more severe than normal to expedite deterioration to afford some measure of probable life at normal conditions.

Accelerator: 1) A substance that increases the speed of a chemical reaction. 2) Something to increase velocity.

Acceptable (nuclear power): Demonstrated to be adequate by the safety analysis of the station.

Acceptance test: Made to demonstrate the degree of compliance with specified requirements.

Accepted: Approval for a specific installation or arrangement of equipment or materials.

Accessible: Capable of being removed or exposed without damaging the building structure or finish, or not permanently closed in by the structure or finish of the building.

ACSR (aluminum, conductor, steel reinforced): A bare composite of aluminum and steel wires, usually aluminum around steel.

Active power: In a 3-phase symmetrical circuit: $p = 3vi \cos \theta$; in a 1-phase, 2-wire circuit, $p = vi \cos \theta$.

Actuated equipment (nuclear power): Component(s) that perform a protective function.

Administrative authority: An organization exercising jurisdiction over the National Electrical Safety Code.

AEC (Atomic Energy Commission): Now defunct; see ERDA and NRC.

AEIC: Association of Edison Illuminating Companies.

Aggregate: Material mixed with cement and water to produce concrete.

Aging: The irreversible change of material properties after exposure to an environment for an interval of time.

AIA: 1) American Institute of Architects. 2) Aircraft Industries Association.

Air cleaner: Device used for removal of air-borne impurities.

Air diffuser: Air distribution outlet designed to direct airflow into desired patterns.

Air entrained concrete: Concrete in which a small amount of air is trapped by addition of a special material to produce greater durability.

Air oven: A lab oven used to heat by convection of hot air.

Al: Aluminum.

Al-Cu: An abbreviation for aluminum and copper, commonly marked on terminals, lugs and other electrical connectors to indicate that the device is suitable for use with either aluminum conductors or copper conductors.

Alive: Energized; having voltage applied.

Alloy: A substance having metallic properties and being composed of elemental metal and one or more chemical elements.

Alternator: A device to produce alternating current.

Alumoweld®: An aluminum clad steel wire by Copperweld Steel Corp.

Ambient temperature: Temperature of fluid (usually air) that surrounds an object on all sides.

American bond: Brickwork pattern consisting of five courses of stretchers followed by one bonding course of headers.

Ammeter: An electric meter used to measure current, calibrated in amperes.

Ampacity: The current-carrying capacity of conductors or equipment, expressed in amperes.

Ampere (A): The basic SI unit measuring the quantity of electricity.

Ampere-turn: The product of amperes times the number of turns in a coil.

Amplification: Procedure of expanding the strength of a signal.

Amplifier: 1) A device that enables an input signal to directly control a larger energy flow. 2) The process of increasing the strength of an input.

Amplitude: The maximum value of a wave.

Analog: Pertaining to data from continuously varying physical quantities.

Angle, roll over (overhead): The sum of the vertical angles between the conductor and the horizontal on both sides of the traveler; excessive roll over angles can cause premature splice failures.

Angular velocity: The average time rate of change of angular position; in electrical circuits = 2 f, and f equals frequency.

ANI (American Nuclear Insurers): A voluntary unincorporated association of companies providing property and liability insurance for US nuclear power plants: formerly NELPIA.

Annealing: The process of preventing or removing objectional stresses in materials by controlled cooling from a heated state; measured by tensile strength.

Annealing, bright: Annealing in a protective environment to prevent discoloration of the surface.

Anode: 1) Positive electrode through which current enters a non-metallic conductor such as an electrolytic cell. 2) The negative pole of a storage battery.

ANSI (American National Standards Institute): An organization that publishes nationally recognized standards.

Antenna: A device for transmission or reception of electromagnetic waves.

Antioxidant: Retards or prevents degradation of materials exposed to oxygen (air) or peroxides.

Antisiphon trap: Trap in a drainage system designed to preserve a water seal by defeating siphonage.

Aperture seal (nuclear): A seal between containment aperture and the electrical penetration assembly.

Appliance: Equipment designed for a particular purpose which utilizes electricity to produce heat, light, mechanical motion, etc., usually complete in itself, generally other than industrial, normally in standard sizes or types.

Approved: 1) Acceptable to the authority having legal enforcement. 2) Per OSHA: A product that has been tested to standards and found suitable for general application, subject to limitations outlined in the nationally recognized testing lab's listing.

Apron: Piece of horizonal wood trim under the sill of the interior casing of a window.

Areaway: Open space below the ground level immediately outside a building. It is enclosed by substantial walls.

Arc: A flow of current across an insulating medium.

Arc furnace: Heats by heavy current flow through the material to be heated.

Arcing time: The time elapsing from the severance of the circuit to the final interruption of current flow.

Arc resistance: The time required for an arc to establish a conductive path in or across a material.

Armature: 1) Rotating machine: the member in which alternating voltage is generated. 2) Electromagnet: the member which is moved by magnetic force.

Armor: Mechanical protector for cables; usually a helical winding of metal tape, formed so that each convolution locks mechanically upon the previous one (interlocked armor); may be a formed metal tube or a helical wrap of wires.

Arrester: Wire screen secured to the top of an incinerator to confine sparks and other products of burning.

Ashlar: Squared and dressed stones used for facing a masonry wall; short upright wood pieces extending from the attic floor to the rafters forming a dwarf wall.

Askarel: A synthetic insulating oil which is nonflammable but very toxic—being replaced by silicone oils.

ASME: American Society of Mechanical Engineers.

Associated circuits (nuclear power): Nonclass 1E circuits that share power supplies or are not physically separated from Class 1E circuits.

ASTM (American Society for Testing and Materials): A group writing standards for testing materials, and specifications for materials.

Asymmetrical: Not identical on both sides of a central line; unsymmetrical.

Attachment plug or cap: The male connector for electrical cords.

Attenuation: A decrease in energy magnitude during transmission.

Audible: Capable of being heard by humans.

Auditable data: Technical information which is documented and organized so as to be readily understandable and traceable to independently verify inferences of conclusions based on these records.

Autoclave: A heated pressure vessel used to bond, cure, seal, or used for environmental testing.

Automatic: Operating by own mechanism when actuated by some impersonal influence: nonmanual: self-acting.

Automatic transfer equipment: A device to transfer a load from one power source to another, usually from normal to emergency source and back.

Autotransformer: Any transformer where primary and secondary connections are made to a single cell.

Auxiliary: A device or equipment which aids the main device or equipment.

AWG (American Wire Gage): The standard for measuring wires in America.

BX: A nickname for armored cable (wires with a spiral-wound, flexible steel outer jacketing); although used generically, BX is a registered tradename of the General Electric Company.

Backfill: Loose earth placed outside foundation walls for filling and grading.

Back pressure: Pressure in the low side of a refrigerating system; also called suction pressure or low side pressure.

Ballast: A device designed to stabilize current flow.

Balloon framing: System of small house framing; two by fours extending two stories with inch by quarter ledger strips notched into the studs to support the second-story floor beams.

Bank: An installed grouping of a number of units of the same type of electrical equipment; such as "a bank of transformers" or "a bank of capacitors" or a "meter bank," etc.

Bar: A long solid product having one cross-sectional dimension of 0.375 inch or more.

Bare (conductor): Not insulated; nor coated.

Bargeboard: Ornamented board covering the roof boards and projecting over the slope of the roof.

Barometer: Instrument for measuring atmospheric pressure.

Barrier: A partition; such as an insulating board to separate bus bars of opposite voltages.

Base: One of the regions or terminals of a transistor.

Base ambient temperature: The temperature of a cable group when there is no load on any cable of the group or on the duct bank containing the group.

Base load: The minimum load over a period of time.

Batten: Narrow wood strips used to cover joints.

Batter: Slope of the exposed face of a retaining wall.

Battery: A device which changes chemical to electrical energy, used to store electricity.

Bead: Narrow projecting molding with a rounded surface; or in plastering, a metal strip embedded in plaster at the projecting corner of a wall.

Beam: A horizontal member of wood, reinforced concrete, steel, or other material used to span the space between posts, columns, girders, or over an opening in a wall.

Bearing plate: Steel plate placed under one end of a beam or truss for load distribution.

Bearing wall: Wall supporting a load other than its own weight.

Bed: Place or material in which stone or brick is laid; horizontal surface of positioned stone; lower surface of brick.

Bedding: A layer of material to act as a cushion or interconnection between two elements of a device, such as the jute layer between the sheath and wire armor in a submarine cable; sometimes incorrectly used to refer to extruded insulation shields.

Belt: The outer protective nonmetallic covering of cable; its jacket.

Belted type cable: A multiple conductor cable having a layer of insulation over the core conductor assembly.

Bench mark: Point of reference from which measurements are made.

Bessel function: A mathematical solution to a differential equation which is used to solve changes in conductor resistance and mutual inductance between conductors with respect to frequency changes due to skin and proximity effects.

Bias vacuum tube: Difference of potential between control grid and cathode. Transistor—difference of potential between base and emitter and base and collector. Magnetic amplifier—level of flux density in the magnetic amplifier core under no-signal condition.

BIL (Basic Impulse Level): A reference impulse insulation strength.

Bimetal strip: Temperature regulating or indicating device that works on the principle that two dissimilar metals with unequal expansion rates, welded together, will bend as temperature changes.

Binder: Material used to hold assembly together.

Birdcage: The undesired unwinding of a stranded cable.

Blackbody: A hypothetical body that absorbs, without reflection, all of the electromagnetic radiation incident on its surface.

Blister: A defect in metal, on or near the surface, resulting from the expansion of gas in the subsurface zone. Very small blisters may be called "pinheads" or "pepper blisters."

Block bridging: Solid wood members nailed between joists to stiffen a floor.

Blueprint: See definition of drawing.

Bobbins: Metal spools, small metal reels.

Boiler: Closed container in which a liquid may be heated and vaporized.

Boiling point: Temperature at which a liquid boils or generates bubbles of vapor when heated.

Bond: A mechanical connection between metallic parts of an electrical system, such as between a neutral wire and a meter enclosure or service conduit to the enclosure for the service equipment with a bonding locknut or bushing; the junction of welded parts; the adhesive for abrasive grains in grinding wheels.

Bonding bushing: A special conduit bushing equipped with a conductor terminal to take a bonding jumper; also has a screw or other sharp device to bite into the enclosure wall to bond the conduit to the enclosure without a jumper when there are no concentric knockouts left in the wall of the enclosure.

Bonding jumper: A bare or insulated conductor used to assure the required electrical conductivity between metal parts required to be electrically connected. Frequently used from a bonding bushing to the service equipment enclosure to provide a path around concentric knockouts in an enclosure wall, also used to bond one raceway to another.

Bonding locknut: A threaded locknut for use on the end of a conduit terminal, but a locknut equipped with a screw through its lip. When the locknut is installed, the screw is tightened so its end bites into the wall of the enclosure close to the edge of the knockout.

Braid: An interwoven cylindrical covering of fiber or wire.

Branch circuit: That portion of a wiring system extending beyond the final overcurrent device protecting a circuit.

Braze: The joining together of two metal pieces, without melting them, using heat and diffusion of a jointing alloy of capillary thickness.

Breadboard: Laboratory idiom for an experimental circuit.

Breakdown: The abrupt change of resistance from high to low, allowing current flow: an initial rolling or drawing operation.

Breaker strip: Thin strips of material placed between phase conductors and the grounding conductor in flat parallel portable cables; the breaker strips provide extra mechanical and electrical protection.

Breakout: The point at which conductor(s) are taken out of a multiconductor assembly.

Bridge: A circuit which measures by balancing four impedances through which the same current flows:

Wheatstone—resistance
Kelvin—low resistance
Schering—capacitance, dissipation factor, dielectric constant
Wien—capacitance, dissipation factor

British thermal unit (Btu): Quantity of heat required to raise the temperature of 1 pound of water 1 degree Fahrenheit.

Brittle point: The highest temperature at which a chilled strip of polymer will crack when it is held at one end and impacted at the other end.

Brush: A conductor between the stationary and rotating parts of a machine, usually of carbon.

Buck: Rough wood door frame placed on a wall partition to which the door moldings are attached; completely fabricated steel door frame set in a wall or partition to receive the door.

Buff: To lightly abrade.

Bug: A crimped or bolted type of electrical connector for splicing wires or cables together. Also used as a verb: "bugged." Example: The wires were bugged together.

Bull-cutters: A larger, long-handled tool for cutting the larger sizes of wire and cable, up to MCM sizes.

Buna: A synthetic rubber insulation of styrene-butadiene; was known as GR-S, now as SBR.

Burner: Device in which combustion of fuel takes place.

Bus: The conductor(s) serving as a common connection for two or more circuits.

Busbars: The conductive bars used as the main current supplying elements of panelboards or switchboards; also the conductive bars duct: an assembly of bus bars within an enclosure which is designed for ease of installation, have no fixed electrical characteristics, and allowing power to be taken off conveniently, usually without circuit interruption.

Butane: Liquid hydrocarbon commonly used as fuel for heating purposes.

Buttress: Projecting structure built against a wall to give it greater strength.

BWR (boiling water reactor-nuclear power): A basic nuclear power fission reactor in which steam is used to transfer the energy from the reactor.

Bypass: Passage at one side of or around a regular passage.

CB: Pronounced "see bee," an expression used to refer to "circuit breaker," taken from the initial letters C and B.

CT: Pronounced "see tee," refers to current transformer, taken from the initial letters C and T.

Cable: An assembly of two or more wires which may be insulated or bare.

Cable, aerial: An assembly of one or more conductors and a supporting messenger.

Cable, armored: A cable having armor (see armor).

Cable, belted: A multiconductor cable having a layer of insulation over the assembled insulated conductors.

Cable, bore-hole: The term given vertical riser cables in mines.

Cable clamp: A device used to clamp around a cable to transmit mechanical strain to all elements of the cable.

Cable, coaxial: A cable used for high frequency, consisting of two cylindrical conductors with a common axis separated by a dielectric; normally the outer conductor is operated at ground potential for shielding.

Cable, control: Used to supply voltage (usually ON or OFF).

Cable, duplex: A twisted pair of cables.

Cable, festoon: A cable draped in accordion fashion from sliding or rolling hangers, usually used to feed moving equipment such as bridge cranes.

Cable, hand: A mining cable used to connect equipment to a reel truck.

Cable, parkway: Designed for direct burial with heavy mechanical protection of jute, lead, and steel wires.

Cable, portable: Used to transmit power to mobile equipment.

Cable, power: Used to supply current (power).

Cable, pressure: A cable having a pressurized fluid (gas or oil) as part of the insulation; paper and oil are the most common fluids.

Cable, ribbon: A flat multiconductor cable.

Cable, service drop: The cable from the utility line to the customer's property.

Cable, signal: Used to transmit data.

Cable, spacer: An aerial distribution cable made of covered conductors held by insulated spacers; designed for wooded areas.

Cable, spread room: A room adjacent to a control room to facilitate routing of cables in trays away from the control panels.

Cable, submarine: Designed for crossing under navigable bodies of water; having heavy mechanical protection against anchors, floating debris and moisture.

Cable tray: A rigid structure to support cables: a type of raceway: normally having the appearance of a ladder and open at the top to facilitate changes.

Cable, tray: A multiconductor having a nonmetallic jacket, designed for use in cable trays; (not to be confused with type TC cable for which the jacket must also be flame retardant).

Cable, triplexed d: Helical assembly of 3 insulated conductors and sometimes a bare grounding conductor.

Cable, unit: A cable having pairs of cables stranded into groups (units) of a given quantity, then these groups form the core.

Cable, vertical riser: Cables utilized in circuits of considerable elevation change; usually incorporate additional components for tensile strength.

Cabling: Helically wrapping together of two or more insulated wires.

Caisson: Sunken panel in a ceiling, contributing to a pattern.

Calender: A machine which mixes and makes slabs of polymers by squeezing heated, viscous material between two counter rotating rollers.

Calibrate: Compare with a standard.

Calorie: Heat required to raise temperature of 1 gram of water 1 degree centigrade.

Cambric: A fine weave of linen or cotton cloth used as insulation.

Candela (cd): The basic SI unit for luminous intensity: the candela is defined as the luminous intensity of 1/600,000 of a square metre of a blackbody at the temperature of freezing platinum.

Cant strip: Beveled strip placed in the angle between the roof and an abutting wall to avoid a sharp bend in the roofing material; strip placed under the lowest row of tiles on a roof to give it the same slope as the rows above it.

Cantilever: Projecting beam or member supported at only one end.

Capacitance: The storage of electricity in a capacitor; the opposition to voltage change: units: farad.

Capacitor: An apparatus consisting of two conducting surfaces separated by an insulating material. It stores energy, blocks the flow of direct current, and permits the flow of alternating current to a degree depending on the capacitance and frequency.

Capillary action: The traveling of liquid along a small interstice due to surface tension.

Capstan: A rotating drum used to pull cables or ropes by friction; the cables are wrapped around the drum.

Carbon black: A black pigment produced by the incomplete burning of natural gas or oil; used as a filler.

Carbon dioxide: Compound of carbon and oxygen that is sometimes used as a refrigerant.

Cascade: The output of one device connected to the input of another.

Catepuller: Two endless belts which squeeze and pull a cable by friction.

Cathode: 1) The negative electrode through which current leaves a nonmetallic conductor, such as an electrolytic cell. 2) The positive pole of a storage battery. 3) Vacuum tube—the electrode that emits electrons.

Cathode-ray tube: The electronic tube which has a screen upon which a beam of electrons from the cathode can be made to create images; for example; the television picture tube.

Cathodic protection: Reduction or prevention of corrosion by making the metal to be protected the cathode in a direct current circuit.

Cavity wall: Wall built of solid masonry units arranged to provide airspace within the wall.

C-C: Center to center.

CCA (Customer Complaint Analysis): A formal investigation of a cable defect or failure.

CEE (International Commission on Rules for the Approval of Electrical Equipment): Controls the standards for electrical products for sale in Europe; analogous to UL in USA.

Centigrade scale: Temperature scale used in metric system. Freezing point of water is 0 degrees; boiling point is 100 degrees.

CFR (Code of Federal Regulations): The general and permanent rules published in the Federal Register by the executive departments and agencies of the Federal Government. The Code is divided into 50 Titles which represent broad areas; Titles are divided into Chapters which usually bear the name of the issuing agency, e.g. Title 30—Mineral

Resources, Chapter 1 = MESA; Title 29—Labor, Chapter XVII—OSHA; Title 10—Energy, Chapter I = NRC.

CFR Title 10, Chapter I (nuclear power): The regulations of the Federal Nuclear Regulatory Commission: a) Standards for protection against radiation. b) Licensing procedures of production and utilization facilities. c) Operators licenses. d) Special nuclear materials. e) Nuclear material packaging for transport. f) Reactor site criteria.

Cgs: Centimetre, gram, second.

Chamfer: Bevel edge surface area produced by cutting away the external angle formed by two faces of stone or lumber.

Charge: The quantity of positive or negative ions in or on an object; unit: coulomb.

Chase: Recess in inner face of masonry wall providing space for pipes and/or ducts.

Check valve: A device that permits fluid flow only in one direction.

Chemical resistance test: Checking performance of materials immersed in different chemicals; loss of strength and dimensional change are measured.

Chimney effect: Tendency of air or gas to rise when heated.

Choke coil: A coil used to limit the flow of alternating current while permitting direct current to pass.

Circuit: A closed path through which current flows from a generator, through various components, and back to the generator.

Circuit breaker: A resettable fuse-like device designed to protect a circuit against overloading.

Circuit foot: One foot of circuit; i.e., if one has a 3-conductor circuit, then each lineal foot of circuit would have 3 circuit feet.

Circular mil: The non-SI unit for measuring the cross-sectional area of a conductor.

CL: Center line.

Class 1E (nuclear power): The safety electrical systems that are essential to emergency reactor shutdown, cooling, and containment.

Class 2 (nuclear power): Items important to reactor operation but not essential to safe shutdown or isolation.

Clearance: The vertical space between a cable and its conduit.

Clearing time: The time from sensing an overcurrent to circuit interruption.

Closing die: A die used to position the individual conductors during cabling.

Coated wire: Wire given a thin coating of another metal such as tin, lead, nickel, etc.; coating by dipping or planting; coating for protection, or to improve its properties.

Coaxial cable: A cable consisting of two conductors concentric with and insulated from each other.

Code: Short for National Electrical Code.

Code installation: An installation that conforms to the local code and/or the national code for safe and efficient installation.

Coefficient of expansion: The change in dimension due to change in temperature.

Coefficient of friction: The ratio of the tangential force needed to start or maintain uniform relative motion between two contacting surfaces to the perpendicular force holding them in contact.

Coil: A wire or cable wound in a series of closed loops.

Cold bend: A test to determine cable or wire characteristics at low temperatures.

Cold cathode: A cathode that does not depend on heat for electron emission.

Cold joint: Improper solder connection due to insufficient heat.

Cold welding: Solid-phase welding using pressure without heat.

Cold work: Permanent strain produced by an external force (such as wire drawing) in a metal below its recrystallization temperature.

Collector: The part of a transistor that collects electrons.

Color code: Identifying conductors by the use of color.

Come along: A cable grip (usually of tubular basket-weave construction which tightens its grip on the cable as it is pulled) with a pulling "eye" on one end for attaching to a pull-rope for pulling conductors into conduit or other raceway.

Comfort zone: Area on psychrometric chart that shows conditions of temperature, humidity, and sometimes air movement in which most people are comfortable.

Common failure mode (nuclear power): An event causing redundant equipment to be inoperable.

Commutator: Device used on electric motors or generators to maintain a unidirectional current.

Compax® die: A wire drawing die made by GE of sintered diamond.

Comples number: A mathematical expression (a + bi) in which 'a' and 'b' are real, i2 = 1; useful in analyzing certain vectors, especially of electrical fields: j is substituted for i in formuale for electrical circuits—i,e. (a + jb).

Compound fill: An insulation which is poured into place while hot.

Compressibility: A density ratio determined under finite testing conditions.

Compression lug or splice: Installed by compressing the connector onto the strand, hopefully into a cold weld.

Compressor: The pump of a refrigerating mechanism that draws a vacuum or low pressure on the cooling side of a refrigerant cycle and squeezes or compresses the gas into the high pressure or condensing side of the cycle.

Computer: An electronic apparatus: 1) For rapidly solving complex and involved problems, usually mathematical or logical. 2) For storing large amounts of data.

Concealed: Rendered inaccessible by the structure or finish of the building. Wires in concealed raceways are considered concealed, even though they may become accessible by withdrawing them.

Concentricity: The measurement of the center of the conductor with respect to the center of the insulation.

Conductance: The ability of material to carry an electric current.

Conductor: Any substance that allows energy flow through it with the transfer being made by physical contact but excluding net mass flow.

Conductor, bare: Having no covering or insulation whatsoever.

Conductor, covered: A conductor having one or more layers of nonconducting materials that are not recognized as insulation under the National Electric Code.

Conductor, insulated: A conductor covered with material recognized as insulation.

Conductor load: The mechanical loads on an aerial conductor—wind, weight, ice, etc.

Conductor, plain: Consisting of only one metal.

Conductor, segmental: Having sections isolated, one from the other and connected in parallel; used to reduce ac resistance.

Conductor, solid: A single wire.

Conductor, stranded: Assembly of several wires, usually twisted or braided.

Conductor stress control: The conducting layer applied to make the conductor a smooth surface in intimate contact with the insulation; formerly called extruded strand shield (ESS).

Conduit: A tubular raceway.

Conduit fill: Amount of cross-sectional area used in a raceway.

Conduit, rigid metal: Conduit made of Schedule 40 pipe, normally 10 foot lengths.

Configuration, cradled: The geometric pattern which cables will take in a conduit when the cables are pulled in parallel and the ratio of the conduit ID to the 1/C cable OD is greater than 3.0.

Configuration, triangular: The geometric pattern which cables will take in a conduit when the cables are triplexed or are pulled in parallel with the ratio of the conduit ID to the 1/C cable OD less than 2.5.

Connection: That part of a circuit which has negligible impedance and which joins components or devices.

Connection, delta: Interconnection of 3 electrical equipment windings in delta (triangular) fashion.

Connection (nuclear power): A cable terminal, splice or seal at the interface of the cable and equipment.

Connection, star: Interconnection of 3 electrical equipment windings in star (wye) fashion.

Connector: A device used to physically and electrically connect two or more conductors.

Connector, pressure: A connector applied using pressure to form a cold weld between the conductor and the connector.

Connector, reducing: Used to join two different size conductors.

Constant current: A type of power system in which the same amount of current flows through each utilization equipment, used for simplicity in street lighting circuits.

Constant voltage: The common type of power in which all loads are connected in parallel, different amounts of current flow thru each load.

Contact: A device designed for repetitive connections.

Contactor: A type of relay.

Containment: (nuclear power): The safety barrier designed to prevent the release of radioactive material in case of reactor accident.

Continuity: The state of being whole, unbroken.

Continuous load: 1) NEC—in operation three hours or more. 2) Nuclear power—8760 hours/year (scheduled maintenance outages permitted).

Continuous vulcanization (CV): A system utilizing heat and pressure to vulcanize insulation after extrusion onto wire or cable; the curing tube may be in a horizontal or vertical pole.

Control: Automatic or manual device used to stop, start, and/or regulate flow of gas, liquid, and/or electricity.

Control, temperature: A thermostatic device that automatically stops and starts a motor, the operation of which is based on temperature changes.

Controller: A device or group of devices that serves to govern in some predetermined manner the electric power delivered to the apparatus to which it is connected.

Convection: The transfer of heat to a fluid by conduction as the fluid moves past the heat source.

Cook annealer: An annealer using heavy electrical current through the conductor as the heat source.

Cook buncher: A buncher using controlled diameter and wire position.

Cooling tower: Device that cools water by water evaporation in air. Water is cooled to the wet bulb temperature of air.

Coordination: The selection of system components to prevent the failure of the whole system due to cascading; limiting system failure by activation of the fewest overcurrent devices, hopefully to one.

Co-polymer: A polymer having two "repeating units".

Copper: A word used by itself to refer to copper conductors. Examples: "A circuit of 500 MCM copper" or "the copper cost of the circuit." It is a good conductor of electricity, easily formed, easily connected to itself and other metals used in electrical wiring.

Copper Development Association: A manufacturer's group to promote the use of copper.

Copper, electrolytic: Copper of high purity, refined by electrolysis, used for electrical conductors.

Copper loss: The energy dissipated in the copper conductors of a circuit, due to heat loss of I^2R produced by current flow through the conductor. Term is sometimes used to refer to the same type of loss in aluminum circuit conductors.

Copperweld®: This is the trade name for a conductor composed of a steel core with a heavy copper coating. It is used where strength is needed such as large, long, vertical risers; most commonly used for ground rods.

Cord: A small flexible conductor assembly, usually jacketed.

Cord set: A cord having a wiring connector on one or more ends.

Core: The portion of a foundry mold that shapes the interior of a hollow casting.

Core (cable): The portion of an insulated cable under a protective covering.

Cored hole: At the time of casting, a sand core is placed in the mold so that the metal flows around it. When the casting is cold, the sand core is broken away, leaving the hole.

Corona: A low energy electrical discharge caused by ionization of a gas by an electric field.

Corrosion: The deterioration of a substance (usually a metal) because of a reaction with its environment.

Coulomb: The derived SI unit for quantity of electricity or electrical charge: One coulomb equals one ampere-second.

Counterbore: A tool that enlarges an already-machined round hole to a certain depth. The pilot of the tool fits in the smaller hole, and the larger part counterbores or makes the end of the hole larger.

Counter emf: The voltage opposing the applied voltage and the current in a coil; caused by a flow of current in the coil; also known as back emf.

Coupling: The means by which signals are transferred from one circuit to another.

Coupon: A piece of metal for testing, of specified size; a piece of metal from which a test specimen may be prepared.

CPE (chlorinated polyethylene): A plastic for jackets.

Cramp: Iron rod with ends bent to a right angle; used to hold blocks of stone together.

Crawl space: Shallow space between the first tier of beams and the ground (no basement).

Crazing: Fine cracks which may extend in a network on or under the surface of a material; usually occurs in the presence of an organic liquid or vapor.

Cross head: The mechanism on an extruder where the material is applied; it holds the die, guider and core tube; usually just called "head".

Cross-link: To cure by linking molecules together in a polymer—either by using chemical cross-linking agents or radiation.

Cross-linked polyethylene: Thermosetting polyethylene which has better physical properties than plain polyethylene: used as an insulation having good physical properties.

Cross talk: Undesired pickup of signals by a second circuit.

CRT: Cathode Ray Tube.

Crystal: A solid composed of atoms, ions, or molecules arranged in a pattern which is periodic in three dimensions.

CU: Copper.

Cure: To change the properties of a polymeric system into a more stable, usable condition by the use of heat, radiation, or reaction with chemical additives.

Current (I): The time rate of flow of electric charges; Unit: ampere.

Current, charging: The current needed to bring the cable up to voltage; determined by capacitance of the cable; after withdrawal of voltage, the charging current returns to the circuit; the charging current will be 90° out of phase with the voltage.

Current density: The current per unit cross-sectional area.

Current-induced: Current in a conductor due to the application of a time-varying electromagnetic field.

Current, leakage: That small amount of current which flows through insulation whenever a voltage is present and heats the insulation because of the insulation's resistance; the leakage current is in phase with the voltage, and is a power loss.

Current limiting: A characteristic of short-circuit protective devices, such as fuses, by which the device operates so fast on high short circuit currents that less than a quarter wave of the alternating cycle is permitted to flow before the circuit is opened, thereby limiting the thermal and magnetic energy to a certain maximum value, regardless of the current available.

Cut-in: The connection of electrical service to a building, from the power company line to the service equipment, e.g., "the building was cut-in" or "the power company cut-in the service.

Cut-in-card: The certificate of approval issued by the electrical inspection authority to the electrical contractor, to be given to the power company as evidence that the building electrical system is safe for connection or "cut-in" by the power company.

Cutout: A fuse holder which may be used to isolate part of a circuit.

Cutout box: A surface mounting enclosure with a cover equipped with swinging doors, used to enclose fuses.

Cutover: Changing from one reel to another without stopping the manufacturing process.

Cut resistance: The ability of a material to withstand mechanical pressure without rupture or becoming ineffective.

Cycle: 1) An interval of space or time in which one set of events or phenomena is completed. 2) A set of operations that are repeated regularly in the same sequence. 3) When a system in a given state goes thru a number of different processes and finally returns to its initial state.

Cyclic aging: A test on a closed loop of cable having voltage applied, and induced current applied in cycles to cause expansion and contraction; simulates cable operating in a dry environment.

Cyclotron: A device for accelerating positive ions by causing them to move in semicircular paths in a magnetic field.

Damper: Valve for controlling air flow.

Damping: The dissipation of energy with time or distance.

DBE (Design Basis Event) (nuclear power): Postulated abnormal events used to establish the performance requirements of the structures, systems and components.

Dead: 1) Not having electrical charge. 2) Not having voltage applied.

Dead-end: A mechanical terminating device on a building or pole to provide support at the end of an overhead electric circuit. A dead-end is also the term used to refer to the last pole in the pole line. The pole at which the electric circuiting is brought down the pole to go underground or to the building served.

Dead-front: A switchboard or panel or other electrical apparatus without "live" (energized) terminals or parts exposed on front where personnel might make contact.

Dead man: Reinforced concrete anchor set in earth and tied to the retaining wall for stability.

Deadman's switch: A switch necessitating a positive action by the operator to keep the system or equipment running or energized.

Debug: To examine or test a procedure, routine, or equipment for the purpose of detecting and correcting errors especially during start-up.

Decay (nuclear power): The transmutation of a nucleus to a stable energy condition.

Defeater: A means to deactivate a safety interlock system.

Defense in depth (nuclear power): A basic design philosophy to keep nuclear power plants safe during normal operations and the worst imagined accidents. There are 3 levels of defense: 1) Accident prevention, quality assurance, redundancy, inspection testing. 2) Protection devices and systems. 3) Safety systems to function in case 1 and 2 fail.

Deflection: Deviation of the central axis of a beam from normal when the beam is loaded.

Deflection plate: The part of a certain type of electron tube that provides an electrical field to produce deflection of an electron beam.

Demand: 1) The measure of the maximum load of a utility's customer over a short period of time. 2) The load integrated over a specified time interval.

Demand factor: For an electrical system or feeder circuit, this is a ratio of the amount of connected load (in kVA or amperes) which will be operating at the same time to the total amount of connected load on the circuit. An 80% demand factor, for instance, indicates that only 80% of the connected load on a circuit will ever be operating at the same time. Conductor capacity can be based on that amount of load.

Demonstration (nuclear power): A course of reasoning showing that a certain result is a consequence of assumed premises: an explanation or illustration.

Density: Closeness of texture or consistency.

Derating: The intentional reduction of stress/strength ratio in the application of a material; usually for the purpose of reducing the occurrence of a stress-related failure.

Derating factor: A factor used to reduce ampacity when the cable is used in environments other than the standard.

Detection: The process of obtaining the separation of the modulation, component from the received signal.

Device: An item intended to carry, or help carry, but not utilize electrical energy.

Dew point: The temperature at which vapor starts to condense (liquify) from a gas-vapor mixture at constant pressure.

Die: 1) Wire: a metal device having a conical hole which is used to reduce the diameter of wire which is drawn (pulled) through the die or series of dies. 2) Extruder: the fixed part of the mold.

Dielectric: An insulator or a term referring to the insulation between the plates of a capacitor.

Dielectric absorption: The storage of charges within an insulation; evidenced by the decrease of current flow after the application of dc voltage.

Dielectric dispersion: The change in relative capacitance due to change in frequency.

Dielectric heating: The heating of an insulating material by ac induced internal losses; normally frequencies above 10mHz are used.

Dielectric loss: The time rate at which electrical energy is transformed into heat in a dielectric when it is subjected to a changing electric field.

Dielectric phase angle: The phase angle between the sinusoidal ac voltage applied to a dielectric and the component of the current having the same period.

Dielectric strength: The maximum voltage which an insulation can withstand without breaking down; usually expressed as a gradient—vpm (volts per mil).

Diode: A device having two electrodes, the cathode and the plate or anode—used as a rectifier and detector.

Direct current (dc): 1) Electricity which flows only in one direction. 2) The type of electricity produced by a battery and dc generators.

Direction of lay: The lateral direction, designated as left-hand or right-hand, in which the elements of a cable run over the top of the cable as they recede from an observer looking along the axis of the cable.

Disconnect: A switch for disconnecting an electrical circuit or load (motor, transformer, panel) from the conductors which supply power to it, e. g., "He pulled the motor disconnect," meaning he opened the disconnect switch to the motor.

Disconnecting means: A device, a group of devices, or other means whereby the conductors of a circuit can be disconnected from their supply source.

Dispersion: Holding fine particles in suspension throughout a second substance.

Displacement current: An expression for the effective current flow across a capacitor.

Dissipation factor: Energy lost when voltage is applied across an insulation because of capacitance: the cotangent of the phase angle between voltage and current in a reactive component; because the shift is so great, we use the complement (angle) of the angle θ which is used for power factor; dissipation factor = tan = cot θ: is quite sensitive to contamination and deterioration of insulation: also known as power factor (of dielectrics).

Distortion: Unfaithful reproduction of signals.

Distribution, statistical analysis: A statistical method used to analyze data by correlating data to a theoretical curve to a) Test validity of data. b) Predict performance at conditions different from those used to produce the data: The normal distribution curve is most common.

Diversity factor: The ratio of the sum of load demands to a system demand.

DOAL: Diameter Overall.

DOC: Diameter Over Conductor: note that for cables having a stress control, the diameter over the stress control layer becomes DOC.

Documents (nuclear power): Pertaining to Class 1E equipment and systems.

DOI: Diameter Over Insulation.

DOJ: Diameter Over Jacket.

Donkey: A motor-driven power machine, on legs, used for threading and/or cutting conduit.

DOS: Diameter Over Insulation Shield.

DOSC: Diameter Over Stress Control.

Dose, radiation: The amount of energy per unit mass of material deposited at each point of an object undergoing radiation.

Double-strength glass: One-eighth inch thick sheet glass (single strength glass is 1/10 inch thick).

Draft indicator: An instrument used to indicate or measure chimney draft or combustion gas movement.

Drain wire: A bonding wire laid parallel to and touching shields.

Drawing: Reducing wire diameter by pulling through dies.

Drawing, block diagram: A simplified drawing of a system showing major items as blocks; normally used to show how the system works and the relationship between major items.

Drawing, line schematic (diagram): Shows how a circuit works.

Drawing, plot or layout: Shows the "floor plan."

Drawing, wiring diagram: Shows how the devices are interconnected.

Drill: A circular tool used for machining a hole.

Drip: Projecting horizontal course sloped outward to throw water away from a building.

Drip loop: An intentional sag placed in service entrance conductors where they connect to the utility service drop conductors on overhead services; the drop loop will conduct any rain water to a point lower than the service head, to prevent moisture being forced into the service conductors by hydrostatic pressure and running through the service head into the service conduit or cable.

Drum: The part of a cable reel on which the cable is wound.

Dry: Not normally subjected to moisture.

Dry bulb: An instrument with a sensitive element that measures ambient (moving) air temperature.

Drywall: Interior wall construction consisting of plasterboard, wood paneling, or plywood nailed directly to the studs without application of plaster.

Dual extrusion: Extruding two materials simultaneously using two extruders feeding a common cross head.

Duct: A tube or channel through which air is conveyed or moved.

Duct bank: Several underground conduits grouped together.

Ductility: The ability of a material to deform plastically before fracturing.

Dumbbell: A die-cut specimen of uniform thickness used for testing tensile and elongation of materials.

Durometer: An instrument to measure hardness of a rubber-like material.

Duty, continuous: A service requirement that demands operation at a substantially constant load for an indefinitely long time.

Duty, intermittent: A service requirement that demands operation for alternate intervals of load and no load, load and rest, or load, no load, and rest.

Duty, periodic: A type of intermittent duty in which the load conditions regularly reoccur.

Duty, short-time: A requirement of service that demands operations at loads and for intervals of time, both which may be subject to wide variation.

Dwarf partition: Partition that ends short of the ceiling.

Dwell: A planned delay in a timed control program.

Dynamic: A state in which one or more quantities exhibit appreciable change within an arbitrarily short time interval.

Dynamometer: A device for measuring power output or power input of a mechanism.

EC: Electrical Conductor of Aluminum.

Ecdentricity: 1) A measure of the entering of an item within a circular area. 2) The percentage ratio of the difference between the maximum and minimum thickness to the minimum thickness of an annular area.

ECCS (Emergency Core Cooling System): (nuclear power): A system to flood the fueled portion of the reactor and remove the residual heat produced by radioactive decay.

Eddy currents: Circulating currents induced in conducting materials by varying magnetic fields; usually considered undesirable because they represent loss of energy and cause heating.

Edison base: The standard screw base used for ordinary lamps.

EEI: Edison Electric Institute.

Effective temperature: Overall effect on a person of air temperature, humidity, and air movement.

Efficiency: The ratio of the output to the input.

Elasticity: That property of recovering original size and shape after deformation.

Elastomer: A material which, at room temperature, stretches under low stress to at least twice its length and snaps back to the original length upon release of stress.

Elbow: A short conduit which is bent.

Electric defrosting: Use of electric resistance heating coils to melt ice and frost off evaporators during defrosting.

Electric heating: House heating system in which heat from electrical resistance units is used to heat rooms.

Electric water valve: Solenoid type (electrically operated) valve used to turn water flow on and off.

Electricity: Relating to the flow or presence of charged particles; a fundamental physical force or energy.

Electrocution: Death caused by electrical current through the heart, usually in excess of 50 ma.

Electrode: A conductor through which current transfers to another material.

Electrolysis: The production of chemical changes by the passage of current from an electrode to an electrolyte or vice versa.

Electrolyte: A liquid or solid that conducts electricity by the flow of ions.

Electrolytic condenser-capacitor: Plate or surface capable of storing small electrical charges. Common electrolytic condensers are formed by rolling thin sheets of foil between insulating materials. Condenser capacity is expressed in microfarads.

Electromagnet: A device consisting of a ferromagnetic core and a coil that produces appreciable magnetic effects only when an electric current exists in the coil.

Electromotive force (emf) voltage: Electrical force that causes current (free electrons) to flow or move in an electrical circuit. Unit of measurement is the volt.

Electron: The subatomic particle that carries the unit negative charge of electricity.

Electronegative gas: A type of insulating gas used in pressure cables; such as SF6.

Electron emission: The release of electrons from the surface of a material into surrounding space due to heat, light, high voltage, or other causes.

Electronics: The science dealing with the development and application of devices and systems involving the flow of electrons in vacuum, gaseous media, and semi-conductors.

Electro-osmosis: The movement of fluids through diaphragms because of electric current.

Electroplating: Depositing a metal in an adherent form upon an object using electrolysis.

Electropneumatic: An electrically controlled pneumatic device.

Electrostatics: Electrical charges at rest in the frame of reference.

Electrotherapy: The use of electricity in treatment of disease.

Electrothermics: Direct transformations of electric and heat energy.

Electrotinning: Depositing tin on an object.

Elevation: Drawing showing the projection of a building on a vertical plane.

Elongation: 1) The fractional increase in length of a material stressed in tension. 2) The amount of stretch of a material in a given length before breaking.

EMA (Electrical Moisture Absorption): A water tank test during which the sample cables are subjected to voltage while the water is maintained at rated temperature; the immersion time is long, with the object being to accelerate failure due to moisture in the insulation; simulates buried cable.

EMI: Electromagnetic interference.

Emitter: The part of a transistor that emits electrons.

Emulsifying agent: A material that increases the stability of an emulsion.

Emulsion: The colloidal suspension of one liquid in another liquid, such as oil in water for lubrication.

Enameled wire: Wire insulated with a thin baked-on varnish enamel, used in coils to allow the maximum number of turns in a given space.

Enclosed: Surrounded by a case that will prevent anyone from accidentally touching live parts.

Energy: The ability to do work; such as heat, light, electrical, mechanical, etc.

Engine: An apparatus which converts heat to mechanical energy.

Environment: 1) The universe within which a system

must operate. 2) All the elements over which the designer has no control and that affect a system or its inputs.

EPA (Environmental Protection Agency): The federal regulatory agency responsible for keeping and improving the quality of our living environment—mainly air and water.

Epitaxial: A very significant thin film type of deposit for making certain devices in microcircuits involving a realignment of molecules.

EPRI (Electric Power Research Institute): An organization to develop and manage a technology for improving electric power production, distribution and utilization; sponsored by electric utilities.

Equilibrium: Properties are time constant.

Equipment: A general term including material, fittings, devices appliances, fixtures, apparatus, and the like used as part of, or in connection with, an electrical installation.

Equipotential: Having the same voltage at all points.

Equivalent circuit: An arrangement of circuit elements that has characteristics over a range of interest electrically equivalent to those of a different circuit or device.

ERDA (Energy Research & Development Administration): Federal agency (replacing part of AEC) for research and relating to energy—new sources, better efficiency, etc.

Erosion: Destruction by abrasive action of fluids.

Etching: Revealing structural details of a metal surface using chemical or electrolytic action.

ETL: Electrical Testing Laboratory.

Evaporation: A term applied to the changing of a liquid to a gas, heat is absorbed in this process.

Evaporator: Part of a refrigerating mechanism in which the refrigerant vaporizes and absorbs heat.

Excitation losses: Losses in a transformer or electrical machine because of voltage.

Excite: To initiate or develop a magnetic field.

Expansion bolt: Bolt with a casing arranged to wedge the bolt into a masonry wall to provide an anchor.

Expansion joint: Joint between two adjoining concrete members arranged to permit expansion and contraction with temperature changes.

Expansion, thermal: The fractional change in unit length per unit temperature change.

Expansion valve: A device in a refrigerating system that maintains a pressure difference between the high side and low side and is operated by pressure.

Explosion proof: Designed and constructed to withstand an internal explosion without creating an external explosion or fire.

Exponential: Pertaining to the mathematical expression; y = aebx.

Exposed (as applied to live parts): Live parts that a person could inadvertently touch or approach nearer than a safe distance. This term is applied to parts not suitably guarded, isolated, insulated.

Exposed (as applied to wiring method): Not concealed; externally operable; capable of being operated without exposing the operator to contact with live parts.

Extender: A substance added to a plastic to reduce the amount of the primary resin required per unit area.

Extraction: The transfer of a material from a substance to a liquid in contact with the substance.

Extrude: To form materials to a given cross section by forcing through a die.

Extruder types: a) Strip—uses strips of compound. b) Powder/pellet—uses compound in powder or pellet form.

Eyelet: Something used on printed circuit boards to make reliable connections from one side of the board to the other.

Facade: Main front of a building.

Face: An operation that machines the sides or ends of the piece.

Face brick: Brick selected for appearance in an exposed wall.

Face of a gear: That portion of the tooth curve above the pitch circle and measured across the rim of the gear from one end of the tooth to the other.

Facsimile: The remote reproduction of graphic material: an exact copy.

Factor of safety: Radio of ultimate strength of material to maximum permissible stress in use.

Factorial experiment: Having more than one factor as a controlled variable in one experiment; produces much data per experiment, but the results are complex to analyze.

Fail-safe control: A device that opens a circuit when the sensing element fails to operate.

Failure: Termination of the ability of an item to perform the required function.

Fan: A radial or axial flow device used for moving or producing artificial currents of air.

FAO: This symbol on a mechanical drawing means that the piece is machined or finished all over.

Farad: The basic unit of capacitance: 1 farad equals one coulomb per volt.

Fatigue: The weakening or breakdown of a material due to cyclic stress.

Fatigue strength: The maximum stress that can be sustained for a specified number of cycles without failure, the stress being completely reversed within each cycle unless otherwise stated.

Fault: An abnormal connection in a circuit.

Fault, arcing: A fault having high impedance causing arcing.

Fault, bolting: A fault of very low impedance.

Fault, ground: A fault to ground.

Feedback: The process of transferring energy from the output circuit of a device back to its input.

Feeder: A circuit, such as conductors in conduit or a busway run, which carries a large block of power from the service equipment to a sub-feeder panel or a branch circuit panel or to some point at which the block or power is broken down into smaller circuits.

Ferranti effect: When the voltage is greater than the source voltage in an ac cable or transmission line.

Fiber optics: Transmission of energy by light through glass fibers.

Fibrillation: A continued, uncoordinated activity in the fibers of the heart, diaphragm, or other muscles.

Fiddle: A small, hand-operated drill.

Fidelity: The degree to which a system accurately reproduces an input.

Field: The effect produced in surrounding space by an electrically charged object, by electrons in motion, or by a magnet.

Field, electrostatic: The region near a charged object.

Filament: A cathode in the form of a metal wire in an electron tube.

Filled tape: A fabric tape having interstices, but not necessarily the surface, filled with a compound to prevent wicking, improve strength, make conductive, etc.

Filler: A cheap and relatively inert substance added to plastic or rubber to make it less costly, and improve physical properties.

Filler, cable: Materials used to fill voids and spaces in a cable construction: normally to give a smooth outer configuration, and also may serve as flame retardants, etc.

Fillet: The rounded corner or portion that joins two surfaces which are at an angle to each other.

Film: A rectangular product having thickness of 0.010 inch thick or less.

Filter: A porous article through which a gas or liquid is passed to separate out matter in suspension; a circuit or devices that pass one frequency or frequency band while blocking others, or vice versa.

Final: The final inspection of an electrical installation, e.g., "The contractor got the final on the job."

Final tests: Those performed on the completed cable (after manufacturing.)

Fines: Fill material such as rocks having 1/8 inch as the largest dimension.

Finish plaster: Final or white coat of plaster.

Firebrick: Brick made to withstand high temperatures that is used for lining chimneys, incinerators, and similar structures.

Fireproof wood: Chemically treated wood; fire-resistive, used where incombustible materials are required.

Fire-rated doors: Doors designed to resist standard fire tests and labeled for identification.

Fire-resistance rating: The time in hours, the material or construction will withstand fire exposure as determined by certain standards.

Fire-shield cable: Material or devices to prevent fire spread between raceways.

Fire-stop: A barrier to prevent fire spread.

Fish: To fish wire or cable means to pull it through conduit, raceway or other confined spaces, like walls or ceilings.

Fish tape: A flexible metal tape for fishing through conduits or other raceway to pull in wires or cables; also made in non-metallic form of "rigid rope" for hand fishing of raceways.

Fission: (nuclear power): The splitting of an atom into two fragments, by bombarding its nucleus with particles releasing high kinetic energy and two or three neutrons along with radiation; the most important type of fission is that caused by neutrons because it can be self-sustaining due to chain reactions; the newly released neutrons can cause other fissions to occur.

Fitting: An accessory such as a locknut, bushing, or other part of a wiring system that is intended primarily to perform a mechanical rather than an electrical function.

Five hundred thousands: Referring to size of conductors by their MCM rating, e.g., "Two hundred and fifty thousands" is a number of 250 MCM conductors; "Twin three hundred thousands" is two conductors of 300,000 circular mil size.

Flag: A visual indicator for event happenings such as the activation and reclosing of an automatic circuit breaker.

Flame-retardant: 1) Does not support or convey flame. 2) An additive for rubber or plastic that enhances its flame resistance.

Flange: The circular disks on a reel to support the drum and keep the cable on the drum.

Flapper, valve: The type of valve used in refrigeration compressors that allow gaseous refrigerants to flow in only one direction.

Flashover: A momentary electrical interconnection around or over the surface of an insulator.

Flashpoint: A lowest temperature at which a combustible substance ignites in air when exposed to flame.

Flat: Of uniform thickness; eliminates the drops of beams and girders.

Flat wire: A rectangular wire having 0.188 inch thickness or less, 1¼ inch width or less.

Flemish bond: Pattern of bonding in brickwork consisting of alternate headers and stretchers in the same course.

Flex: Common term used to refer to flexible metallic conduit.

Flexural strength: The strength of a material in bending, expressed as the tensile stress of the outermost fibers of a bent test sample at the instant of failure.

Flitch beam: Built-up beam consisting of a steel plate sandwiched between wood members and bolted.

Floating: Not having a distinct reference level with respect to voltage measurements.

Float valve: Type of valve that is operated by a sphere or pan which floats on a liquid surface and controls the level of liquid.

Flooding: Act of filling a space with a liquid.

Flow meter: Instrument used to measure velocity or volume of fluid movement.

Flux: 1) The rate of flow of energy across or through a surface. 2) A substance used to promote or facilitate soldering or welding by removing surface oxides.

Foamed insulation: Insulation made sponge-like using foaming or blowing agents to create the cells.

Foil: Metal film.

Footing: Structural unit used to distribute loads to the bearing materials.

Forced convection: Movement of fluid by mechanical force such as fans or pumps.

Foundation: Composed of footings, piers, foundation walls (basement walls), and any special underground construction necessary to properly support the structure.

FPM: Feet per minute.

FR1: See VW1.

Frequency: The number of complete cycles an alternating electric current, sound wave, or vibrating object undergoes per second.

Friction pile: Pile with supporting capacity produced by friction with the soil in contact with the pile.

Friction tape: An insulating tape made of asphalt impregnated cloth; used on 600V cables.

Frost line: Deepest level below grade to which frost penetrates in a geographic area.

Fuel cell: A cell that can continually change chemical energy to electrical energy.

Full braid: One made of a single material as opposed to one of a mixture of materials.

Function: A quantity whose value depends upon the value of another quantity.

Furring: Thin wood, brick, or metal applied to joists, studs, or wall to form a level surface (as for attaching wallboard) or airspace.

Fuse: A protecting device which opens a circuit when the fusible element is severed by heating, due to overcurrent passing through. Rating: voltage, normal current, maximum let-thru current, time delay of interruption.

Fuse, dual element: A fuse having two fuse characteristics; the usual combination is having an overcurrent limit and a time delay before activation.

Fuse, nonrenewable or one-time: A fuse which must be replaced after it interrupts a circuit.

Fuse, renewable link: A fuse which may be reused after current interruption by replacing the meltable link.

Fusible plug: A plug or fitting made with a metal of a known low melting temperature; used as a safety device to release pressures in case of fire.

Fusion (nuclear power): This is the joining or fusing of two light nuclei such as those of hydrogen or helium; such fusion reaction is strongly exothermic and radiating. Nuclear fusion requires an external force to overcome the natural electrical repulsion of the two nuclei. Science is striving to develop power reactors in which a controlled fusion can take place. Unlimited explosive power is achieved by using a fission bomb to make the fusion bomb.

Gain: The ratio of output to input power, voltage, or current, respectively.

Galvanometer: An instrument for indicating or measuring a small electrical current by means of a mechanical motion derived from electromagnetic or dynamic forces.

Gambrel roof: Roof with its slope broken by an obtuse angle.

Garage: A building or portion of a building in which one or more self-propelled vehicles carrying volatile, flammable liquid for fuel or power are kept.

Garden bond: Bond formed by inserting headers at wide intervals.

Gas: Vapor phase or state of a substance.

Gas filled pipe cable: See pressure cable.

Gas pocket: A cavity caused by entrapped gas.

Gate: A device that makes an electronic circuit operable for a short time.

Gauge: 1) Dimension expressed in terms of a system of arbitrary reference numbers; dimensions expressed in decimals are preferred. 2) To measure.

Gem box: The most common rectangular outlet box used to hold wall switches and receptacle outlets installed recessed in walls; made in wide variety on constructions, 2 in. wide by 3 in. high by various depths, without clamps for conduit and with or without clamps for cable (armored for nonmetallic sheathed), single gang boxes which can be ganged together for more than one device.

Generator: 1) A rotating machine to convert from mechanical to electrical energy. 2) Automotive-mechanical to direct current. 3) General-apparatus, equipment, etc., to convert or change energy from one form to another.

Geometric factor: A parameter used and determined solely by the relative dimensions and configuration of the conductors and insulation of a cable.

Girder: A large beam made of wood, steel, or reinforced concrete.

Girt: Heavy timber framed into corner posts as support for the building.

Government anchor: A V-shaped anchor usually made of ½ inch round bars to secure the steel beam to masonry.

Grade beam: Horizontal, reinforced concrete beam between two supporting piers at or below ground supporting a wall or structure.

Graded insulation: Combining insulations in a manner to improve the electric field distribution across the combination.

Gradient: The rate of change of a variable magnitude.

Grain: An individual crystal in a polycrystalline metal or alloy.

Gray: The derived SI unit for absorbed radiation dose: one gray equals one joule per kilogram.

Greenfield: Another name used to refer to flexible metal conduit.

Grid: An electrode having one or more openings for the passage of electrons or ions.

Grid leak: A resistor of high ohmic value connected between the control grid and the cathode in a grid-leak capacitor detector circuit and used for automatic biasing.

Grillage: Steel framework in a foundation designed to spread a concentrated load over a wider area; generally enclosed in concrete.

Grille: An ornamental or louvered opening placed at the end of an air passageway.

Groined ceiling: Arched ceiling consisting of two intersecting curved planes.

Grommet: A plastic, metal or rubber doughnut-shaped protector for wires or tubing as they pass through a hole in an object.

Ground: A large conducting body (as the earth) used as a common return for an electric circuit and as an arbitrary zero of potential.

Ground check: A pilot wire in portable cables to monitor the grounding circuit.

Ground coil: A heat exchanger buried in the ground that may be used either as an evaporator or a condenser.

Grounded conductor: A system or circuit conductor that is intentionally grounded.

Grounded: Connected to earth.

Grounding: The device or conductor connected to ground designed to conduct only in abnormal conditions.

Grounding conductor: A conductor used to connect metal equipment enclosures and/or the system grounded conductor to a grounding electrode, such as the ground wire run to the water pipe at a service; also may be a bare or insulated conductor used to ground motor frames, panel boxes and other metal equipment enclosures used throughout an electrical system. In most conduit systems, the conduit is used as the ground conductor.

Ground Fault Interrupter (GFI): A protective device that detects abnormal current flowing to ground and then interrupts the circuit.

Grounds: Narrow strips of wood nailed to walls as guides to plastering and as a nailing base for interior trim.

Group ambient temperature: The no-load temperature of a cable group with all other cables or ducts loaded.

Guard: 1) A conductor situated to conduct interference to its source and prevent its influence upon the desired signal. 2) A mechanical barrier against physical contact.

Guider: The adjustable part of the mold of an extruder.

Gusset: A plate or bracket for strengthening an angle in framework.

Gutter: The space provided along the sides and at the top and bottom of enclosures for switches, panels, and other apparatus, to provide for arranging conductors which terminate at the lugs or terminals of the enclosed equipment. Gutter is also used to refer to a rectangular sheet metal enclosure with removable side, used for splicing and tapping wires at distribution centers and motor control layouts.

Guy: A tension wire connected to a tall structure and another fixed object to add strength to the structure.

H beam: Steel beam with wider flanges than an I beam.

Half hard: A relative measure of conductor temper.

Half lap, joint: Joint formed by cutting away half the thickness of each piece.

Half wave: Rectifying only half of a sinusoidal ac supply.

Half effect: The changing of current density in a conductor due to a magnetic field extraneous to the conductor.

Hand shake: Requiring mutual events prior to change.

Handhole: A small box in a raceway used to facilitate cable installation into which workmen reach but do not enter.

Handy box: The commonly used, single-gang outlet box used for surface mounting to enclose wall switches or receptacles, on concrete or cinder block construction of industrial and commercial buildings, non-gangable, also made for recessed mounting, also known as "utility boxes."

Hard drawn: A relative measure of temper; drawn to obtain maximum strength.

Hardness: Resistance to plastic deformation usually by deformation: stiffness or temper: resistance to scratching, abrasion or cutting.

Harmonic: An oscillation whose frequency is an integral multiple of the fundamental frequency.

Harness: A group of conductors laced or bundled in a given configuration, usually with many breakouts.

Hat: A special pallet for transporting long rubber strips or coils of wire; the pallets look like a hat.

Hazardous: Ignitable vapors, dust, or fibers that may cause fire or explosion.

HDP: High density polyethylene.

Header: Brick laid with an end exposed in the wall; wood beam set between two trimmers and carrying the tail beams.

Heat: A fundamental physical force or energy relating to temperature.

Heat dissipation: The flow of heat from a hot body to a cooler body by: 1) convection 2) radiation 3) conduction.

Heat exchanger: A device used to transfer heat from a warm or hot surface to a cold or cooler surface. Evaporators and condensers are heat exchangers.

Heat load: Amount of heat, measured in Btu, that is removed during a period of 24 hours.

Heat pump: A compression cycle system used to supply or remove heat to or from a temperature-controlled space.

Heat sink: A part used to absorb heat from another device.

Heat transfer: Movement of heat from one body or substance to another. Heat may be transferred by radiation, conduction, convection, or a combination of these.

Heat treatment: Heating and cooling a solid metal or alloy to obtain desired properties or conditions; excluding heating for hot work.

Heating valve: Amount of heat that may be obtained by burning a fuel; usually expressed in Btu per pound or gallon.

Heavy water (nuclear power): Heavy water, D^2O, contains deuterium which are hydrogen atoms having twice the ordinary mass.

Helix: The path followed when winding a wire or strip around a tube at a constant angle.

Henry: The derived SI unit for inductance: one henry equals one weber per ampere.

Hermetic motor: A motor designed to operate within refrigeration fluid.

Hertz: The derived SI unit for frequency: one hertz equals one cycle per second.

Hickey: 1) A conduit bending tool. 2) A box fitting for hanging lighting fixtures.

High-hat: A ceiling recessed incandescent lighting fixture of round cross-section, looking like a man's high hat in the shape of its construction.

High pressure cutout: Electrical control switch operated by the high side pressure that automatically opens an electrical circuit if too high head pressure or condensing pressure is reached.

High side: Parts of a refrigerating system that are under condensing or high side pressure.

Hi-pot test: A high-potential test in which equipment insulation is subjected to voltage level higher than that for which it is rated to find any weak spots or deficiencies in the insulation.

HMP, HMPE: High molecular weight polyethylene.

Hole: A mobile vacancy in the electron structure of semi-conductors that acts like a positive electron charge with mass.

Home run: That part of a branch circuit from the panelboard housing the branch circuit fuse or CB and the first junction box at which the branch circuit is spliced to lighting or receptacle devices or to conductors which continue the branch circuit to the next outlet or junction box. The term "home run" is usually reserved to multi-outlet lighting and appliance circuits.

Horsepower: The non-SI unit for power: 1 hp = 1 HP = 746 w (electric) = 9800 w (boiler).

Hot: Energized with electricity.

Hot dip: Coating by dipping into a molten bath.

Hot gas bypass: Piping system in a refrigerating unit that moves hot refrigerant gas from a condenser into the low pressure side.

Hot leg: A circuit conductor which normally operates at a voltage above ground; the phase wires or energized circuit wires other than a grounded neutral wire or grounded phase leg.

Hot junction: That part of the thermoelectric circuit which released heat.

Hot modulus: Stress at 100% elongation after 5 minutes of conditioning at a given temperature (normally 130° C).

Hot stick: A long insulated stick having a hook at one end which is used to open energized switches, etc.

Hot wire: A resistance wire in an electrical relay that expands when heated and contracts when cooled.

HTGR (Hi-Temp, Gas-Cooled Reactor) (nuclear power): A basic nuclear power fission reactor in which the reactor heats a gas; the gas exchanges its heat with a secondary loop to produce steam for the turbine.

Hub: 1) A fitting to attach threaded conduit to boxes. 2) The central part of a cylinder into which a shaft may be inserted. 3) A reference point used for overhead line layout.

Hum: Interference from ac power, normally of low frequency and audible.

Humidity: Moisture, dampness. Relative humidity is the ratio of the quantity of vapor present in the air to the greatest amount possible at a given temperature.

Hydrometer: Floating instrument used to measure specific gravity of a liquid. Specific gravity is the ratio of the density of a material to the density of a substance accepted as a standard.

Hydronic: Type of heating system that circulates a heated fluid, usually water, through baseboard coils. Circulation pump is usually controlled by a thermostat.

Hygrometer: An instrument used to measure the degree of moisture in the atmosphere.

Hygroscopic: Readily absorbing and retaining moisture.

Hypot®: Registered trade name by Associated Research, Inc. for their hi-pot tester.

Hysteresis: The time lag exhibited by a body in reacting to changes in forces affecting it; an internal friction.

Hypalon®: The Dupont trade name for chlorosulfonated polyethylene.

I beam: Rolled steel beam or built-up beam of an I section.

IACS: (International Annealed Copper Standard): Refined copper for electrical conductors: 100% conductivity at 20° C.

IBEW: International Brotherhood of Electrical Workers.

IC: Pronounced "eye see": Refers to interrupting capacity of any device required to break current (switch, circuit breaker, fuse, etc.), taken from the initial letters I and C, is the amount of current which the device can interrupt without damage to itself.

ID: Inside diameter.

Identified: Marked to be recognized as grounded.

IEC: International Electrochemical Commission.

IEEE: Institute of Electrical and Electronics Engineers.

IIR (Isobutylene Isoprene Rubber): Butyl synthetic rubber.

Ignition transformer: A transformer designed to provide a high voltage current.

Impedance: (A): The opposition to current flow in an ac circuit; impedance includes resistance (R), capacitive reactance (xc) and inductive reactance (XL); unit—ohm.

Impedance matching: Matching source and load impedance for optimum energy transfer with minimum distortion.

Impulse: A surge of unidirectional polarity.

Inching: Momentary activation of machinery used for inspection or maintenance.

Incombustible material: Material that will not ignite or actively support combustion in a surrounding temperature of 1200 degrees Fahrenheit during an exposure of five minutes; also, material that will not melt when the temperature of the material is maintained at 900 degrees Fahrenheit for at least five minutes.

Indoor: Not suitable for exposure to the weather.

Inductance: The creation of a voltage due to a time-varying current; the opposition to current change, causing current changes to lag behind voltage changes: Units— henry.

Induction heater: The heating of a conducting material in a varying electromagnetic field due to the material's internal losses.

Induction machine: An asynchronous ac machine to change phase or frequency by converting energy—from electrical to mechanical, then from mechanical to electrical.

Inductor: A device having winding(s) with or without a magnetic core for creating inductance in a circuit.

Infrared lamp: An electrical device that emits infrared rays—invisible rays just beyond red in the visible spectrum.

Infrared radiation: Radiant energy given off by heated bodies which transmits heat and will pass through glass.

Ink: The material used for legends and color coding.

In phase: The condition existing when waves pass through their maximum and minimum values of like polarity at the same instant.

Instantaneous value: The value of a variable quantity at a given instant.

Intrinsically safe: Incapable of releasing sufficient electrical or thermal energy under normal or abnormal conditions to cause ignition of a specific hazardous atmospheric mixture in its most ignitable concentration.

Instrument: A device for measuring the value of the quantity under observation.

Insulated: Separated from other conducting surfaces by a substance permanently offering a high resistance to the passage of energy through the substance.

Insulation, electrical: A medium in which it is possible to maintain an electrical field with little supply of energy from additional sources; the energy required to produce the electric field is fully recoverable only in a complete vacuum (the ideal dielectric) when the field or applied voltage is removed: used to a) save space b) enhance safety c) improve appearance.

Insulation, class rating: A temperature rating descriptive of classes of insulations for which various tests are made to distinguish the materials; not related necessarily to operating temperatures.

Insulation fall-in: The filling of strand interstices, especially the inner interstices, which may contribute to connection failures.

Insulation level (cable): The thickness of insulation for circuits having ground fault detectors which interrupt fault currents within 1) 1 minute = 100% level 2) 1 hour = 133% level 3) Over 1 hour = 173% level.

Insulation, thermal: Substance used to retard or slow the flow of heat through a wall or partition.

Intercalated tapes: Two or more tapes of different materials helically wound and overlapping on a cable to separate the materials.

Interface: 1) A shared boundary. 2) (nuclear power): a junction between Class 1E and other equipment of systems.

Interference: Extraneous signals or power which are undesired.

Integrated circuit: A circuit in which different types of devices such as resistors, capacitors, and transistors are made from a single piece of material and then connected to form a circuit.

Integrator: Any device producing an output proportionate to the integral of one variable with respect to a second variable; the second is usually time.

Interconnected system: Operating with two or more power systems connected thru tie lines.

Interlock: A safety device to insure that a piece of apparatus will not operate until certain conditions have been satisfied.

Interpolate: To estimate an intermediate between two values in a sequence.

Interrupting time: The sum of the opening time and arcing time of a circuit opening device.

Interstice: The space or void between assembled conductors and within the overall circumference of the assembly.

Inverter: An item which changes dc to ac.

Ion: An electrically charged atom or radical.

Ionization: 1) The process or the result of any process by which a neutral atom or molecule acquires charge. 2) A breakdown that occurs in gaseous parts of an insulation when the dielectric stress exceeds a critical value without initiating a complete breakdown of the insulation system; ionization is harmful to living tissue, and is detectable and measurable; may be evidenced by corona.

Ionization factor: This is the difference between percent dissipation factors at two specified values of electrical stress; the lower of the two stresses is usually so selected that the effect of the ionization on dissipation factor at this stress is negligible.

IPCEA (Insulated Power Cable Engineers Association): The association of cable manufacturing engineers who make nationally recognized specifications and tests for cables.

IR (Insulation resistance): The measurement of the dc resistance of insulating material; can be either volume of surface resistivity; extremely temperature sensitive.

IR drop: The voltage drop across a resistance due to the flow of current through the resistor.

IRK (Insulation dc resistance constant): A system to classify materials according to their resistance on a 1000 foot basis at 15.5°C (60°F).

Irradiation, atomic: Bombardment with a variety of subatomic particles; usually caused changes in physical properties.

ISO: International Organization for Standardization who have put together the "SI" units that are now the international standards for measuring.

Isolated: Not readily accessible to persons unless special means for access are used.

Isolating: Referring to switches, this means that the switch is not a loadbreak type and must only be opened when no current is flowing in the circuit. This term also refers to transformers (an isolating transformer) used to provide magnetic isolation of one circuit from another, thereby breaking a metallic conductive path between the circuits.

Isotope: Atoms of a given element, each having different mass from the other because of different quantities of subatomic particles in the nucleus; isotopes are useful because several are naturally radiating and thus can become radiation sources for medical treatments or researching labs; a common isotope is Cobalt 60.

I^2t: Relating to the heating effect of a current (amps-squared) for a specified time (seconds), under specified conditions.

Jack: A plug-in type terminal.

Jacket: A non-metallic polymeric close fitting protective covering over cable insulation; the cable may have one or more conductors.

Jacket, conducting: An electrically conducting polymeric covering over an insulation.

Jan: Joint army and navy specification.

Jamb: Upright member forming the side of a door or window opening.

Jamming: The wedging of a cable such that it can no longer be moved during installation.

JB: Pronounced "jay bee," refers to any junction box, taken from the initial letters J and B.

Joule: The derived SI unit for energy, work, quantity of heat: one joule equals one newton-metre.

Jumper: A short length of conductor, usually a temporary connection.

Junction: A connection of two or more conductors.

Junction box: Group of electrical terminals housed in a protective box or container.

Jute: A fibrous natural material used as filler or bedding.

ka: KiloAmpere.

kc: Kilocycle, use kiloHertz.

kcmil: 1,000 circular mils. 250 kcmil is a conductor with a cross-sectional area of 250,000 circular mils.

kelvin (K): The basic SI unit of temperature: 1/273.16 of thermodynamic temperature of the triple point of water.

Kel: Kilogram.

kHz: KiloHertz.

Kilogram (kg): The basic SI unit for mass; an arbitrary unit represented by an artifact kept in Paris, France.

Kilometer: A metric linear measurement: 1000 meters.

Kilowatt: Unit of electrical power equal to 1000 watts.

Kilowatt-ft: The product of load in kilowatts and the circuits distance over which a load is carried in feet; used to compute voltage drop.

Kinetic energy: Energy by virtue of motion.

Kirchoff's Laws: 1) The algebraic sum of the currents at any point in a circuit is zero. 2) The algebraic sum of the product of the current and the impedance in each conductor in a circuit is equal to the electromotive force in the circuit.

Knockout: A portion of an enclosure designed to be readily removed for installation of a raceway.

KO: Pronounced "kay oh," a knockout, the partially cut opening in boxes, panel cabinets and other enclosures, which can be easily knocked out with a screw driver and hammer to provide a clean hole for connecting conduit, cable or some fittings.

KVA: Kilovolts times Ampere.

LA: Lightning arrestor.

Lacquer: A protective coating or finish that dries to form a film by evaporation of a volatile constituent.

Lally column: Concrete-filled cylindrical steel structural column.

Lamp: A device to convert electrical energy to radiant energy, normally visible light: usually only 10-20% is converted to light.

Laminated core: An assembly of steel sheets for use as an element of magnetic circuit; the assembly has the property of reducing eddy-current losses.

Laminated wood: Wood built up of piles or laminations that have been joined either with glue or with mechanical fasteners. The piles usually are too thick to be classified as veneer, and the grain of all piles is parallel.

Labeled: Items having trademark of nationally recognized testing lab.

Lagging: The wood covering for a reel.

Lap: The relative position of applied tape edges; "closed butt lap"—tapes just touching: "open butt" or "negative lap"—tapes not touching: "positive lap" or "lap"—tapes overlapping.

Latent heat: Heat given off or absorbed in a process (as vaporization or fusion) other than a change in temperature.

Law of charges: Like charges repel, unlike charges attract.

Law of magnetism: Like poles repel, unlike poles attract.

Lay: The axial length of one turn of the helix of any element in a cable.

Lay direction: Direction of helical lay when viewed from the end of the cable.

Lay length: Distance along the axis for one turn of a helical element.

Lead (leed): A short connecting wire brought out from a device or apparatus.

Lead squeeze: The amount of compression of a cable by a lead sheath.

Leading: Applying a lead sheath.

Leakage: Undesirable conduction of current.

Leakage distance: The shortest distance along an insulation surface between conductors.

Leg: A portion of a circuit.

Legend, embossed: Molded letters and numbers in the jacket surface, letters may be raised or embedded.

Lenz' Law: "In all cases the induced current is in such a direction as to oppose the motion which generates it."

Lighting outlet: An outlet intended for the direct connection of a lamp holder, lighting fixture, or pendant cord terminating or a lamp holder.

Lightning arrestor: A device designed to protect circuits and apparatus from high transient voltage by diverting the over-voltage to ground.

LIM (Laboratory Inspection Manual): A summary of specified values for quality testing.

Limit: A boundary of a controlled variable.

Limit control: Control used to open or close electrical circuits as temperature or pressure limits are reached.

Limiter: A device in which some characteristic of the output is automatically prevented from exceeding in predetermined value.

Line: A circuit between two points: ropes used during overhead construction.

Line, bull: A rope for large loads.

Line, finger: A rope attached to a device on a pole when a device is hung, so further conductor installation can be done from the ground.

Line, pilot: A small rope strung first.

Line, tag: A rope to guide devices being hoisted.

Linear: Arranged in a line.

Linearity: When the effect is directly proportional to the cause.

Liquid absorbent: A chemical in liquid form that has the property to absorb moisture.

Liquid line: The tube that carries liquid refrigerant from the condenser or liquid receiver to the refrigerant control mechanism.

Liquid receiver: Cylinder connected to a condenser outlet for storage of liquid refrigerant in a system.

Lissajous Figure: A special case of an s-y plot in which the signals applied to both axes are sinusoidal functions: useful for determining phase and harmonic relationships.

Listed: Items in a list published by a nationally recognized independent lab that makes periodic tests.

Liter: Metric unit of volume that equals 61.03 cubic inches.

Live: Energized.

Live-front: Any panel or other switching and protection assembly, such as switchboard or motor control center, which has exposed electrically energized parts on its front, presenting the possibility of contact by personnel.

Live load: Any load on a structure other than a dead load includes the weight of persons occupying the building and freestanding material.

LMFBR (liquid metal, fast-breeder reactor) (nuclear power): A basic nuclear power fission reactor in which a metal (sodium) is heated by the reactor: this heats a second metal loop which transfer its heat to a third loop to produce steam for the turbine: this reactor can also produce other nuclear fuel at the same time.

LMP: Low Molecular Weight Polyethylene.

Load: 1) A device that receives power. 2) The power delivered to such a device.

Load-break: Referring to switches or other control devices, this phrase means that the device is capable of safely interrupting load current—to distinguish such devices from other disconnect devices which are not rated for breaking load current and must be opened only after the load current has been broken by some other switching device.

Load center: An assembly of circuit breakers or switches.

Load factor: The ratio of the average to the peak load over a period.

Load losses: Those losses incidental to providing power.

LOCA (Loss of Coolant Accident) (nuclear power): The test to simulate nuclear reactor accident exhibited by high radiation, high temperature, etc.

Location, damp: A location subject to a moderate amount of moisture such as some basements, barns, cold-storage, warehouses, and the like.

Location, dry: A location not normally subject to dampness or wetness; a location classified as dry may be temporarily subject to dampness or wetness.

Location, wet: A location subject to saturation with water or other liquids.

Lock seam: Joining of two sheets of metal consisting of a folded, pressed, and soldered joint.

Locked rotor: When the circuits of a motor are energized but the rotor is not turning.

Lockout: To keep a circuit locked open.

Logarithm: The exponent that indicates the power to which a number is raised to produce a given number.

Logarithmic: Pertaining to the function $y = \log x$.

Longwall machine (mining): A machine used to undercut coal at relatively long working spaces.

Looping-in: Avoiding splices by looping wire thru device connections instead of cutting the wire.

Loss: Power expended without doing useful work.

Lug: A device for terminating a conductor to facilitate the mechanical connection.

Lumen: The derived SI unit for luminous flux.

Lus: The derived SI unit for illuminance.

LV: Low voltage.

LWBR (nuclear): Light water breeder reactor.

Machine: An item to transmit and modify force or motion, normally to do work.

Magnet: A body that produces a magnetic field external to itself; magnets attract iron particles.

Magnetic field: 1) A magnetic field is said to exist at a point if a force over and above any electrostatic force is exerted on a moving charge at the point. 2) The force field established by ac through a conductor, especially a coiled conductor.

Magnetic pole: Those portions of the magnet toward which the external magnetic induction appears to converge (south) or diverge (north).

Manhole: A subsurface chamber, large enough for a man, to facilitate cable installation splices, etc., in a duct bank.

Manual: Operated by mechanical force applied directly by personal intervention.

Marker: A tape or colored thread in a cable which identifies the cable manufacturer.

Mass: The property that determines the acceleration the body will have when acted upon by a given force: Unit = qkilogram.

Mat: A concrete base for heavy electrical apparatus, such as transformers, motors, generators, etc., sometimes the term includes the concrete base, a bed of crushed stone around it and an enclosing chain-link fence.

Matrix: A multi-dimensional array of items.

Matte surface: A surface from which reflection is predominately diffused.

MCM: An expression referring to conductors of sizes from 250 MCM which stands for Thousand Circular Mils up to 2000 MCM.

Mean: An intermediate value: arithmetic—sum of values divided by the quantity of the values: the average.

Mechanical water absorption: A check of how much water will be absorbed by material in warm water for seven days (mg/sq. in. surface).

Medium hard: A relative measure of conductor temper.

Megger: The term used to identify a test instrument for measuring the insulation resistance of conductors and other electrical equipment; specifically, a megohm (million ohms) meter; but this is a registered trade name of the James Biddle Co.

Megohmmeter: An instrument for measuring extremely high resistance.

Melt index: The extrusion rate of a material through a specified orifice at specified conditions.

Melting time: That time required for an overcurrent to sever a fuse.

Messenger: The supporting member of an aerial cable.

Metal clad (MC): The cable core is enclosed in a flexible metal covering.

Metal-clad switchgear: Switchgear having each power circuit device in its own metal enclosed compartment.

Meter: An instrument designed to measure; metric unit of linear measurement equal to 39.37 inches.

Meter pan: A shallow metal enclosure with a round opening, through which a kilowatt hour meter is mounted, as the usual meter for measuring the amount of energy consumed by a particular building or other electrical system.

MFT: Thousands of feet.

Mho: Reciprocal of ohm.

MI cable: Mineral insulated, metal sheathed cable.

Mica: A silicate which separates into layers and has high insulation resistance, dielectric strength and heat resistance.

Micrometer (mike): A tool for measuring linear dimensions accurately to 0.001 inch or to 0.01 mm.

Micro structure: The structure of polished and etched metals as revealed by a microscope at a magnification of more than ten diameters.

Microwave: Radio waves of frequencies above one gigahertz.

Mil: A unit used in measuring the diameter of wire, equal to 0.001 inch (25.4 micrometers).

MIL: Military specification.

Mil scale: The heavy oxide layer formed during hot fabrication or heat treatment of metals.

mm: Millimeter: 1 meter ÷ 1000.

MM: Mining machine.

Minimum average: The specified average insulation or jacket thickness.

Minimum at a point: Specifications that permit the thickness at one point to be less than the average.

Mks: Meter, kilogram, second.

Modem: Equipment that connects data transmitting/

receiving equipment to telephone lines: a word contraction of modulator-demodulator.

Modulation: The varying of a "carrier" wave characteristic by a characteristic of a second "modulating" wave.

Modulus of electricity: The ratio of stress (force) to strain (deformation) in a material that is elastically deformed.

Moisture-repellent: So constructed or treated that moisture will not penetrate.

Moisture-resistance: So constructed or treated that moisture will not readily injure.

Molded case breaker: A circuit breaker enclosed in an insulating housing.

Mole (mol): The basic SI unit for amount of substance. One mole is the amount of substance of a system that contains as many elementary entities as there are atoms in 0.012 kilograms of carbon 12.

Molecule: The group of atoms which constitutes the smallest particle in which a compound or material can exist separately.

Motor: An apparatus to convert from electrical to mechanical energy.

Motor, capacitor: A single-phase induction motor with an auxiliary starting winding connected in series with a condenser for better starting characteristics.

Motor control: Device to start and/or stop a motor at certain temperature or pressure conditions.

Motor control center: A grouping of motor controls such as starters.

Motor effect: Movement of adjacent conductors by magnetic forces due to currents in the conductors.

Mouse: Any weighted line used for dropping down between finished walls to attach to cable to pull the cable up; a type of vertical fishing between walls.

MPT: Male pipe thread.

MPX: Multiplexer.

MTW: Machine tool wire.

Multioutlet assembly: A type of surface or flush raceway designed to hold conductors and attachment plug receptacles.

Multiple barrier protection (nuclear power): The keeping of radioactive fission products from the public by placing multiple barriers around the reactor: for BWR these are fuel cladding, the reactor vessel, and the containment building.

Multiplex: To interleave or simultaneously transmit two or more messages on a single channel.

Mutual inductance: The condition of voltage in a second conductor because of a change in current in another adjacent conductor.

Mw: Megawatt: 106 watts.

Mylar®: DuPont trade name for a polyester film whose generic name is oriented polyethylene terephthalate; used for insulation, binding tapes.

N/A: 1) not available 2) not applicable.

National Electrical Code (NEC): A national consensus standard for the installation of electrical systems.

Natural convection: Movement of a fluid or air caused by temperature change.

NBR: Nitrite-butadiene rubber: synthetic rubber.

NBS: National Bureau of Standards.

NC: Normally closed.

Negative: Connected to the negative terminal of a power supply.

NEMA: National Electrical Manufacturers Association.

Neoprene: An oil resistant synthetic rubber used for jackets; originally a DuPont trade name, now a generic term for polychloroprene.

Network: An aggregation of interconnected conductors consisting of feeders, mains, and services.

Network limiter: A current limiting fuse for protecting a single conductor.

Neutral: The element of a circuit from which other voltages are referenced with respect to magnitude and time displacement in steady state conditions.

Neutral block: The neutral terminal block in a panelboard, meter enclosure, gutter or other enclosure in which circuit conductors are terminated or subdivided.

Neutral wire: A circuit conductor which is common to the other conductors of the circuit, having the same voltage between it and each of the other circuit wires and usually operating grounded; such as the neutral of 3 wire, single-phase, or 3-phase 4-wire wye systems.

Neutron: Subatomic particle contained in the nucleus of an atom: electrical neutral.

Newton: The derived SI unit for force: the force which will give one kilogram mass an acceleration of one meter per second.

NFPA (National Fire Protection Association): An organization to promote the science and improve the methods of fire protection which sponsors various codes, including the National Code.

Nineteen hundred box: A commonly used term to refer to any 2-gang 4-inch square outlet box used for two wiring devices or for one wiring device with a single-gang cover where the number of wires requires this box capacity.

Nipple: A threaded pipe or conduit of less than two feet length.

NO: Normally open.

Node: A junction of two or more branches of a network.

Nominal: Relating to a designated size that may vary from the actual.

Nominal rating: The maximum constant load which may be increased for a specified amount for two hours without exceeding temperature limits specified from the previous steady state temperature conditions: usually 25 or 50 percent increase is used.

Nomograph: A chart or diagram with which equations

can be solved graphically by placing a straightedge on two known values and reading where the straightedge crosses the scale of the unknown value.

Nonautomatic: Used to describe an action requiring personal intervention for its control.

Noncode installation: A system installed where there are no local, state, or national codes in force.

Normal charge: The thermal element charge that is part liquid and part gas under all operating conditions.

NPT: National tapered pipe thread.

NR: 1) Nonreturnable reel; a reel designed for one-time use only. 2) Natural rubber.

NRC (Nuclear Regulatory Commission): The federal agency for atomic element usage; formerly AEC.

NSD (neutral supported drop): A type of service cable.

Nylon®: This is the DuPont trade name for polyhexamethylene-adipamide which is the thermoplastic used as insulation and jacketing material.

OC: Overcurrent.

OD: Outside diameter.

OF: Oxygen free.

Offgassing: Percentage of a gas released during combustion.

Ohm: The derived SI unit for electrical resistance or impedance: one ohm equals one volt per ampere.

Ohmmeter: An instrument for measuring resistance in ohms.

Ohm's Law: Mathematical relationship between voltage, current, and resistance in an electric circuit.

OI (Official Interpretation): An interpretation of the National Electrical Code made to help resolve a specific problem between an inspector and an installer.

Oil-: The prefix designating the operation of a device submerged in oil to cool or quench or insulate.

Oil-proof: The accumulation of oil or vapors will not prevent safe successful operation.

Oil-tight: Construction preventing the entrance of oil or vapors not under pressure.

OL (nuclear): Operating License.

OL: Overload.

Open: A circuit which is energized by not allowing useful current to flow.

Opening time: The period between which an activation signal is initiated and switch contacts part.

Optimization: The procedure used in the design of a system to maximize or minimize some performance index.

Organic: Matter originating in plant or animal life, or composed of chemicals of that origin.

Orifice: Accurate size opening for controlling fluid flow.

Oscillation: The variation, usually with time, of the magnitude of a quantity which is alternately greater and smaller than a reference.

Oscillator: A device that produces an alternating or pulsating current or voltage electronically.

Oscillograph: An instrument primarily for producing a graph of rapidly varying electrical quantities.

Oscilloscope: An instrument primarily for making visible rapidly varying electrical quantities: oscilloscopes function similarly to TV sets.

OSHA (Occupational Safety and Health Act): Federal Law #91-596 of 1970 charging all employers engaged in business affecting interstate commerce to be responsible for providing a safe working place: it is administered by the Department of Labor: the OSHA regulations are published in Title 29, Chapter XVII, Part 1910 of the CFR and the Federal Register.

Osmosis: The diffusion of fluids through membranes.

Ought sizes: An expression referring to conductors of sizes No. 1/0, 2/0, 3/0 or 4/0.

Outage: When a component is not available to perform its intended function.

Outdoor: Designed for use out-of-doors.

Outgassing: Dissipation of gas from a material.

Outlet: A point on the wiring system at which current is taken to supply utilization equipment.

Outline lighting: An arrangement of incandescent lamps or gaseous tubes to outline and call attention to certain features, such as the shape of a building or the decoration of a window.

Output: 1) The energy delivered by a circuit or device. 2) The terminals for such delivery.

Oven: An enclosure and associated sensors and heaters for maintaining components at a controlled and usually constant temperature.

Overcurrent protection: De-energizing a circuit whenever the current exceeds a predetermined value; the usual devices are fuses, circuit breakers, or magnetic relays.

Overload: Load greater than the load for which the system or mechanism was intended.

Overvoltage (cable): Voltages above normal operating voltages, usually due to: a) switching loads on/off. b) lighting. c) single phasing.

Oxidize: 1) To combine with oxygen. 2) To remove one or more electrons. 3) To dehydrogenate.

Oxygen bomb test: Aging a rubber sample in a pressure container in a pure oxygen environment: not deteriorated by ozone.

Oxygen index: A test to rate flammability of materials in a mixture of oxygen and nitrogen.

Ozone: An active molecule of oxygen which may attack insulation: produced by corona in air.

Ozone resistance tests: A test for rubber, not used now because the synthetic rubbers do not deteriorate from ozone.

P102: A cable legend signifying acceptance for listing by Pennsylvanic Department of Mines (POM).

Pad: A coil of tape.

Pad-Mounted: A shortened expression for "pad-mount transformer," which is completely enclosed transformer mounted outdoors on a concrete pad, without need for a surrounding chain-link fence around the metal, box-like transformer enclosure.

Pan: A sheet metal enclosure for a watt-hour meter, commonly called a "meter pan."

Panel: A unit for one or more sections of flat material suitable for mounting electrical devices.

Panelboard: A single panel or group of panel units designed for assembly in the form of a single panel; includes buses and may come with or without switches and/or automatic overcurrent protective devices for the control of light, heat, or power circuits of individual as well as aggregate capacity. It is designed to be placed in a cabinet or cutout box that is in or against a wall or partition and is accessible only from the front.

Paper-lead cable: One having oil impregnated paper insulation and a lead sheath.

Parallel: Connections of two or more devices between the same two terminals of a circuit.

Parameter: A variable given a constant value for a specific process or purpose.

Pascal: The derived SI unit for pressure or stress: one pascal equals one newton per square meter.

Patch: To connect circuits together temporarily.

Payoff: The equipment to guide the feeding of wire.

PB: Pushbutton; pull box.

PE: 1) Polyethylene 2) Professional engineer.

Peak value: The largest instantaneous value of a variable.

Penciling: The tapering of insulation to relieve electrical stress at a splice or termination.

Penetration (nuclear): A fitting to seal pipes or cables thru the containment vessel wall.

Pentode: An electron tube with five electrodes or elements.

Period: The minimum interval during which the same characteristics of a periodic phenomenon recur.

Permalloy: An alloy of nickel and iron that is easily magnetized and demagnetized.

Permeability: 1) The passage or diffusion of a vapor, liquid or solid through a barrier without physically or chemically affecting either. 2) The rate of such passage.

Permissible mine equipment: Equipment that has been formally accepted by the Federal agency.

Per-unit quantity: The ratio of the actual value to an arbitrary base value of a quantity.

PES: Power Engineering Society of IEEE.

PF: Power factor.

pH: An expression of the degree of acidity or alkalinity of a substance: on the scale of 1-10, acid is under 7, neutral is 7, alkaline is over 7.

Phase: The fractional part "t/p" of the period through which a quantity has advanced, relative to an arbitrary origin.

Phase angle: The measure of the progression of a periodic wave in time or space from a chosen instant or position.

Phase conductor: The conductors other than the neutral.

Phase leg: One of the phase conductors (an ungrounded or "hot" conductor) of a polyphase electrical system.

Phase out: A procedure by which the individual phases of a polyphase circuit or system are identified; such as to "phase out" a 3-phase circuit for a motor in order to identify phase A, phase B and phase C to know how to connect them to the motor to get the correct phase rotation so the motor will rotate in the desired direction.

Phase sequence: The order in which the successive members of a periodic wave set reach their positive maximum values: a) zero phase sequence—no phase shift: b) plus/minus phase sequence—normal phase shift.

Phase shift: The absolute magnitude of the difference between two phase angles.

Phasor quantity: A complex algebraic expression for sinusoidal wave.

Photocell: A device in which the current-voltage characteristic is a function of incident radiation (light).

Photoelectric control: A control sensitive to incident light.

Photoelectricity: A physical action wherein an electrical flow is generated by light waves.

Photon: An elementary quantity of radiant energy (quantum).

Pi ***(π):*** The ratio of the circumference of a circle to its diameter.

Pick: The grouping or band of parallel threads in a braid.

Pickle: A solution or process to loosen or remove corrosion products from a metal.

Pickup value: The minimum input that will cause a device to complete a designated action.

Picocoulomb: 10-12 coulombs.

Piezoelectric effect: Some materials become electrically polarized when they are mechanically strained: the direction and magnitude of the polarization depends upon the nature, amount and the direction of the strain: in such materials the reversal is also true in that a strain results from the application of an electric field.

Pigtail: A flexible conductor attached to an apparatus for connection to a circuit.

PILC cable: Paper insulated, lead covered.

Pilot lamp: A lamp that indicadtes the condition of an associated circuit.

Pilot wire: An auxiliary insulated conductor in a power cable used for control or data.

Pitch diameter: The diameter through the center of a layer in a concentric layup of a cable or strand.

Pitting: Small cavities in a metal surface.

Plasma: A gas made up of charged particles.

Plastic: Pliable and capable of being shaped by pressure.

Plastic deformation: Permanent change in dimensions of an object under load.

Plasticizer: Agent added to plastic to improve flow and processability and to reduce brittleness.

Plate: A rectangular product having thickness of 0.25 inch or more.

Plating: Forming an adherent layer of metal on an object.

Plenum: Chamber or space forming a part of an air conditioning system.

Plowing: Burying cable in a split in the earth made by a blade.

Plug: A male connector for insertion into an outlet or jack.

Plugging: Braking an induction motor by reversing the phase sequence of the power to the motor.

Polarity: 1) Distinguishing one conductor or terminal from another. 2) Identifying how devices are to be connected such as + or -.

Polarization Index: Ratio of insulation resistance measured after 10 minutes to the measure at 1 minute with voltage continuously applied.

Pole: 1) That portion of a device associated exclusively with one electrically separated conducting path of the main circuit of device. 2) A supporting circular column.

Poly: Polyethylene.

Polychloroprene: Generic name for neoprene.

Polycrystalline: Pertaining to a solid having many crystals.

Polyethylene: A thermoplastic insulation having excellent electrical properties, good chemical resistance (useful as jacketing), good mechanical properties with the exception of temperature rating.

Polymer: A high-molecular-weight compound whose structure can usually be represented by a repeated small unit.

Polyphase circuits: ac circuits having two or more interrelated voltages, usually of equal amplitudes, phase differences, and periods, etc: if a neutral conductor exists, the voltages referenced to the neutral are equal in amplitude and phase: the most common version is that of 3-phase, equal in amplitude with phases 120° apart.

Polypropylene: A thermoplastic insulation similar to polyethylene, but with slightly better properties.

Polytetrafluoroethylene (PTFE): A thermally stable (-90 to + 250°C) insulation having good electrical and physical properties even at high frequencies.

Porcelain: Ceramic chinalike coating applied to steel surfaces.

Portable: Designed to be movable from one place to another, not necessarily while in operation.

Positive: Connected to the positive terminal of a power supply.

Potential: The difference in voltage between two points of a circuit. Frequently, one is assumed to be ground (zero potential).

Potential energy: Energy of a body or system with respect to the position of the body or the arrangement of the particles of the system.

Potentiometer: An instrument for measuring an unknown voltage or potential difference by balancing it, wholly or in part, by a known potential difference produced by the flow of known currents in a network of circuits of known electrical constants.

Pothead: A terminator for high-voltage circuit conductor to keep moisture out of the insulation and to protect the cable end, along with providing a suitable stress relief cone for shielded-type conductors.

Power: 1) Work per unit of time. 2) The time rate of transferring energy: as an adjective, the word "power" is descriptive of the energy used to perform useful work: pound-feet per second, watts.

Power, active: In a 3-phase symmetrical circuit: $p = 3$ vi cos θ; in a 1-phase, 2 wire circuit, $p =$ vi cos θ.

Power, apparent: The product of rms volts times rms amperes.

Power element: Sensitive element of a temperature-operated control.

Power factor: Correction coefficient for ac power necessary because of changing current and voltage values.

Power loss (cable): Losses due to internal cable impedance, mainly I^2R: the losses cause heating.

Power pool: A group of power systems operating as an interconnected system.

P-P: Peak to peak.

Precast concrete: Concrete units (such as piles or vaults) cast away from the construction site and set in place.

Precious metal: Gold, silver, or platinum.

Premolded: A splice or termination manufactured of polymers, ready for field application.

Pressure: An energy impact on a unit area; force or thrust exerted on a surface.

Pressure motor control: A device that opens and closes an electrical circuit as pressures change.

Primary: Normally referring to the part of a device or equipment connected to the power supply circuit.

Primary control: Device that directly controls operation of heating, air-conditioning, ventilation and similar systems.

Printed circuit: A board having interconnecting wiring printed on its surface and designed for mounting of electronic components.

Process: Path of succession of states through which a system passes.

Program, computer: The ordered listing of sequence of events designed to direct the computer to accomplish a task.

Propagation: The travel of waves through or along a medium.

Property: An observable characteristic.

Protected enclosure: Having all openings protected with screening, etc.

Protector, circuit: An electrical device that will open an electrical circuit if excessive electrical conditions occur.

Proton: The hydrogen atom nucleus; it is electrically positive.

Prototype: The first full size working model.

Proximity effect: The distortion of current density due to magnetic fields; increased by conductor diameter, close spacing, frequency, and magnetic materials such as steel conduit or beams.

PSAR (Preliminary Safety Analyses Report) (nuclear): Construction permit.

PSI: Pound force per square inch.

PT: Potential transformer.

Pull box: A sheet metal box-like enclosure used in conduit runs, either single conduits or multiple conduits, to facilitate pulling in of cables from point to point in long runs or to provide installation of conduit support bushings needed to support the weight of long riser cables or to provide for turns in multiple-conduit runs.

Pull-down: Localized reduction of conductor diameter by longitudinal stress.

Pulling compound (lubricant): A substance applied to the surface of a cable to reduce the coefficient of friction during installation.

Pulling eye: A device attached to a cable to facilitate field connection of pulling ropes.

Pulsating function: A periodic function whose average value over a period is not zero.

Pulse: A brief excursion of a quantity from normal.

Pumped storage (hydro power): The storage of power by pumping a reservoir full of water during off-peak, then depleting the water to generate when needed.

Puncture: Where breakdown occurs in an insulation.

Purge: To clean.

Push button: A switch activated by buttons.

PVC (polyvinyl chloride): A thermoplastic insulation and jacket compound.

PWR (pressurized water reactor-nuclear power): A basic nuclear fission reaction in which water is used to transfer energy from the reactor; the water exchanges its heat with a secondary loop to produce steam for the turbine.

Pyroconductivity: Electric conductivity that develops with changing temperature, and notably upon fusion, in solids that are practically non-conductive at atmospheric temperatures.

Pyrometer: Thermometer that measures the radiation from a heated body.

QA (Quality Assurance): All the planned and systematic actions to provide confidence that a structure, system, or component will perform satisfactorily.

QAP (nuclear): Quality assurance policy.

Quadruplexed: Twisting of four conductors together.

Qualified life (nuclear power): The period of time for which satisfactory performance can be demonstrated for a specific set of service conditions.

Qualified person: A person familiar with construction, operation and hazards.

Quick-: A device that has a high contact speed independent of the operator; example—quick-make or quick-break.

Raceway: Any channel designed expressly for holding wire, cables, or bars and used solely for that purpose.

Rack (cable): A device to support cables.

Radar: A radio detecting and ranging system.

Radial feeder: A feeder connected to a single source.

Radian (rad): A supplementary SI unit for plane angles; the plane angle with its vertex at the center of a circle that is subtended by an arc equal in length to the radius.

Radiant energy: Energy traveling in the form of electromagnetic waves.

Radiant heating: Heating system in which warm or hot surfaces are used to radiate heat into the space to be conditioned.

Radiation: The process of emitting radiant energy in the form of waves or particles.

Radiation, blackbody: Energy given off by an ideal radiating surface at any temperature.

Radiation, nuclear: The release of particles and rays during disintegration or decay of atom's nucleus; these rays cause ionization; they are: alpha particles, beta particles, gamma rays.

Radius, bending: The radii around which cables are pulled.

Radius, training: The radii to which cables are bent by hand positioning, not while the cables are under tension.

Rail clamp: A device to connect cable to a track rail.

Rainshield: An inverted funnel to increase the creepage over a stress cone.

Raintight: So constructed or protected that exposure to a beating rain will not result in the entrance of water.

Raked joint: Joint formed in brickwork by raking out some of the mortar an even distance from the face of the wall.

Ram: Random access memory.

Rated: Indicating the limits of operating characteristics for application under specified conditions.

Rating, temperature (cable): The highest conductor temperature attained in any part of the circuit during a) normal operation b) emergency overload c) shot circuit.

Rating, voltage: The thickness of insulation necessary to confine voltage to a cable conductor after withstanding the rigors of cable installation and normal operating environment.

REA (Rural Electrification Administration): A federally supported program to provide electrical utilities in rural areas.

Reactance: 1) The imaginary part of impedance. 2) The opposition to ac due to capacitance (Xc) and inductance (XL).

Reactor: A device to introduce capacitive or inductive reactance into a circuit.

Reactor, nuclear: An assembly designed for a sustained nuclear chain reaction; the chain reaction occurs when the mass of the fuel reaches a critical value having enough free neutrons or enough heat to sustain fusion.

Real time: The actual time during which a physical process transpires.

Reamer: A finishing tool that finishes a circular hole more accurately than a drill.

Receptacle: A contact device installed at an outlet for the connection of an attachment plug and flexible cord to supply portable equipment.

Reciprocating: Action in which the motion is back and forth in a straight line.

Recognized component: An item to be used as a subcomponent and tested for safety by UL; is UL's trademark for recognized component.

Recorder: A device that makes a permanent record, usually visual, of varying signals.

Rectifiers: Devices used to change alternating current to unidirectional current.

Rectify: To change from ac to dc.

Red-leg: That phase conductor of a 3-phase, 4-wire, delta-connected system that is not connected to the single-phase power supply; the conductor with the highest voltage above ground, which must be identified (per NEC) and is commonly painted red to provide such identification.

Redraw: Drawing of wire through consecutive dies.

Reducing joint: A splice of two different size conductors.

Reduction: The gain of electrons by a constituent of a chemical reaction.

Redundancy: The use of auxiliary items to perform the same functions for the purpose of improving reliability and safety.

Reel: A drum having flanges on the ends; reels are used for wire/cable storage.

Reflective insulation: Thin sheets of metal or foil on paper set in the exterior walls of a building to reflect radiant energy.

Refrigerant: Substance used in refrigerating mechanism to absorb heat in an evaporator coil and to release heat in a condenser as the substance goes from a gaseous state back to a liquid state.

Register: Combination grille and damper assembly covering on an air opening or end of an air duct.

Regulation: The maximum amount that a power supply output will change as a result of the specified change in line voltage, output load, temperature, or time.

Reinforced jacket: A cable jacket having reinforcing fiber between layers.

Relative capacitance: The ratio of a material's capacitance to that of a vacuum of the same configuration; will vary with frequency and temperature.

Relative humidity: Ratio of amount of water vapor present in air to greatest amount possible at same temperature.

Relay: A device designed to abruptly change a circuit because of a specified control input.

Relay, overcurrent: A relay designed to open a circuit when current in excess of a particular setting flows through the sensor.

Reliability: The probability that a device will function without failure over a specified time period or amount of usage.

Relief valve: Safety device to permit the escape of steam or hot water subjected to excessive pressures or temperatures.

Remote-control circuits: The control of a circuit through relays, etc.

Repeatability: The closeness of agreement among repeated measurements of the same variable under the same conditions.

Reproducibility: The ability of a system or element to maintain its output or input over a relatively long period of time.

Reservoir, thermal: A body to which and from which heat can be transferred indefinitely without a change in the temperature of the reservoir.

Residual elements: Elements present in an alloy in small quantities, but not added intentionally.

Residual stress: Stress present in a body that is free of external forces or thermal gradients.

Resin: The polymeric base of all jacketing, insulating, etc. compounds, both rubber and plastic.

Resistance: The opposition in a conductor to current; the real part of impedance.

Resistance furnace: A furnace which heats by the flow current against ohmic resistance internal to the furnace.

Resistance, thermal: The opposition to heat flow; for cables it is expressed by degrees centigrade per watt per foot of cable.

Resistance welding: Welding by pressure and heat when the work piece's resistance in an electric circuit produces the heat.

Resistivity: A material characteristic opposing the flow of energy through the material; expressed as a constant for each material: it is affected by temper, temperature, contamination, alloying, coating, etc.

Resistor: A device whose primary purpose is to introduce resistance.

Resistor, bleeder: 1) Used to drain current after a device is de-energized. 2) To improve voltage regulation. 3) To protect against voltage surges.

Resolution: The degree to which nearly equal values of a quantity can be discriminated.

Resolver: A device whose input and output is a vector quantity.

Resonance: In a circuit containing both inductance and capacitance.

Resonating: The maximizing or minimizing of the amplitude or other characteristics provided the maximum or minimum is of interest.

Response: A quantitative expression of the output as a function of the input under conditions that must be explicitly stated.

Restrike: A resumption of current between contacts during an opening operation after an interval of zero current of ¼ cycle at normal frequency or longer.

Reverse lay: Reversing the direction of lay about every five feet during cabling of aerial cable to facilitate field connections.

Reversible process: Can be reversed and leaves no change in system or surroundings.

RF: Radio frequency: 10kGz to GHz.

RFI: Radio frequency interference.

Rheology: The science of the flow and deformation of matter.

Rheostat: A variable resistor, which can be varied while energized, normally one used in a power circuit.

Ring-out: 1) A circular section of insulation or jacket. 2) The continuity testing of a conductor.

Ripple: The ac component from a dc power supply arising from sources within the power supply.

Riser; A vertical run of conductors in conduit or busway, for carrying electrical power from one level to another in a building.

Riser valve: Device used to manually control flow of refrigerant in vertical piping.

RMS (Root-mean-square): The square root of the average of the square of the function taken throughout the period.

Rock duster: A machine to distribute rock dust over coal to prevent dust explosions.

Rod: The shape of solidified metal convenient for wire drawing, usually 5⁄16 inch or larger.

Rolling: Reducing the cross-sectional area of metal stock or otherwise shaping metal products using rotating rolls.

ROM: Read only memory.

Romex: General Cable's trade name for type NM cable; but it is used generically by electrical men to refer to any nonmetallic sheathed cable.

Rope job: An installation of non-metallic sheathed cable.

Rotor: Rotating part of a mechanism.

Rough inspection: The first inspection made of an electrical installation after the conductors, boxes and other equipment have been installed in a building under construction.

Roughing in: The first stage of an electrical installation, when the raceway, cable, wires, boxes and other equipment are installed; that electrical work which must be done before any finishing or cover-up phases of building construction can be undertaken.

Round off: To delete the least significant digits of a numeral and adjust the remaining by given rules.

RPM: Revolutions per minute.

Rubber, chlorosulfonated polyethylene (CP): A synthetic rubber insulation and jacket compound developed by DuPont as Hypalon®.

Rubber, ethylene propylene: A synthetic rubber insulation having excellent electrical properties.

Running board: A device to permit stringing more than one conductor simultaneously.

Sacrificial protection: Prevention of corrosion by coupling a metal to an electrochemically more active metal which is sacrificed.

SAE: Society of Automotive Engineers.

Safety conductor: A safety sling used during overhead line construction.

Safety control: A device that will stop the refrigerating unit if unsafe pressures and/or temperatures are reached.

Safety factor: The ratio of the maximum stress which something can withstand to the estimated stress which it can withstand.

Safety motor control: Electrical device used to open a circuit if the temperature, pressure, and/or the current flow exceed safe conditions.

Safety plug: Device that will release the contents of a container above normal pressure conditions and before rupture pressures are reached.

Sag: The difference in elevation of a suspended conductor.

Sag, apparent: Sag between two points at 60°F and no wind.

Sag, final: Sag under specified conditions after the conductor has been externally loaded, then the load removed.

Sag, initial: Sag prior to external loading.

Sag, maximum: Sag at midpoint between two supports.

Sag section: Conductor between two snubs.

Sag snub: Where a conductor is held fixed and the other end moved to adjust sag.

Sag, total: Under ice loading.

Sampling: A small quantity taken as a sample for inspection or analysis.

Saturation: The condition existing in a circuit when an increase in the driving signal does not produce any further change in the resultant effect.

Scalar: A quantity (as mass or time) that has a magnitude described by a real number and no direction.

Scan: To examine sequentially, part by part.

Scavenger pump: Mechanism used to remove fluid from sump or containers.

Scintillation: The optical photons emitted as a result of the incidence of ionization radiation.

Scope: Slang for oscilloscope.

Scram (nuclear): The rapid shutdown of a nuclear reactor.

Screen pack: A metal screen used for straining.

SE: Service entrance.

Sealed: Preventing entrance.

Sealed motor compressor: A mechanical compressor consisting of a compressor and a motor, both of which are enclosed in the same sealed housing, with no external shaft or shaft seals, and with the motor operating in the refrigerant atmosphere.

Sealing compound: The material poured into an electrical fitting to seal and prevent the passage of vapors.

Secondary: The second circuit of a device or equipment, which is not normally connected to the supply circuit.

Seebeck Effect: The generation of a voltage by a temperature difference between the junctions in a circuit composed of two homogeneous electrical conductors of dissimilar composition: or in a nonhomogeneous conductor the voltage produced by a temperature gradient in a nonhomogeneous region.

Self inductance: Magnetic field induced in the conductor carrying the current.

Semiconductor: A material that has electrical properties of current flow between a conductor and an insulator.

Sensible heat: Heat that causes a change in temperature of a substance.

Sensor: A material or device that goes through a physical change or an electronic characteristic change as conditions change.

Separable insulated connector: An insulated device to facilitate power cable connections and separations.

Separator: Material used to maintain physical spacing between elements in cables, such as: a layer of tape to prevent jacket sticking to individual conductors.

Sequence controls: Devices that act in series or in time order.

Service: The equipment used to connect to the conductors run from the utility line, including metering, switching and protective devices; also the electric power delivered to the premises, rated in voltage and amperes, such as a "100-amp, 480 volt service."

Service cable: The service conductors made up in the form of a cable.

Service conductors: The supply conductors that extend from the street main or transformers to the service equipment of the premises being supplied.

Service drop: Run of cables from the power company's aerial power lines to the point of connection to a customer's premises.

Service entrance: The point at which power is supplied to a building, including the equipment used for this purpose (service main switch or panel or switchboard, metering devices, overcurrent protective devices, conductors for connecting to the power company's conductors and raceways for such conductors).

Service equipment: The necessary equipment, usually consisting of a circuit breaker or switch and fuses and their accessories, located near the point of entrance of supply conductors to a building and intended to constitute the main control and cutoff means for the supply to the building.

Service lateral: The underground service conductors between the street main, including any risers at a pole or other structure or from transformers, and the first point of connection to the service-entrance conductors in a terminal box, meter, or other enclosure with adequate space, inside or outside the building wall. Where there is no terminal box, meter, or other enclosure with adequate space, the point of connection is the entrance point of the service conductors into the building.

Service raceway: The rigid metal conduit, electrical metallic tubing, or other raceway that encloses the service-entrance conductors.

Service valve: A device to be attached to a system that provides an opening for gauges and/or charging lines.

Serving: A layer of helically applied material.

Servomechanism: A feedback control system in which at least one of the system signals represents mechanical motion.

Setting (of circuit breaker): The value of the current at which the circuit breaker is set to trip.

Shaded pole motor: A small dc motor used for light start loads that has no brushes or commutator.

Shaft furnace: A furnace used for pouring wire bars from continuous melting of cathodes.

Shall: Mandatory requirement of a Code.

Shaving: Removing about 0.001 inch of metal surface.

Shear: The lateral displacement in a body due to an external force causing sliding action.

Sheath: A metallic close fitting protective covering.

Shield: The conducting barrier against electromagnetic fields.

Shield, braid: A shield of interwoven small wires.

Shield, insulation: An electrically conducting layer to provide a smooth surface in intimate contact with the insulation outer surface; used to eliminate electrostatic charges external to the shield, and to provide a fixed known path to ground.

Shield, tape: The insulation shielding system whose current carrying component is thin metallic tapes, now normally used in conjunction with a conducting layer of tapes or extruded polymer.

Shim: Thin piece of material used to bring members to an even or level bearing.

Shore feeder: From ship to shore feeder.

Shore hardness: A measure of the hardness of a plastic.

Short-circuit: An often unintended low-resistance path through which current flows around, rather than through, a component or circuit.

Short cycling: Refrigerating system that starts and stops more frequently than it should.

Short time (nuclear power): Operation for 2 hours out of a 24-hour period.

Short-time overload rating (nuclear): The limiting overload current that one third (must be at least three) of the conductors in an assembly thru a penetration can carry, with all other conductors fully loaded.

Shrinkable tubing: A tubing which may be reduced in size by applying heat or solvents.

Shroud: Housing over a condenser or evaporator.

Shunt: A device having appreciable resistance or impedance connected in parallel across other devices or apparatus to divert some of the current: appreciable voltage exists across the shunt and appreciable current may exist in it.

Sidewall load: The normal force exerted on a cable under tension at a bend; quite often called sidewall pressure.

Signal: A detectable physical quantity or impulse (as a voltage, current, or magnetic field strength) by which messages or information can be transmitted.

Signal circuit: Any electrical circuit supplying energy to an appliance that gives a recognizable signal.

Silica gel: Chemical compound used as a drier.

Silicon controlled rectifier (SCR): Electronic semiconductor that contains silicon.

Sill: Horizontal timber forming the lowest member of a wood frame house; lowest member of a window frame.

Sine wave, ac: Wave form of single frequency alternating current; wave whose displacement is the sine of the angle proportional to time or distance.

Single-phase circuit: An ac circuit having one source voltage supplied over two conductors.

Single-phase motor: Electric motor that operates on single-phase alternating current.

Single-phasing: The abnormal operation of a three phase machine when its supply is changed by accidental opening of one conductor.

Sintering: Forming articles from fusible powders by pressing the powder just under its melting point.

Skin effect: The tendency of current to crowd toward the outer surface of a conductor: increases with conductor diameter and frequency.

Slip: The difference between the speed of a rotating magnetic field and the speed of its rotor.

Slippercoat: A surface lubricant factory applied to a cable to facilitate pulling, and to prevent jacket sticking.

Sliver: A defect consisting of a very thin elongated piece of metal attached by only one end to the parent metal into whose surface it has been rolled.

Slot: A channel opening in the stator or rotor of a rotating machine for ventilation and the insertion of windings.

Soap: Slang for pulling compound.

Soffit: Underside of a stair, arch, or cornice.

Soft drawn: 1) A relative measure of the tensile strength of a conductor. 2) Wire which has been annealed to remove the effects of cold working. 3) Drawn to a low tensile.

Solar cell: The direct conversion of electromagnetic radiation into electricity; certain combinations of transparent conducting films separated by thin layers of semiconducting materials.

Solar heat: Heat from visible and invisible energy waves from the sun.

Solder: To braze with tin alloy.

Solenoid: Electric conductor wound as a helix with a small pitch: coil.

Solidly grounded: No intentional impedance in the grounding circuit.

Solid state: A device, circuit, or system which does not depend upon physical movement of solids, liquids, gases or plasma.

Solid type PI cable: A pressure cable without constant pressure controls.

Soluble oil: Specially prepared oil whose water emulsion is used as a curing grinding or drawing lubricant.

Solution: 1) Homogenous mixture of two or more components. 2) Solving a problem.

SP: Single pole.

Space heater: A heater for occupied spaces.

Span: A conductor between two consecutive supports.

Spark: A brilliantly luminous flow of electricity of short duration that characterizes an electrical breakdown.

Spark gap: Any short air space between two conductors.

Spark test: A voltage withstand test on a cable while in production with the cable moving: it is a simple way to test long lengths of cable.

SPDT: Single pole double throw.

Specific heat: Ratio of the quantity of heat required to raise the temperature of a body 1 degree to that required to raise the temperature of an equal mass of water 1 degree.

Specs: Abbreviation for the word "specifications", which is the written precise description of the scope and details of an electrical installation and the equipment to be used in the system.

Spectrum: The distribution of the amplitude (and sometimes phase) of the components of the wave as a function of frequency.

Spike: A pulse having great magnitude.

Splice: The electrical and mechanical connection between two pieces of cable.

Splice tube: The movable section of vulcanizing tube at the extruder.

Split fitting: A conduit fitting which may be installed after the wires have been installed.

Split-phase motor: Motor with two stator windings. Winding in use while starting is disconnected by a centrifugal switch after the motor attains speed, then the motor operates on the other winding.

Split system: Refrigeration or air conditioning installation that places the condensing unit outside or remote from the evaporator. It is also applicable to heat pump installations.

Split-wire: A way of wiring a duplex plug outlet (a receptacle) with a 3-wire, 120/240 volt single-phase circuit so that one hot leg and the neutral feeds one of the receptacle outlets and the other hot leg and the common neutral feeds the other receptacle outlet. This gives the capacity of two separate circuits to the duplex receptacle, the split-wired receptacle.

Spray cooling: Method of refrigerating by spraying refrigerant inside the evaporator or by spraying refrigerated water.

SPST: Single pole single throw.

Spurious response: Any response other than the desired response of an electric transducer or device.

Squirrel cage motor: An induction motor having the primary winding (usually the stator) connected to the power and a current is induced in the secondary cage winding (usually the rotor).

SSR: Solid state relay.

Stability factor: Percent change in dissipation factor with respect to time.

Stack: Any vertical line of soil, waste, or vent piping.

Standard conditions: Temperature of 68 degrees Fahrenheit, pressure of 29.92 inches of mercury, and relative humidity of 30 percent.

Standard deviation: 1) A measure of data from the average. 2) The root mean square of the individual deviations from the average.

Standard reference position: The nonoperated or deenergized condition.

Standing wave: A wave in which, for any component of the field, the ratio of its instantaneous value at one point to that at any other point, does not vary with time.

Standoff: An insulated support.

Starter: 1) An electric controller for accelerating a motor from rest to normal speed and to stop the motor. 2) A device used to start an electric discharge lamp.

Starting relay: An electrical device that connects and/or disconnects the starting winding of an electric motor.

Starting winding: Winding in an electric motor used only during the brief period when the motor is starting.

Static: Interference caused by electrical disturbances in the atmosphere.

Stator: The portion of a rotating machine that includes and supports the stationary active parts.

Steady state: When a characteristic exhibits only negligible change over a long period of time.

Steam: Water in vapor state.

Steam heating: Heating system in which steam from a boiler is conducted to radiators in a space to be heated.

Steradian: The supplemental SI unit for solid angle: the three dimensional angle having its vertex at the center of a sphere and including the area of the spherical surface equal to that of a square with sides equal in length to the radius of the sphere.

Strain: 1) A change in characteristic resulting from external forces. 2) To screen foreign materials from a substance.

Strand: A group of wires, usually twisted or braided.

Strand, annular: A concentric conductor over a core: used for large conductors (1000 MCM @ 60 Hertz) to make use of skin effect: core may be of rope, or twisted I-beam.

Strand, bunch: A substrand for a rope-lay conductor: the wires in the substrand are stranded simultaneously with the same direction: bunched conductors flex easily and with little stress.

Strand, class: A system to indicate the type of stranding: the postscripts are alpha.

Strand, combination: A concentric strand having the outlet layer of different size: done to provide smoother outer surface: wires are sized with +5% tolerance from nominal.

Strand, compact: A concentric stranding made to a specified diameter of 8%-10% less than standard by using smaller than normal closing die, and for larger sizes, preshaping the strands for the outer layer(s).

Strand, compressed: The making of a tight stranded conductor by using a small closing die.

Strand, concentric: Having a core surrounded by one or more layers of helically laid wires each of one size, each layer increased by six.

Strand, herringbone lay: When adjacent bunches have opposite direction of lay in a layer of a rope-lay cable.

Strand, regular lay: Rope stranding having left-hand lay within the substrands and right-hand lay for the conductor.

Strand, nonspecular: One having a treated surface to reduce light reflection.

Strand, reverse-lay: A stranding having alternate direction of lay for each layer.

Strand, rope-lay: A conductor having a lay-up of substrands; substrand groups are bunched or concentric.

Strand, sector: A stranded conductor formed into sectors of a circle to reduce the overall diameter of a cable.

Strand, segmental: One having sectors of the stranded conductor formed and insulated one from the other, operated in parallel: used to reduce ac resistance in single conductor cables.

Strand, unilay (unidirectional): Stranding having the same direction of lay for all layers: used to reduce diameter, but is more prone to birdcaging.

Stratification of air: Condition in which there is little or no air movement in the room; air lies in temperature layers.

Stress: 1) An internal force set up within a body to resist or hold it in equilibrium. 2) The externally applied forces.

Stress-relief cone: Mechanical element to relieve the electrical stress at a shield cable termination; used above 2kV.

Striking: The process of establishing an arc or a spark.

Striking distance: The effective distance between two conductors separated by an insulating fluid such as air.

Stringers; Members supporting the treads and risers of a stair.

Strip: To remove insulation or jacket.

Strip cooler: A device to cool strips of compound.

Strut: A compression member other than a column or pedestal.

Studs: Vertically set skeleton members of a partition or wall to which the lath is nailed.

Sub-panel: A panelboard in a residential system which is fed from the service panel; or any panel in any system which is fed from another, or main, panel supplied by a circuit from another panel.

Substation: An assembly of devices and apparatus to monitor, control, transform or modify electrical power.

Superconductors: Materials whose resistance and magnetic permeability are infinitesimal at absolute zero (-273°C).

Supervised circuit: A closed circuit having a current-responsive device to indicate a break or ground.

Surge: 1) A sudden increase in voltage and current. 2) Transient condition.

Switch: A device for opening and closing or for changing the connection of a circuit.

Switch, ac general-use snap: A general-use snap switch suitable only for use on alternating current circuits and for controlling the following:

- Resistive and inductive loads (including electric discharge lamps) not exceeding the ampere rating at the voltage involved.
- Tungsten-filament lamp loads not exceeding the ampere rating of the switches at the rated voltage.
- Motor loads not exceeding 80 percent of the ampere rating of the switches at the rated voltage.

Switch, ac-dc general-use snap: A type of general-use snap switch suitable for use on either direct or alternating-current circuits and for controlling the following:

- Resistive loads not exceeding the ampere rating at the voltage involved.
- Inductive loads not exceeding one-half the ampere rating at the voltage involved, except that switches having a marked horsepower rating are suitable for controlling motors not exceeding the horsepower rating of the switch at the voltage involved.
- Tungsten-filament lamp loads not exceeding the ampere rating at 125 volts, when marked with the letter T.

Switch, general-use: A switch intended for use in general distribution and branch circuits. It is rated in amperes and is capable of interrupting its rated voltage.

Switch, general-use snap: A type of general-use switch so constructed that it can be installed in flush device boxes or on outlet covers, or otherwise used in conjunction with wiring systems recognized by the National Electrical Code.

Switch, isolating: A switch intended for isolating an electrical circuit from the source of power. It has no interrupting rating and is intended to be operated only after the circuit has been opened by some other means.

Switch, knife: A switch in which the circuit is closed by a moving blade engaging contact clips.

Switch-leg: That part of a circuit run from a lighting outlet box where a luminaire or lampholder is installed down to an outlet box which contains the wall switch that turns the light or other load on or off; it is a control leg of the branch circuit.

Switch, motor-circuit: A switch, rated in horsepower, capable of interrupting the maximum operating overload current of a motor having the same horsepower rating as the switch at the rated voltage.

Switchboard; A large single panel, frame, or assembly of panels having switches, overcurrent, and other protective devices, buses, and usually instruments mounted on the face or back or both. Switchboards are generally accessible from the rear and from the front and are not intended to be installed in cabinets.

Symmetrical: Exhibiting symmetry.

Symmetry: The correspondence in size, form and arrangement of parts on opposite sides of a plane or line or point.

Synchronism: When connected ac systems, machines or a combination operate at the same frequency and when the phase angle displacements between voltages in them are constant, or vary about a steady and stable average value.

Synchronous: Simultaneous in action and in time (in phase).

Synchronous machine: A machine in which the average speed of normal operation is exactly proportional to the frequency of the system to which it is connected.

Synchronous speed: The speed of rotation of the magnetic flux produced by linking the primary winding.

Synchrotron: A device for accelerating charged particles to high energies in a vacuum; the particles are guided by a changing magnetic field while they are accelerated in a closed path.

System: A region of space or quantity of matter undergoing study.

Tachometer: An instrument for measuring revolutions per minute.

Take-off: The procedure by which a listing is made of the numbers and types of electrical components and devices for an installation, taken from the electrical plans, drawings and specs for the job.

Take-up: 1) A device to pull wire or cable. 2) The process of accumulating wire or cable.

Tandem extrusion: Extruding two materials, the second being applied over the first, with the two extruders being just a few inches or feet apart in the process.

Tangent (geometry): A line that touches a curve at a point so that it is closer to the curve in the vicinity of the point than any other line drawn through the point: (trigonometry) in a right triangle it is the ratio of the opposite to the adjacent sides for a given angle.

Tank test: The immersion of a cable in water while making electrical tests; the water is used as a conducting element surrounding the cable.

Tap: A splice connection of a wire to another wire (such as a feeder conductor in an auxiliary gutter) where the smaller conductor runs a short distance (usually only a few feet, but could be up to 25 feet) to supply a panelboard or motor controller or switch. Also called a "tap-off" indicating that energy is being taken from one circuit or piece of equipment to supply another circuit or load; a tool that cuts or machines threads in the side of a round hole.

Tap drill: Drill used to form hole prior to placing threads in hole. The drill is the size of the root diameter of tap threads.

Tape: A relatively narrow, long, thin, flexible fabric, film or mat or combination thereof: helically applied tapes are used for cable insulation, especially at splices: for the first century the primary insulation for cables above 2kV was oil saturated paper tapes.

Target, sag: A visual reference used when sagging.

TC: 1) Thermocouple 2) Time constant 3) Timed closing.

TDR: 1) Time delay relay 2) Time domain reflectometer, pulse-echo (radar) testing of cables; signal travels thru cable until impedance discontinuity is encountered, then part of signal is reflected back; distance to fault can be estimated. Useful for finding faults, broken shields or conductor.

Technical Appeal Board (TAB) of UL: A group to recommend solutions to technical differences between UL and a UL client.

Teflon®: A DuPont trade name for polytetrafluoroethylene which is used as high temperature insulation and has low dissipation factor and low relative capacitance.

Telegraphy: Telecommunication by the use of a signal code.

Telemetering: Measurement with the aid of intermediate means that permits interpretation at a distance from the primary detector.

Telephone: The transmission and reception of sound by electronics.

Temper: A measure of the tensile strength of a conductor; indicative of the amount of annealing or cold working done to the conductor.

Temperature, ambient: The temperature of the surrounding medium, such as air around a cable.

Temperature, coefficient of resistance: The unit change in resistance per degree temperature change.

Temperature, emergency: The temperature to which a cable can be operated for a short length of time, with some loss of useful life.

Temperature humidity index: Actual temperature and humidity of a sample of air compared to air at standard conditions.

Temperature, operating: The temperature at which a device is designed or rated for normal operating conditions; for cables: the maximum temperature for the conductor during normal operation.

Temporary: This single word is used to mean either "temporary service" (which a power company will give to provide electric power in a building under construction) or "temporary inspection" (the inspection that a code-enforcing agency will make of a temporary service prior to inspection of the electrical work in a building under construction.)

Tensile strength: The greatest longitudinal stress a material—such as a conductor—can withstand before rupture or failure while in service.

Tension, final unloaded conductor: The tension after the conductor has been stretched for an appreciable time by loads simulating ice and wind.

Tension, initial conductor: The tension prior to any external load.

Tension, working: The tension which should be used for a portable cable on a power reel; it should not exceed 10% of the cable breaking strength.

Terminal: A device used for connecting cables.

Termination: 1) The connection of a cable. 2) The preparation of shielded cable for connection.

Testlight: Light provided with test leads that is used to test or probe electrical circuits to determine if they are energized.

Test, proof: Made to demonstrate that the item is in satisfactory condition.

Test, voltage breakdown: a) Step method— applying a multiple of rated voltage to a cable for several minutes, then increasing the applied voltage by 20% for the same period until breakdown. b) Applying a voltage at a specified rate until breakdown.

Test, voltage life: Applying a multiple of rated voltages over a long time period until breakdown: time to failure is the parameter measured.

Test, volume resistivity: Measuring the resistance of a material such as the conducting jacket or conductor stress control.

Test, water absorption: Determination of how much water a given volume of material will absorb in a given time period; this test is being superseded by the EMA test.

Therm: Quantity of heat equivalent to 100,000 Btu.

Thermal cutout: An overcurrent protective device containing a heater element in addition to, and affecting, a renewable fusible member that opens the circuit. It is not designed to interrupt short-circuit currents.

Thermal endurance: The relationship between temperature and time of degrading insulation until failure, under specified conditions.

Thermal protector (applied to motors): A protective device that is assembled as an integral part of a motor or motor-compressor and that, when properly applied, protects the motor against dangerous overheating due to overload and failure to start.

Thermal relay (hot wire relay): Electrical control used to actuate a refrigeration system. This system uses a wire to convert electrical energy into heat energy.

Thermal shock: Subjecting something to a rapid, large temperature change.

Thermally protected (as applied to motors): When the words thermally protected appear on the nameplate of a motor or motor-compressor, it means that the motor is provided with a thermal protector.

Thermionic emission: The liberation of electrons or ions from a solid or liquid as a result of its thermal energy.

Thermistor: An electronic device that makes use of the change of resistivity of a semiconductor with change in temperature.

Thermocouple: A device using the Seebeck effect to measure temperature.

Thermodisk defrost control: Electrical switch with bimetal disk that is controlled by electrical energy.

Thermodynamics: Science that deals with the relationships between heat and mechanical energy and their interconversion.

Thermoelectric generator: A device interaction of a heat flow and the charge carriers in an electric circuit, and that requires, for this process, the existence of a temperature difference in the electric circuit.

Thermoelectric heat pump: A device that transfers thermal energy from one body to another by the direct interaction of an electrical current and the heat flow.

Thermometer: Device for measuring temperatures.

Thermoplastic: Materials which when reheated, will become pliable with no change of physical properties.

Thermoset: Materials which may be molded, but when cured, undergo an irreversible chemical and physical property change.

Thermostat: Device responsive to ambient temperature conditions.

Thermostatic expansion valve: A control valve operated by temperature and pressure within an evaporator coil, which controls the flow of refrigerant.

Thermostatic valve: Valve controlled by thermostatic elements.

Three phase circuit: A polyphase circuit of three interrelated voltages for which the phase difference is 120°: the common form of generated power.

Thumper: A device used to locate faults in a cable by the release of power surges from a capacitor, characterized by the audible noise when the cable breaks down.

Thyratron: A gas-filled triode tube that is used in electronic control circuits.

Timers: Mechanism used to control on and off times of an electrical circuit.

Timer-thermostat: Thermostat control that includes a clock mechanism. Unit automatically controls room temperature and changes it according to the time of day.

Tinned: Having a thin coating of pure tin, or tin alloy; the coating may keep rubber from sticking or be used to enhance connection; coatings increase the resistance of the conductor, and may contribute to corrosion by electrolysis.

Toggle: A device having two stable states: i.e.—toggle switch which is used to turn a circuit ON or OFF. A device may also be toggled from slow to fast, etc.

Tolerance: The permissible variation from rated or assigned value.

Topping-off: The finishing touches put to an electrical installation; mounting plates on wall switches, receptacles and other wiring devices; receptacles and other wiring devices; installing fixtures, etc.

Toroid: A coil wound in the form of a doughnut; i.e.—current transformers.

Torquing: Applying a rotating force and measuring or limiting its value.

TPE: Thermal plastic elastomer.

TPR: Thermal plastic rubber.

Tracer: A means of identifying cable.

Transducer: A device by means of which energy can flow from one or more media to another.

Transfer switch: A device for transferring one or more load conductor connections from one power source to another.

Transformer: A static device consisting of winding(s) with or without a tap(s) with or without a magnetic core for introducing mutual coupling by induction between circuits.

Transformer, potential: Designed for use in measuring high voltage: normally the secondary voltage is 120V.

Transformer, power: Designed to transfer electrical power from the primary circuit to the secondary circuit(s) to 1) step-up the secondary voltage at less current or 2) step-down the secondary voltage at more current; with the voltage-current product being constant for either primary or secondary.

Transformer-rectifier: Combination transformer and rectifier in which input in ac may be varied and then rectified into dc.

Transformer, safety isolation: Inserted to provide a nongrounded power supply such that a grounded person accidentally coming in contact with the secondary circuit will not be electrocuted.

Transformer, vault-type: Suitable for occasional submerged operation in water.

Transient: 1) Lasting only a short time; existing briefly; temporary. 2) A temporary component of current existing in a circuit during adjustment to a changed load, different source voltage, or line impulse.

Transistor: An active semiconductor device with three or more terminals.

Transmission: Transfer of electric energy from one location to another through conductors or by radiation or induction fields.

Transmission line: A long electrical circuit.

Transposition: Interchanging position of conductors to neutralize interference.

Traveler: A pulley complete with suspension arm or frame to be attached to overhead line structures during stringing.

Trim: The front cover assembly for a panel, covering all live terminals and the wires in the gutters, but providing openings for the fuse cutouts or circuit breakers mounted in the panel; may include the door for the panel and also a lock.

Triode: A three-electrode electron tube containing an anode, a cathode, and a control electrode.

Triplex: Three cables twisted together.

Trolley wire: Solid conductor designed to resist wear due to rolling or sliding current pickup trolleys.

Trough: Another name for an "auxiliary gutter", which is a sheet metal enclosure of rectangular cross-section, used to supplement wiring spaces at meter centers, distribution centers, switchboards and similar points in wiring systems where splices or taps are made to circuit conductors. The single word "gutter" is also used to refer to this type of enclosure.

Trunk feeder: A feeder connecting two generating stations or a generating station and an important substation.

Trussed: Framed structural pieces consisting of triangles in a single plane for supporting loads over spans.

Tub: An expression sometimes used to refer to large panelboards or control center cabinets, in particular the box-like enclosure without the front trim.

Tube: A hollow long product having uniform wall thickness and uniform cross-section.

Tuning: The adjustment of a circuit or system to secure optimum performance.

Turn: The basic coil element that forms a single conducting loop comprised of one insulated conductor.

Turn ratio: The ratio between the number of turns between windings in a transformer; normally primary to secondary, except for current transformers it is secondary to primary.

Twist test: A test to grade round material for processibility into conductors.

Two-phase: A polyphase ac circuit having two interrelated voltages.

UHF (ultra high frequency): 300 MHz to 3GHz.

UL (Underwriters Laboratories): A nationally known laboratory for testing a product's performance with safety to the user being prime consideration: UL is an independent organization, not controlled by any manufacturer: the best known lab for electrical products.

Ultrasonic: Sounds having frequencies higher than 20 KHz: 20 KHz is at the upper limit of human hearing.

Ultrasonic cleaning: Immersion cleaning aided by ultrasonic waves which cause microagitation.

Ultrasonic detector: A device that detects the ultrasonic noise such as that produced by corona or leaking gas.

Ultraviolet: Radiant energy within the wave length range 10 to 380 nanometers: invisible, filtered by glass, causes teeth to glow, causes suntan.

Undercurrent: Less than normal operating current.

Undervoltage: Less than normal operating voltage.

Ungrounded: Not connected to the earth intentionally.

Universal motor: A motor designed to operate on either ac or dc at about the same speed and output with either.

Urethane foam: Type of insulation that is foamed in between inner an outer walls.

URD (Underground Residential Distribution): A single phase cable usually consisting of an insulated conductor having a bare concentric neutral.

Utilization equipment: Equipment that uses electric energy for mechanical, chemical, heating, lighting, or other useful purposes.

V block: V-shaped groove in a metal block used to hold a shaft.

VA: Volts times amps.

Vacuum; Reduction in pressure below atmospheric pressure.

Vacuum pump: Special high efficiency compressor used for creating high vacuums for testing or drying purposes.

Vacuum switch: A switch with contacts in an evacuated enclosure.

Vacuum tube: A sealed glass enclosure having two or more electrodes between which conduction of electricity may take place. Most vacuum tubes have been replaced with solid-state circuitry.

Valve: Device used for controlling fluid flow.

Valve, expansion: Type of refrigerant control that maintains a pressure difference between high side and low side pressure in a refrigerating mechanism. The valve operates by pressure in the low or suction side.

Valve, solenoid: Valve actuated by magnetic action by means of an electrically energized coil.

Vapor: Word usually used to denote vaporized refrigerant rather than gas.

Vapor barrier: Thin plastic or metal foil sheet used in air conditioned structures to prevent water vapor from penetrating insulating material.

Vapor lock: Condition where liquid is trapped in line because of a bend or improper installation that prevents the vapor from flowing.

Vapor-safe: Constructed so that it may be operated without hazard to its surroundings in hazardous areas in aircraft.

Vapor, saturated: A vapor condition that will result in condensation into liquid droplets as vapor temperature is reduced.

Vapor-tight: So enclosed that vapor will not enter.

Variable speed drive: A motor having an integral coupling device which permits the output speed of the unit to be easily varied thru a continuous range.

Varnish: An insulation that is applied as a liquid and is quite thin.

VD: Voltage drop.

Vector: A mathematical term expressing magnitude and direction.

Ventilation: Circulation of air, system or means of providing fresh air.

Vermiculite: Lightweight inert material resulting from expansion of mica granules at high temperatures that is used as an aggregate in plaster.

Vernier: An auxiliary scale permitting measurements more precise than the main scale.

Viscosity: Internal friction or resistance to flow of a liquid, the constant ratio of shearing stress to rate of shear.

Volatile flammable liquid: A flammable liquid having a flash point below 38oC or whose temperature is above its flash point.

Volt: The derived SI unit for voltage: one volt equals one watt per ampere.

Voltage: The electrical property that provides the energy for current flow; the ratio of the work done to the value of the charge moved when a charge is moved between two points against electrical forces.

Voltage, breakdown: The minimum voltage required to break down an insulation's resistance, allowing a current flow through the insulation, normally at a point.

Voltage, contact: A small voltage which is established whenever two conductors of different materials are brought into contact; due to the difference in work functions or the ease with which electrons can cross the surface boundary in the two directions.

Voltage drop: 1) The loss in voltage between the input to a device and the output from a device due to the internal impedance or resistance of the device. 2) The difference in voltage between two points of a circuit.

Voltage, EHV (extra high voltage): 230-765 kV.

Voltage, induced: A voltage produced in a conductor by a change in magnetic flux linking that path.

Voltage rating of a cable: Phase-to-phase ac voltage when energized by a balanced three phase circuit having a solidly grounded neutral.

Voltage regulator: A device to decrease voltage fluctuations to loads.

Voltage, signal: Voltages to 50 V.

Voltage to ground: The voltage between an energized conductor and earth.

Voltage, UHV (ultra-high voltage): 765 + kV.

Voltage divider: A network consisting of impedance elements connected in series to which a voltage is applied and from which one or more voltages can be obtained across any portion of the network.

Voltmeter: An instrument for measuring voltage.

VOM (volt-ohm-multimeter): A commonly used electrical test instrument to test voltage, current, resistance, and continuity. Instruments are available in both dial and the more popular digital meters.

Vortex tube: Mechanism for cooling or refrigerating that accomplishes a cooling effect by releasing compressed air through a specially designed opening. Air expands in a rapidly spiraling column of air that separates slow moving molecules (cool) from fast moving molecules (hot).

Vulcanize: To cure by chemical reaction that induces extensive changes in the physical properties of a rubber or plastic, brought about by reacting it with sulphur and/or other suitable agents: the change in physical properties include decreased plastic flow, reduced surface tackiness, increased elasticity, much greater tensile strength, and considerably less solubility: the process being hastened by heat and pressure: the method of curing thermosetting materials—rubbers, XLP, etc.

VW-1: A UL rating given single conductor cables as to flame resistant properties, formerly FR-1.

Wall: The thickness of insulation or jacket of cable.

Wall, fire: A dividing wall for the purpose of checking the spread of fire from one part of a building to another.

Waterblocked cable: A multiconductor cable having interstices filled to prevent water flow or wicking.

Water-cooled condenser: Condensing unit that is cooled through the use of water.

Waterproof: Moisture will not interfere with operation.

Water table: A projection of the wall at or near the grade line of a building. It is used to turn the water away from the foundation wall.

Watertight: So constructed that water will not enter.

Watt: The derived SI unit for power, radiant flux: one watt equals one joule per second.

Watt-hour: The number of watts used in one hour.

Watt-hour meter: A meter which measures and registers the integral, with respect to time, of the active power in a circuit.

Wattmeter: An instrument for measuring the magnitude of the active power in a circuit.

Wave: A disturbance that is a function of time or space or both.

Waveform: The geometrical shape as obtained by displaying a characteristic of the wave as a function of some variable when plotted over one primitive period.

Wavelength: The distance measured along the direction of propagation between two points which are in phase on adjacent waves.

Weatherhead: The conduit fitting at a conduit used to allow conductor entry, but prevent weather entry.

Weatherproof: So constructed or protected that exposure to the weather will not interfere with successful operation.

Web: An open braid; central portion of an I beam.

Weight correction factor: The correction factor necessitated by uneven geometric forces placed on cables in conduits; used in computing pulling tensions and sidewall loads.

Weighting: The artificial adjustment of measurements in order to account for factors different from those factors which would be encountered in normal use.

Weld: The joining of materials by fusion or recrystallization across the interface of those materials using heat or pressure or both, with or without filler material.

Welding cable: Very flexible cable used for leads to the rod holders of arc-welders, usually consisting of size 4/0 flexible copper conductors.

Wet bulb: Device used in measurement of relative humidity.

Wet cell battery: A battery having a liquid electrolyte.

Wet locations: Exposed to weather or water spray or buried.

Winding: An assembly of coils designed to act in consort to produce a magnetic flux field or to link a flux field.

Wiped joint: A joint wherein filler metal is applied in liquid form and distributed by mechanical action.

Wire: A slender rod or filament of drawn metal: the term may also refer to insulated wire.

Wire bar: A cast shape which has a square cross section with tapered ends.

Wire, building: That class of wire and cable, usually rated at 600V, which is normally used for interior wire of buildings.

Wire, covered: A wire having a covering not in conformance with NEC standards for material or thickness.

Wire, hookup: Insulated wire for low voltage and current in electronic circuits.

Wire, resistance: Wire having appreciable resistance: used in heating applications such as electric toasters, heaters, etc.

WM: Wattmeter.

Work: Force times distance: pound-feet.

Work function: The minimum energy required to remove an electron from the Fermi level of a material into field-free space: Units—electron volts.

Work hardening: Hardening and embrittlement of metal due to cold working.

Xfmr: Transformer.

X-ray: Penetrating short wavelength electromagnetic radiation created by electron bombardment in high voltage apparatus: produce ionization when they strike certain materials.

Yield strength: The point at which a substance changes from elastic to viscous.

appendix II

ELECTRICAL WIRING SYMBOLS

The purpose of an electrical drawing is to show the complete design and layout of the electrical systems for lighting, power, signal and communication systems, special raceways, and related electrical equipment. In preparing such drawings, the electrical layout is shown through the use of lines, symbols, and notation which should indicate, beyond any question or any doubt, exactly what is required.

Many engineers, designers, and draftsmen use symbols adapted by the United States of America Standards Institute (USASI). However, no definite standard schedule of symbols is always used in its entirety. Consulting engineering firms quite frequently modify these standard symbols to meet their own needs. Therefore, in order to identify the symbols properly, the engineer provides, on one of the drawings or in the written specifications, a list of symbols with a descriptive note for each—clearly indicating the part of the wiring system which it represents.

Figure A2-1 shows a list of electrical symbols which are currently recommended by USASI, while Fig. A2-2 shows another list of symbols which was prepared by the Consulting Engineers Council/U.S. and the Construction Specifications Institute, Inc. Figure A2-3 shows still another list of electrical reference symbols which have been modified for use by one consulting engineering firm.

It should be evident from these symbols that many have the same basic form, but, because of some slight difference, their meaning changes. For example, the outlet symbols in Fig. A2-4 each have the same basic form—a circle—but the addition of a line or an abbreviation gives each an individual meaning. A good procedure to follow in learning symbols is to first learn the basic form and then apply the variations for obtaining different meanings.

The electrical symbols described in the following paragraphs represent a system of electrical notation whose compactness and clarity may be of assistance to electrical engineers, designers, and draftsmen.

The system should be used in place of standard symbols only if there seems to be a decided advantage in doing so.

Some of the symbols are abbreviated idioms, like "WP" for weatherproof and "AFF" for above finished floor. Other symbols are simplified pictographs, like for a double floodlight fixture or for an infrared electric heater with two lamps.

In some cases there are combinations of idioms and pictographs, as in for fusible safety switch, for nonfusible safety switch, and for double-throw safety switch, where the pictograph of a switch enclosure is combined with the abbreviated idioms of F (fusible), N (nonfusible), and DT (double throw), respectively. The numerals indicate the bus bar capacity in amperes.

This list came about as a result of much discussion with consulting engineers, electrical designers, electrical draftsmen, electrical estimators, electricians,

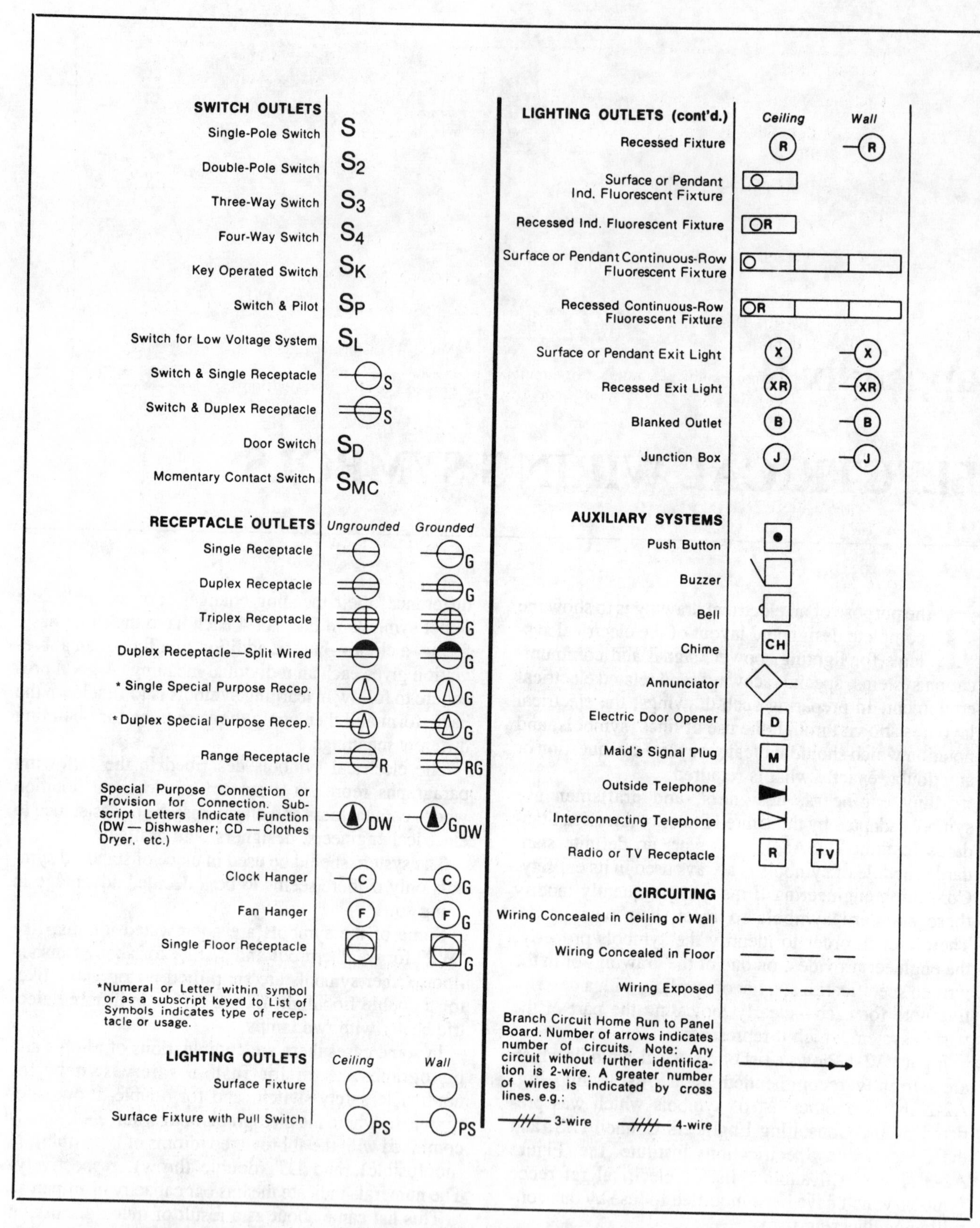

Fig. A2-1: Electrical symbols currently recommended by USASI.

CIRCUITING

Wiring Exposed (not in conduit)	E
Wiring Concealed in Ceiling or Wall	
Wiring Concealed in Floor	
Wiring Existing*	
Wiring Turned Up	
Wiring Turned Down	
Branch Circuit Home Run to Panel Board.	2 1

Number of arrows indicates number of circuits. (A number at each arrow may be used to identify circuit number.)**

BUS DUCTS AND WIREWAYS

Trolley Duct***	T T
Busway (Service, Feeder, or Plug-in)***	B B
Cable Trough Ladder or Channel***	C C
Wireway***	W W

PANELBOARDS, SWITCHBOARDS AND RELATED EQUIPMENT

Flush Mounted Panelboard and Cabinet***	
Surface Mounted Panelboard and Cabinet***	
Switchboard, Power Control Center, Unit Substations (Should be drawn to scale)***	
Flush Mounted Terminal Cabinet (In small scale drawings the TC may be indicated alongside the symbol)***	TC
Surface Mounted Terminal Cabinet (In small scale drawings the TC may be indicated alongside the symbol)***	TC
Pull Box (Identify in relation to Wiring System Section and Size)	
Motor or Other Power Controller (May be a starter or contactor)***	
Externally Operated Disconnection Switch***	
Combination Controller and Disconnection Means***	

POWER EQUIPMENT

Electric Motor (HP as indicated)	¼
Power Transformer	
Pothead (Cable Termination)	
Circuit Element, e.g., Circuit Breaker	CB
Circuit Breaker	
Fusible Element	
Single-Throw Knife Switch	
Double-Throw Knife Switch	
Ground	
Battery	
Contactor	C
Photoelectric Cell	PE
Voltage Cycles, Phase	Ex: 480/60/3
Relay	R
Equipment Connection (as noted)	

*Note: Use heavy-weight line to identify service and feeders. Indicate empty conduit by notation CO (conduit only).
**Note: Any circuit without further identification indicates two-wire circuit. For a greater number of wires, indicate with cross lines. e.g.:

3 wires   4 wires, etc.

Neutral wire may be shown longer. Unless indicated otherwise, the wire size of the circuit is the minimum size required by specification. Identify different functions of wiring system, e.g. signaling system by notation or other means.
***Identify by Notation or Schedule

Fig. A2-2: A list of electrical symbols recommended by the Consulting Engineers Council/US and the Construction Specifications Institute, Inc.

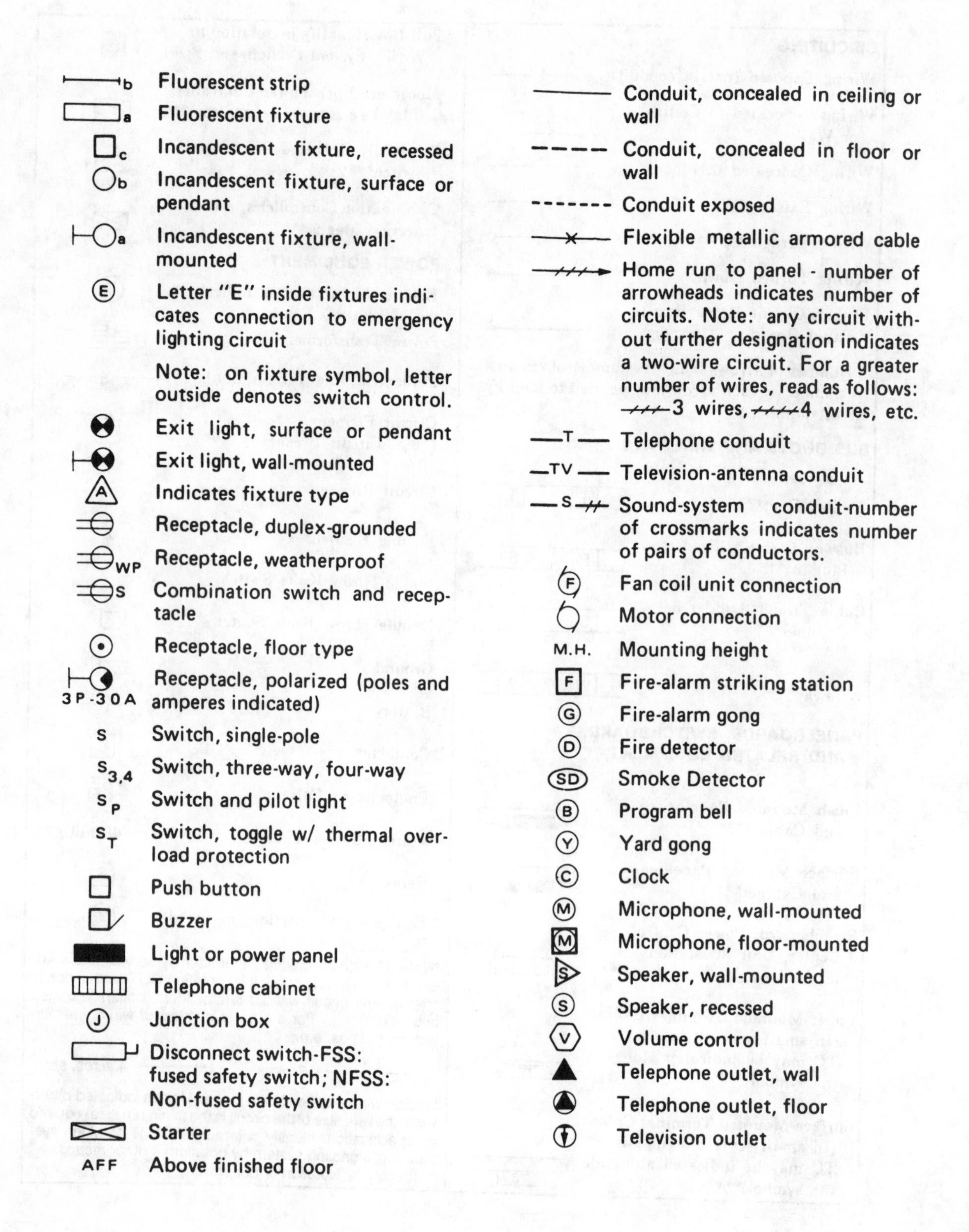

Fig. A2-3: A modified list of electrical symbols used by a consulting engineering firm.

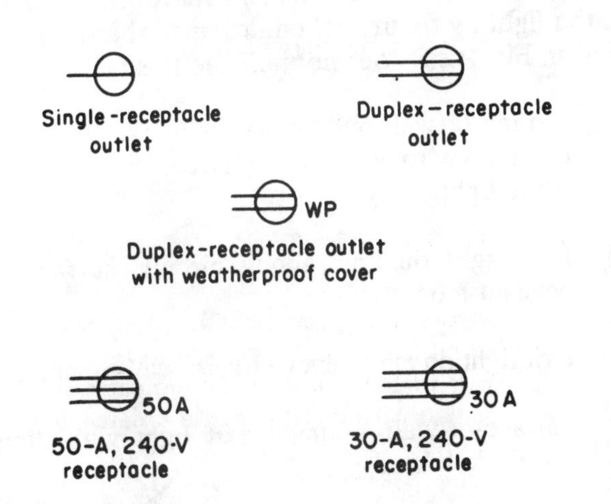

Fig. A2-4: Many electrical symbols have the same basic form but a line or note added gives each an individual meaning.

and others who are required to interpret electrical drawings. It is felt that this list represents a good set of symbols in that they are:

- Easy to draw
- Easily interpreted by workers
- Sufficient for most applications

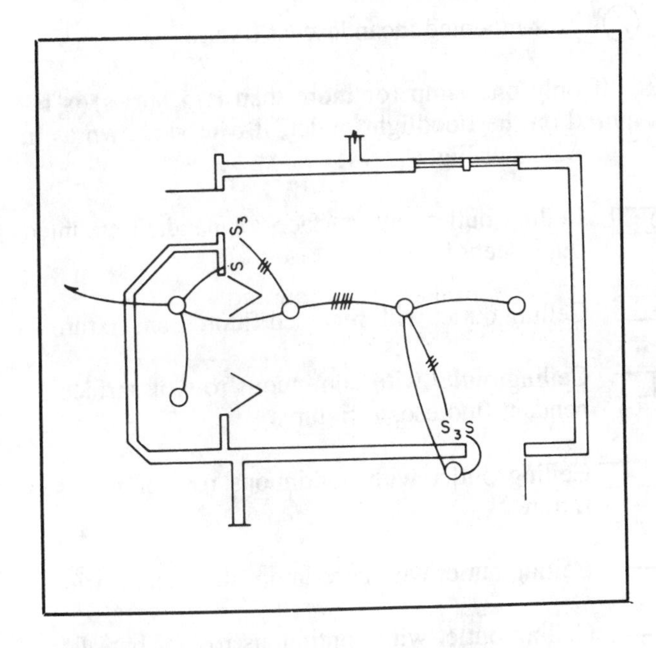

Fig. A2-5: Practical application of surface-mounted incandescent fixtures.

Lighting Outlets

○ Ceiling outlet with surface-mounted incandescent lighting fixture

○ Wall outlet with surface-mounted incandescent lighting fixture

Practical use of these symbols is shown in Fig. A2-5.

◎ Wall outlet with recessed incandescent lighting fixture

◎ Ceiling outlet with recessed incandescent lighting fixture

Practical use of these symbols is shown in Fig. A2-6.

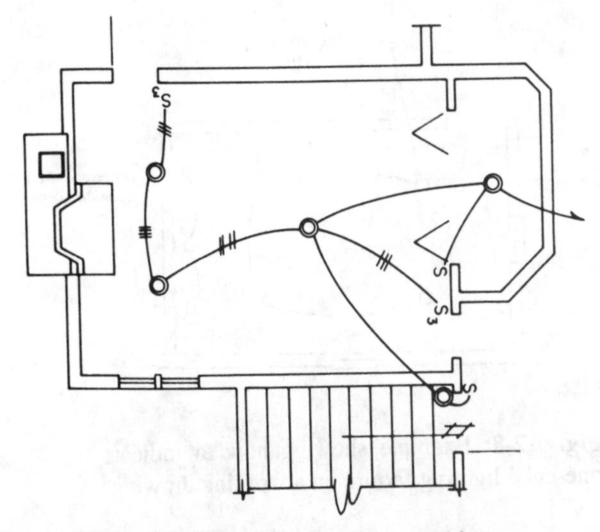

Fig. A2-6: Practical application of recessed incandescent lighting fixtures.

Sometimes these symbols are modified in order to indicate the physical shape of a particular incandescent fixture. The lighting fixture in Fig. A2-7 consists of four 6-in. cubes. This type of lighting fixture may be indicated on the electrical floor plan as shown in Fig. A2-9.

A lighting fixture consisting of one cube may be indicated as shown in Fig. A2-8. All should be drawn as close to scale as possible.

The type of mounting of all fixtures is usually indicated in the lighting fixture schedule shown on the drawings or in the written specifications. The fixture illustrated in Fig. A2-7 is obviously pendant-mounted

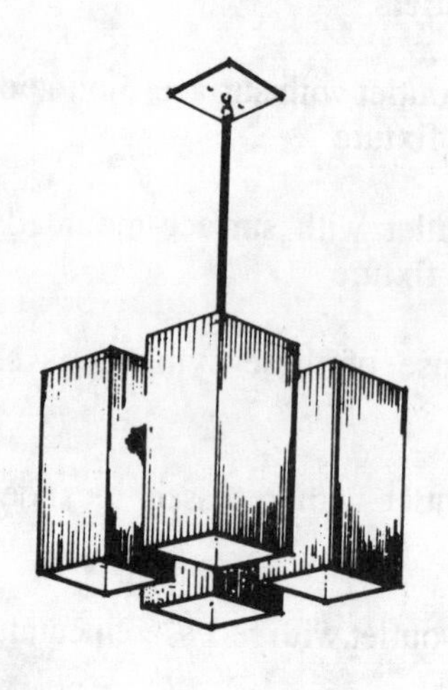

Fig. A2-7: Perspective view of a lighting fixture.

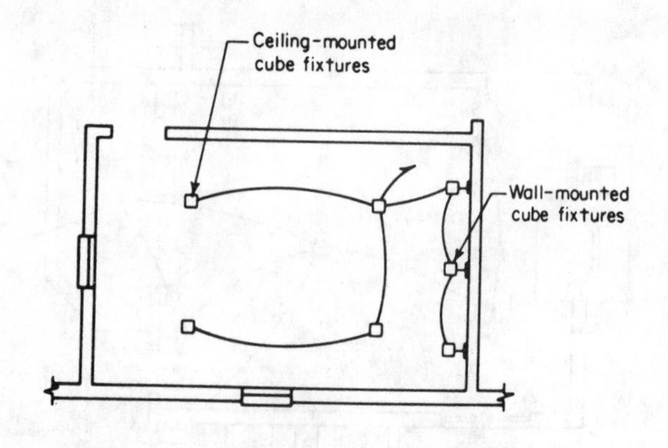

Fig. A2-8: Example showing how to indicate a one-cube lighting fixture on a working drawing.

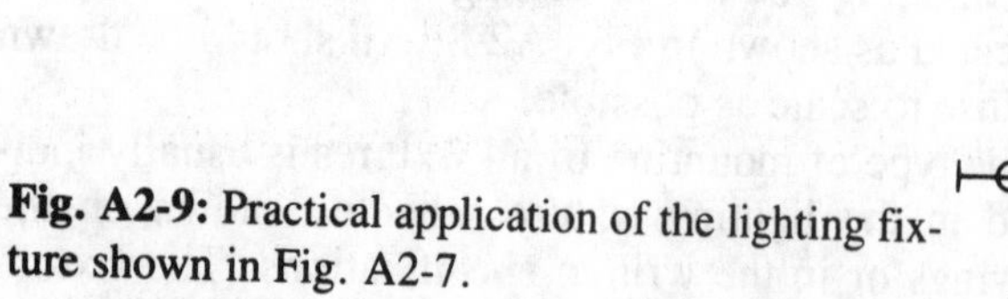

Fig. A2-9: Practical application of the lighting fixture shown in Fig. A2-7.

and should be so indicated in an appropriate column in the lighting fixture schedule, since the floor-plan view in Fig. A2-8 does not indicate this fact.

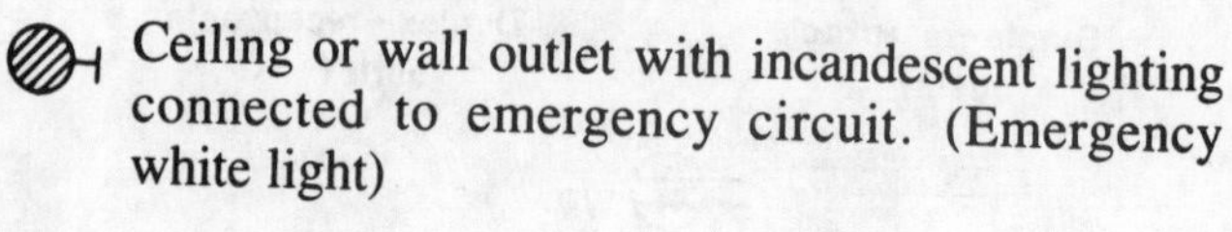

Ceiling or wall outlet with incandescent lighting connected to emergency circuit. (Emergency white light)

Exit light on emergency circuit, surface-or-pendant-mounted

Exit light on emergency circuit, wall-mounted

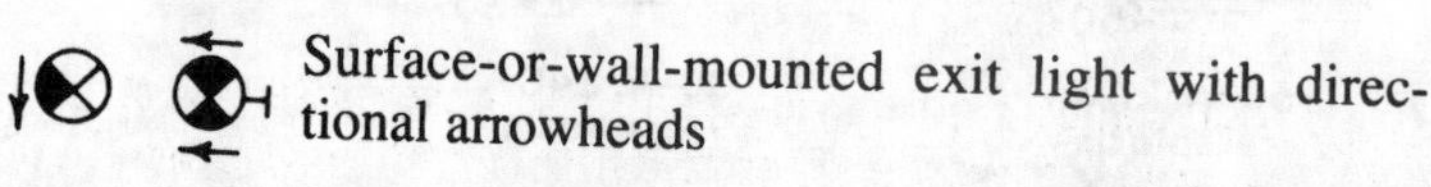

Surface-or-wall-mounted exit light with directional arrowheads

The mounting height of wall-mounted lighting fixtures is sometimes indicated in the symbol lists, especially where most are to be mounted at one height. For example, it might read "... *wall outlet with incandescent fixture mounted 6 ft 6 in. above finished floor to center of outlet box unless otherwise indicated.*" If a few wall-mounted fixtures were to be mounted at 8 ft 0 in. above finished floor, they could be indicated as shown in Fig. A2-10, the letters AFF meaning "above finished floor."

Ground-mounted incandescent uplight

Post-mounted incandescent fixture

If only one lamp (or more than two lamps) is required on the floodlight outlet, it can be shown as in Fig. A2-13 or Fig. A2-14.

Ceiling outlet with surface- or pendant-mounted fluorescent fixture

Ceiling outlet with recessed fluorescent fixture

Ceiling outlet with continuous row of surface or pendant fluorescent fixtures

Ceiling outlet with continuous row of recessed fixtures

Ceiling outlet with bare-lamp fluorescent strip

Ceiling outlet with continuous row of bare-lamp fluorescent-strip lighting

Wall outlet with fluorescent fixture

Fluorescent fixture mounted under cabinet 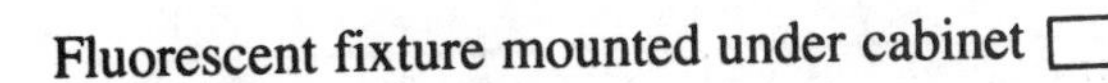

Fig. A2-10: Method of indicating mounting height of a few wall-mounted fixtures.

Modification of the symbols for fluorescent lighting is common. For example, cove lighting with bare-lamp fluorescent strips may be indicated on the drawings as shown in Fig. A2-15.

The lighting layout illustrated in Fig. A2-16 shows practical applications of all fluorescent symbols covered in this writing.

Many electrical drawings do not differentiate between recessed, surface, or pendant-mounted fixtures on the floor plans. Rather, the mounting is indicated either in the lighting fixture schedule or in the written specifications. However, since a major variation in the type of outlet box, outlet supporting means, wiring system arrangement, and outlet connection, plus need of special items such as plaster rings or roughing-in cans, depends upon the way in which a fixture is mounted, the electrician should be able to know the type of mounting in a glance at the drawings. Therefore, the mounting of a lighting fixture should be indicated on the floor plans as well as in the lighting fixture schedule.

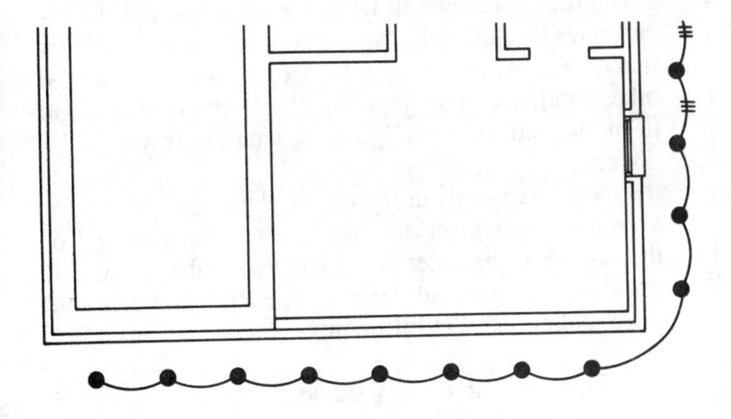

Fig. A2-11: Practical application of ground- and post-mounted incandescent fixtures.

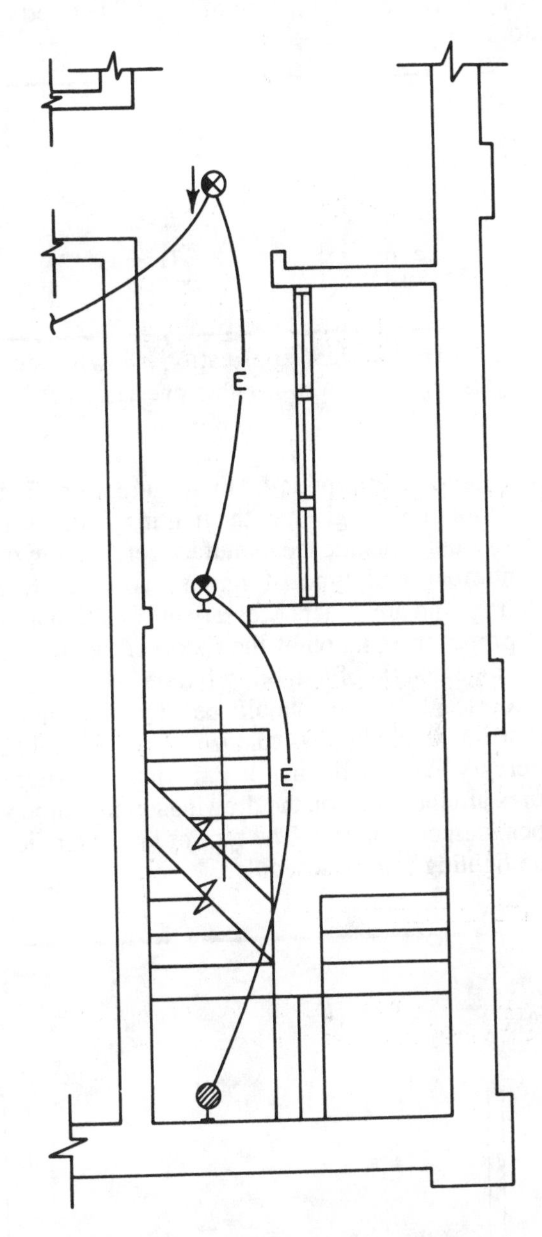

Fig. A2-12: Practical application of exit lights and emergency white incandescent fixtures.

Lighting fixtures are identified as to type of fixture by a numeral placed inside a triangle near each lighting fixture, as shown here and in Fig. A2-17.

The indicated fixture is shown in the symbol lists as follows:

△5 Indicates type of lighting fixture—see schedule

If one type of fixture is used exclusively in one room or area, as shown in Fig. A2-18, the indicator need only appear once with the word "ALL" lettered at the bottom.

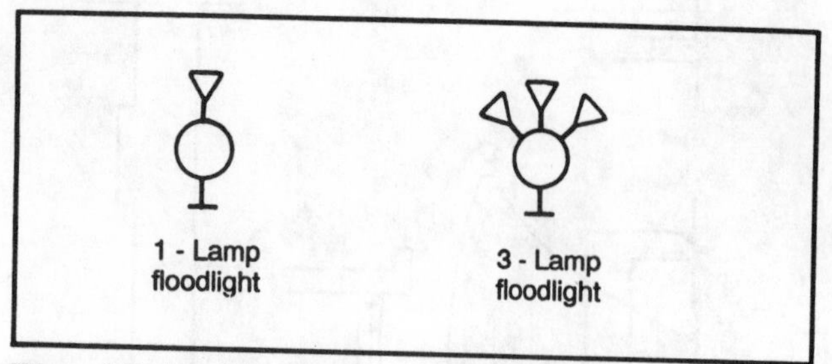

Fig. A2-13: Practical application showing floodlights on the drawings, either one lamp or more than two lamps.

A complete description of the fixture identified by the symbol must be given in the lighting fixture schedule and should include the manufacturer, catalog number, number and type of lamps, voltage, finish, mounting, and any other necessary information needed for a proper installation of the fixtures. Figure A2-19 shows an example of a lighting fixture schedule.

Fluorescent fixtures should be drawn to approximate scale, showing physical size whenever practical.

Mercury vapor and other electric-discharge lighting fixtures are indicated on the drawings in the same way as incandescent fixtures. The type of lamp is indicated in the lighting fixture schedule.

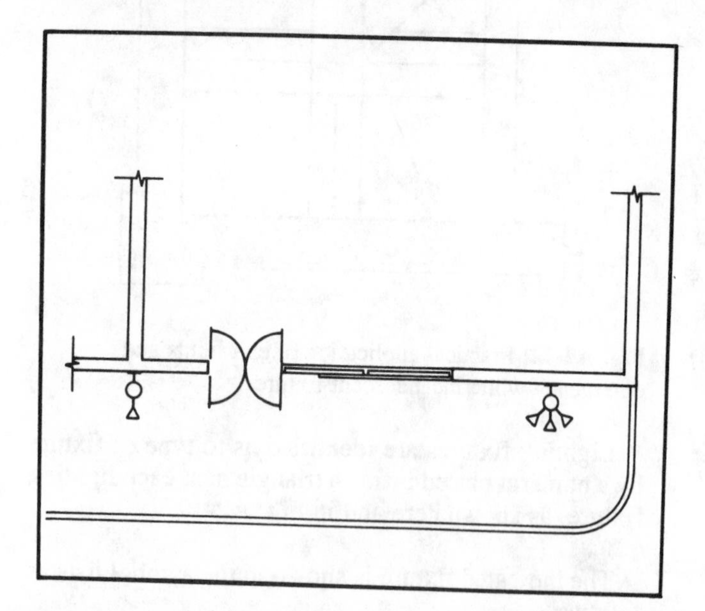

Fig. A2-14: Practical application of floodlights.

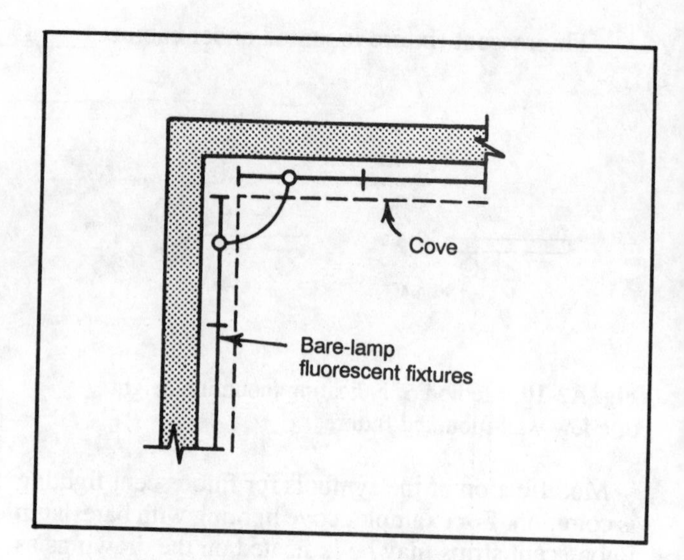

Fig. A2-15: Example of bare-tube fluorescent fixtures shown on a working drawing.

Receptacles and Switches

Every day those who work in the electrical construction industry hear the word "standard." Yes, it would be the ideal situation if all electrical engineers could use one set of standard electrical symbols for all their projects. However, with the present symbols known as "standard" this is not practical.

For example, one consulting engineering firm in Washington, DC, did a large amount of electrical designs for hospitals all over the eastern United States. On one of their jobs, there were over 100 duplex receptacles mounted horizontally in the backsplash of lavatory countertops, while there were over 300 conventional duplex receptacles.

If a draftsman had to letter a note at each of the receptacles located in the backsplash, it is easily seen that much time would be spent on the drawings. On the other hand, a simple symbol with a written explanation in the legend would tell exactly what work was to be done.

One firm recently designed the electrical systems for a group of quick-service restaurants where several of the duplex receptacles were located in the ceiling. If standard symbols had been used, each location would have looked like the following:

⊖ Flush mounted 120V duplex receptacles mounted in the ceiling

Much time was saved by composing a new symbol for the ceiling-mounted receptacles and eliminating the

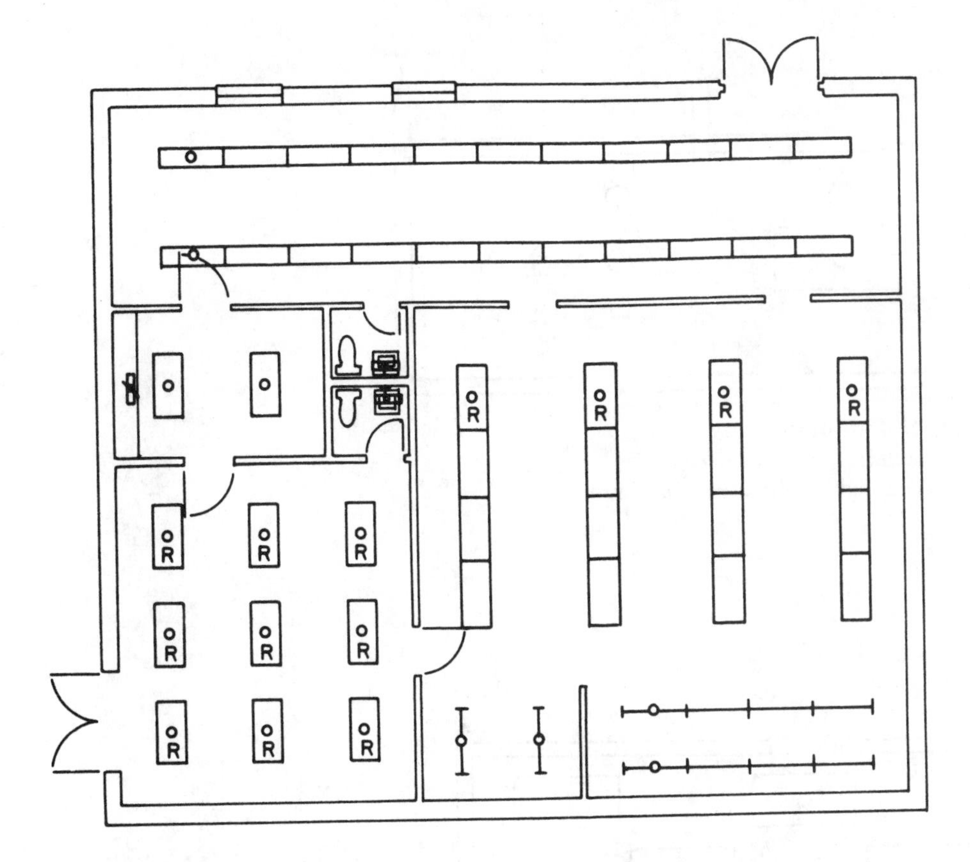

Fig. A2-16: Practical application of all fluorescent lighting symbols covered in this Appendix.

descriptive note at each of the dozen or so receptacles. A note appeared only once in the symbol list as follows:

 Flush-mounted 120V, duplex "twist-lock" receptacle mounted in ceiling with stainless steel plate

The cases are endless, and until a sufficient number of different symbols is available in the standard symbol list, it is highly impractical to use the standard throughout.

If a special outlet is shown on a drawing in only a few locations, then perhaps a descriptive note at each location is best; but if the "special outlet" appears in several locations, then an individual symbol can save much drawing time.

The following are switch and receptacle symbols used in one office on working drawings.

S Single-pole toggle switch mounted 50 inches above finished floor to center of box unless otherwise indicated

S_2 Two-pole toggle switch mounted 50 inches above finished floor to center of box unless otherwise indicated.

S_3 Three-way switch mounted 50 inches above finished floor to center of box unless otherwise indicated.

S_4 Four-way switch mounted 50 inches above finished floor to center of box unless otherwise indicated.

S_L Low-voltage switch to relay mounted 50 inches above finished floor unless otherwise indicated.

S_D Flush-mounted door switch to control closet light.

S_P Switch with pilot light.

Figure A2-20 illustrates some practical applications of switch symbols used on working drawings. Notice that the single-pole switches are used to control lighting

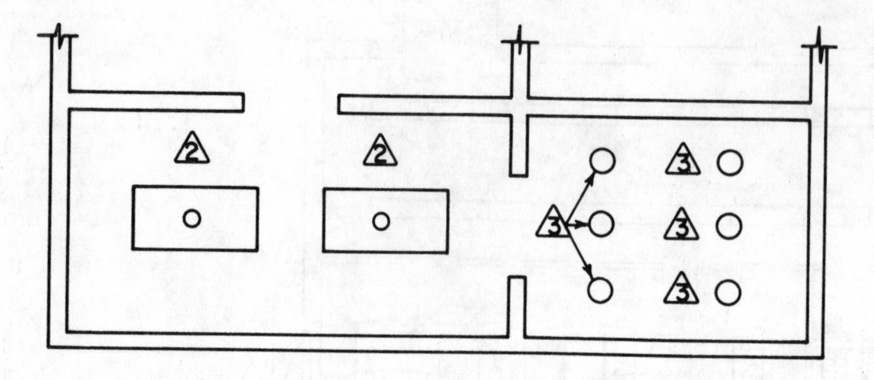

Fig. A2-17: Method of identifying lighting fixtures of different types.

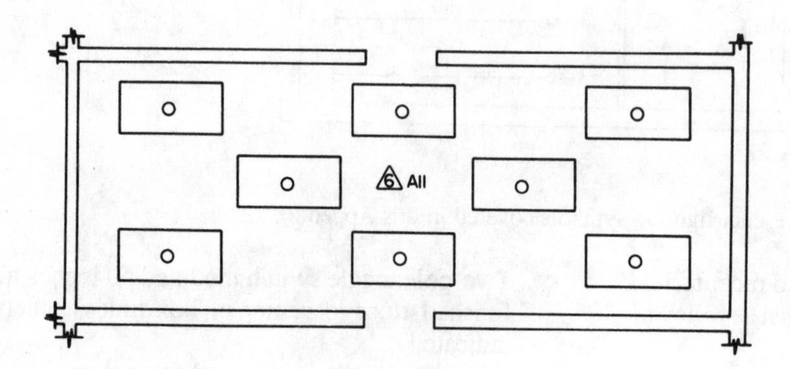

Fig. A2-18: Method of identifying several lighting fixtures of one type.

LIGHTING FIXTURE SCHEDULE						
FIXT. Type	Manufacturer's Description	LAMPS No.	LAMPS Type	Volts	Mounting	Remarks

Fig. A2-19: Typical lighting fixture schedule used on an electrical working drawing to identify lighting fixtures.

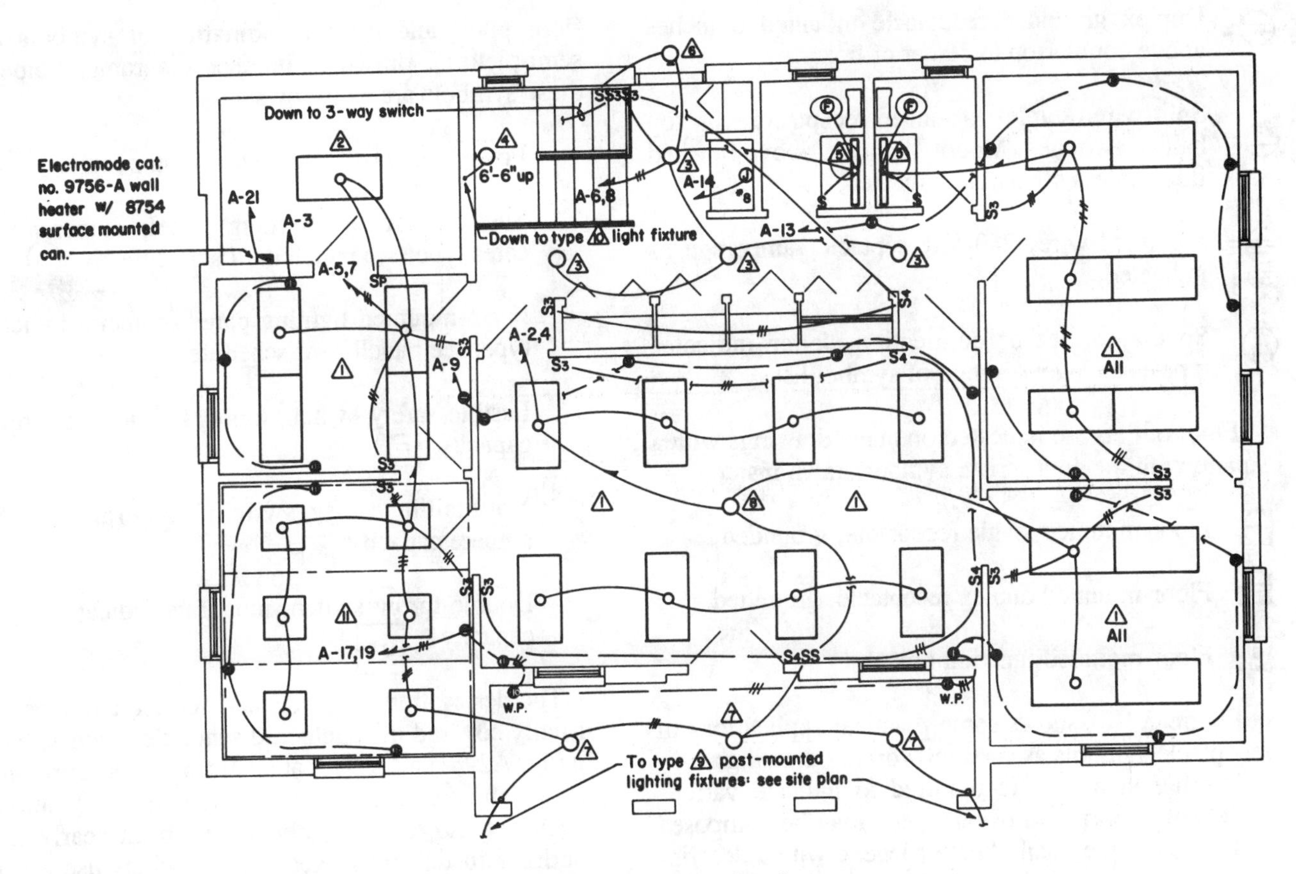

Fig. A2-20: Practical applications of switch symbols used on working drawings.

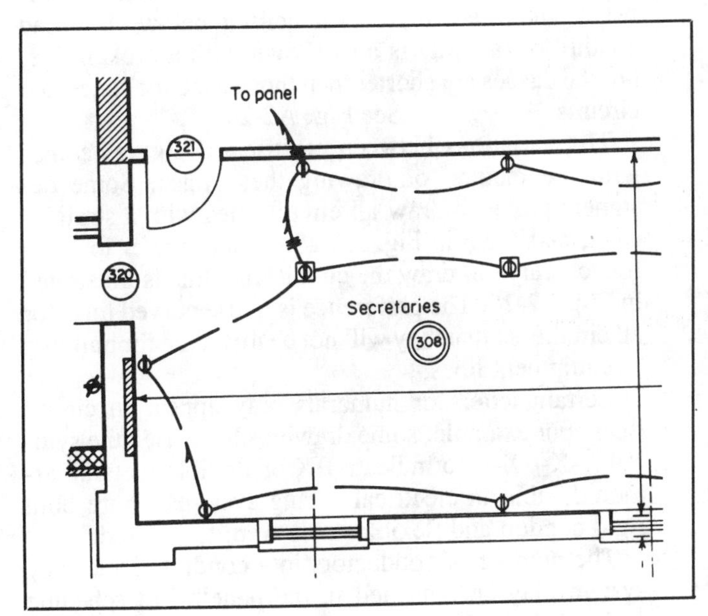

Fig. A2-21: Practical applications of receptacle symbols used on working drawings.

from one point, while the three- and four-way switches, used in combination, control a light or series of lights from two or more points.

The two-pole switch is used to control a series of lights on two separate circuits with only one motion. The switch-pilot light combination is used where it is practical to notice if the item controlled by the switch is energized, such as a light in an attic or closet.

Door switches to control closet lights are quite common in residential construction. When the closet door is opened, the switch button is released and in turn energizes the circuit to the closet light. When the door is closed, the light is de-energized.

Receptacle symbols used on working drawings are numerous. Some consulting engineering firms have used over 50 different symbols for receptacles and power outlets. However, many drawings contain only six different symbols to cover most of the applications. They are as follows:

Duplex grounded receptacle mounted 18 inches above finished floor to center of box unless otherwise indicated.

Duplex grounded receptacle mounted 6 inches above countertop to center of box.

Split-wired duplex grounded receptacle with top half of switched. Mount 18 inches above finished floor unless otherwise indicated.

30a 3-pole, 3-wire, 240-V receptacle, amperage as indicated.

a Special outlet or connection, letter indicates types; see legend at end of symbol list.

Floor outlets are indicated on many drawings with a square with the appropriate symbol drawn inside:

Floor-mounted single receptacle, grounded.

Floor-mounted duplex receptacle, grounded.

Floor-mounted junction box.

Figure A2-21 shows some practical applications of receptacle symbols as used on working drawings.

If other symbols are required to indicate various outlets on working drawings, they may be composed and added to the symbol list or legend with a description of their use. Examples are as follows:

Duplex grounded receptacle with "Twist-Lock" connection.

Whatever the need may be.

Whatever the need may be.

When outlets are located in areas requiring special boxes, covers, etc., they are usually indicated by abbreviations.

Example:

W.P.	Indicates weatherproof cover
E.P.	Indicates explosionproof device, fittings, etc.
EMERG	Indicates outlet on emergency circuit.

Service Equipment

Panelboards, distribution centers, transformers, safety switches, and similar components of the electrical installation are indicated by electrical symbols on floor plans and by a combination of symbols and semi-pictorial drawings in riser diagrams. Some of these symbols are as follows:

Power panel or main distribution panel

Surface-mounted lighting panel; numeral indicates type; see panelboard schedule (A)

Flush-mounted lighting panel; numeral indicates type; see panelboard schedule (A)

Fusible safety switch; numerals indicate ampere capacity F 60A

Nonfusible safety switch; numerals indicate ampere capacity N 30A

Double-throw switch; numerals indicate ampere capacity DT 30A

The description of panels and service equipment is usually covered in panelboard schedules such as those in Fig. A2-22 and A2-23; at other times a description of the panelboard is covered in the written specifications.

Circuit and feeder symbols have been nearly standardized in that most electrical drawings use a solid line ——— to indicate circuits concealed in ceiling or wall and a broken line — — — for conduit or raceways concealed in floor or ceiling below. Exposed conduit or raceway is also shown with a broken line, but the dashes are shorter than those used for concealed circuits – – – . See Fig. A2-24.

The variations between electrical designers comes with the method of drawing these lines. Some designers prefer to draw all circuit lines with a straightedge, as shown in Fig. A2-25. Others prefer to use a French curve to draw the circuit line; this is illustrated in Fig. A2-26. The preference is to use curved lines for all circuits so that they will not be mistaken for building or equipment lines.

Certain letters or numerals may appear in circuit lines. For example, some drawings have used the symbol —X—X— to indicate BX or flexible metallic armored cable in electrical wiring systems where both rigid conduit and flexible cable were to be used.

The number of conductors in a conduit or raceway system may be indicated in the panelboard schedule under the appropriate column, but on the other hand, this information is shown on the floor plans along with the circuits. Most workers find this latter method the easiest to follow. For example, one symbol list con-

Panel ______ 1 & 3 Wire ______ Mounted ______ Ampere Main ______

Location ______ ______ Ampere Bus

CCT No.	Volt - Amperes φ A	φ B	Description	Outlets Ltg	Rec	Cctbkr Pole	Ta	Phase A	B	Cctbkr Ta	Pole	Outlets Rec	Ltg	Description	Volt - Amperes φ A	φ B	CCT No.
1																	2
3																	4
5																	6
7																	8
9																	10
11																	12
13																	14
15																	16
17																	18
33																	34
35																	36
37																	38
39																	40
								Subtotals									
			Total VA/φ Total volt-amperes Line amperes														

Fig. A2-22: One type of panelboard schedule.

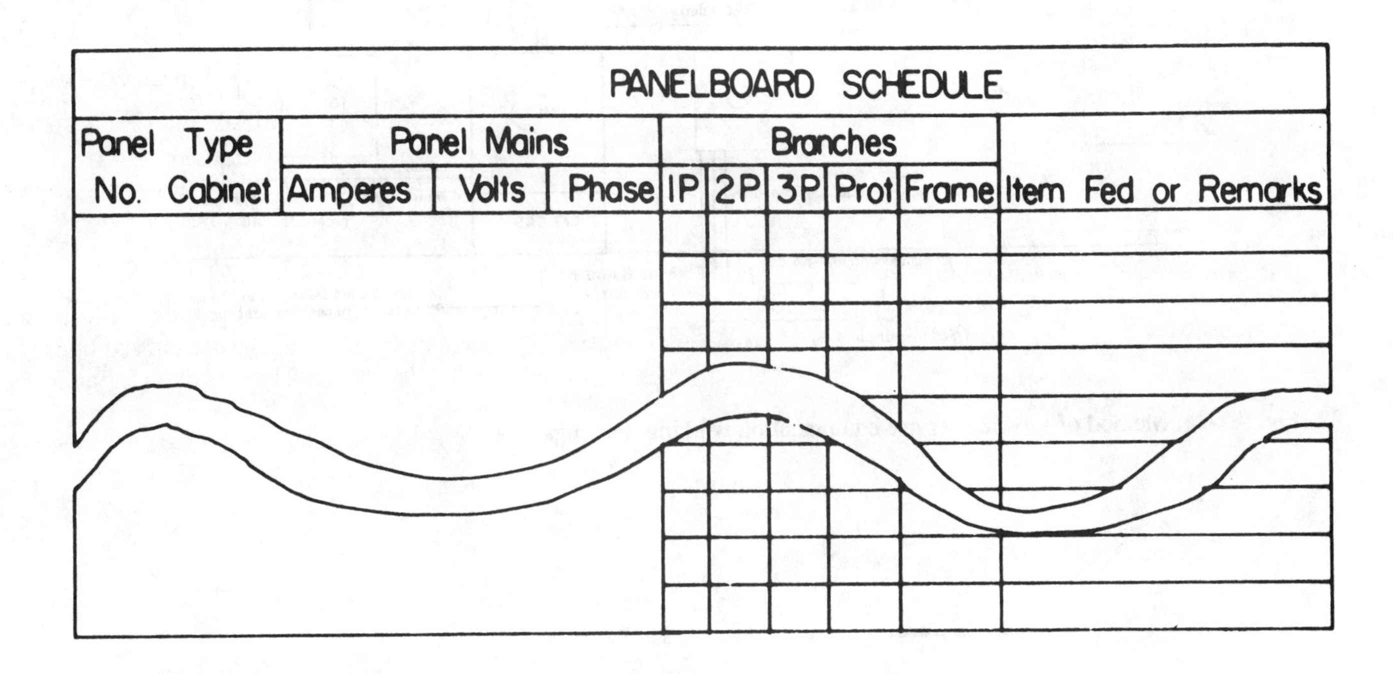

PANELBOARD SCHEDULE										
Panel No.	Type Cabinet	Panel Mains Amperes	Volts	Phase	Branches 1P	2P	3P	Prot	Frame	Item Fed or Remarks

Fig. A2-23: Another type of panelboard schedule.

tains the following symbol to indicate concealed branch circuit wiring:

Branch circuit concealed in ceiling or walls; slash marks indicate number of conductors in run—two conductors are not noted; numerals indicate wire size—No. 12 AWG not noted.

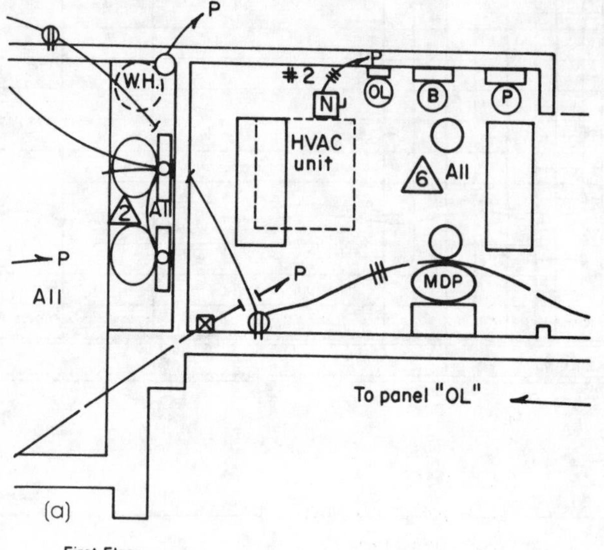

The symbol description says that a solid-line circui with no slash marks or numerals indicates a circui containing two No. 12 AWG conductors. Three slash marks with no numeral indicates three No. 12 AWG conductors, etc. If we want to indicate four No. 10 AWG conductors in a circuit, the symbol is:

Six slashes through the circuit lines on the drawing would indicate six conductors, etc.

Most electrical drawings used the symbol to indicat home runs to panelboard, with the number o arrowheads indicating the number of circuits in the ru and the slash marks indicating the number of conductor in the run. However, since full arrowheads are normall used for call-outs, half-arrowheads are frequently use to indicate branch-circuit home runs to panelboard.

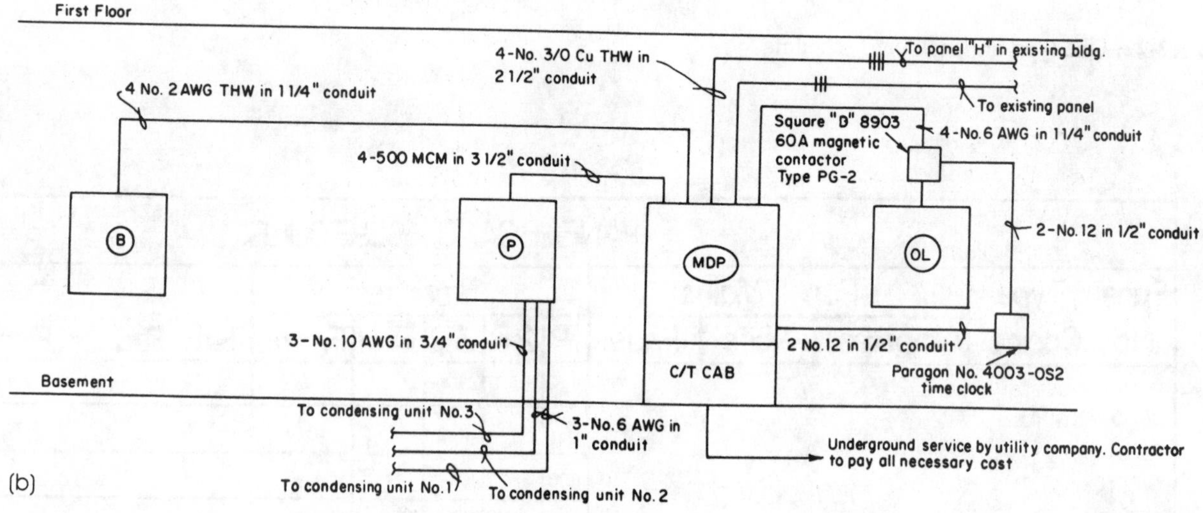

Fig. A2-24: Method of showing service equipment on working drawings.

Fig. A2-25: Circuit drawn with straightedge.

Fig. A2-26: Circuit drawn with French curve.

Index

D

E

F

G

H

I

K

L

M

N

O

P

R

S

Other Practical References

• Electrical Blueprint Reading Revised

Shows how to read and interpret electrical drawings, wiring diagrams, and specifications for constructing electrical systems. Shows how a typical lighting and power layout would appear on a plan, and explains what to do to execute the plan. Describes how to use a panelboard or heating schedule, and includes typical electrical specifications. **208 pages, 8½ x 11, $18.00**

• Electrical Construction Estimator

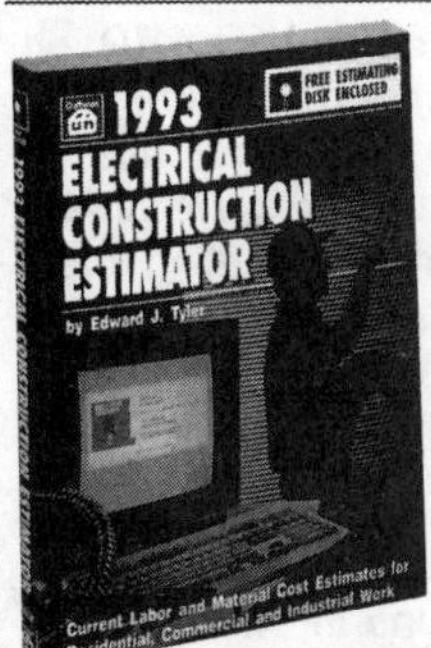

This year's prices for installation of all common electrical work: conduit, wire, boxes, fixtures, switches, outlets, loadcenters, panelboards, raceway, duct, signal systems, and more. Provides material costs, manhours per unit, and total installed cost. Explains what you should know to estimate each part of an electrical system. ***Electrical Estimate Writer is included FREE with the book on a 5¼" high-density (1.2 Mb) disk.*** (Add $10 for extra 5¼" double-density 360K disks or 3½" 720K disks.) **432 pages, 8½ x 11, $28.50. Revised annually.**

• Audiotapes: Estimating Electrical Work

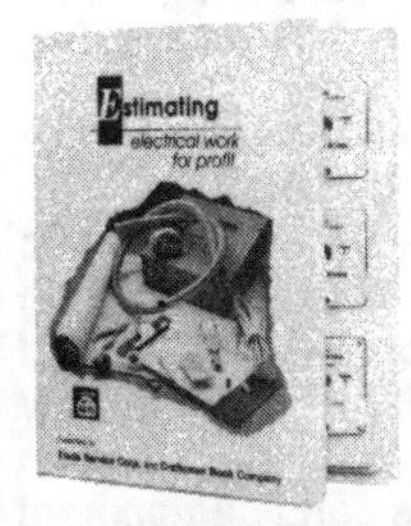

Listen to Trade Service's two-day seminar and study electrical estimating at your own speed for a fraction of the cost of attending the actual seminar. You'll learn what to expect from specifications, how to adjust labor units from a price book to your job, how to make an accurate take-off from the plans, and how to spot hidden costs that other estimators may miss. Includes six 30-minute tapes, a workbook that includes price sheets, specification sheet, bid summary, and estimate recap sheet, blueprints used in the actual seminar, and blank forms for your own use. **$65**

• Residential Wiring

Shows how to install rough and finish wiring in new construction, alterations, and additions. Complete instructions on troubleshooting and repairs. Every subject is referenced to the National Electrical Code, and there's over 24 pages of the most-needed NEC tables to help make your wiring pass inspection — the first time. **352 pages, 5½ x 8½, $18.25**

• Estimating & Bidding for Builders & Remodelers

New and more profitable ways to estimate and bid any type of construction. Shows how to take off labor and material, select the most profitable jobs for your company, estimate with a computer (FREE estimating disk enclosed), fine-tune your markup, and learn from your competition. Includes Estimate Writer, an estimating program with a 30,000-item database on a 5¼" high-density disk when you buy the book. (If your computer can't use high-density disks, add $10 for 5¼" or 3½" double-density disks.) **272 pages, 8½ x 11, $29.75**

• Estimating Electrical Construction

Like taking a class in how to estimate materials and labor for residential and commercial electrical construction. Written by an A.S.P.E. National Estimator of the Year, it teaches you how to use labor units, the plan take-off, and the bid summary to make an accurate estimate, how to deal with suppliers, use pricing sheets, and modify labor units. Provides extensive labor unit tables and blank forms for your next electrical job. **272 pages, 8½ x 11, $19.00**

• Residential Electrical Design

Shows how to draw up an electrical plan from blueprints, including the service entrance; grounding; lighting requirements for kitchen, bedroom and bath; and how to lay them out. Explains how to plan electrical heating systems and what equipment you'll need, how to plan outdoor lighting, and much more. If you're a builder who sometimes has to plan an electrical system, you should have this book. **194 pages, 8½ x 11, $11.50**

• National Construction Estimator

Current building costs for residential, commercial, and industrial construction. Estimated prices for every common building material. Manhours, recommended crew, and labor cost for installation. Includes Estimate Writer, an electronic version of the book on computer disk, with a stand-alone estimating program — free on 5¼" high density (1.2Mb) disk. The National Construction Estimator and Estimate Writer on 1.2Mb disk cost $31.50. (Add $10 if you want Estimate Writer on 5¼" double density 360K disks or 3½" 720K disks.) **592 pages, 8½ x 11, $31.50. Revised annually**

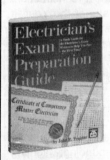

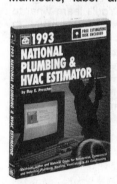

• How to Succeed With Your Own Construction Business

Everything you need to start your own construction business: setting up the paperwork, finding the work, advertising, using contracts, dealing with lenders, estimating, scheduling, finding and keeping good employees, keeping the books, and coping with success. If you're considering starting your own construction business, all the knowledge, tips, and blank forms you need are here. **336 pages, 8½ x 11, $19.50**

• Wood-Frame House Construction

Step-by-step construction details, from the layout of the outer walls, excavation and formwork, to finish carpentry and painting, with clear illustrations and explanations. Everything you need to know about framing, roofing, siding, insulation and vapor barrier, interior finishing, floor coverings, and stairs — complete step-by-step "how to" information on building a frame house. **320 pages, 8½ x 11, $19.75. Revised edition**

• Profits in Buying & Renovating Homes

Step-by-step instructions for selecting, repairing, improving, and selling highly profitable "fixer-uppers." Shows which price ranges offer the highest profit-to-investment ratios, which neighborhoods offer the best return, practical directions for repairs, and tips on dealing with buyers, sellers, and real estate agents. Shows you how to determine your profit before you buy, what "bargains" to avoid, and how to make simple, profitable, inexpensive upgrades. **304 pages, 8½ x 11, $19.75**

• Finish Carpentry

The time-saving methods and proven shortcuts you need to do first class finish work on any job: cornices and rakes, gutters and downspouts, wood shingle roofing, asphalt, asbestos and built-up roofing, prefabricated windows, door bucks and frames, door trim, siding, wallboard, lath and plaster, stairs and railings, cabinets, joinery, and wood flooring. **192 pages, 8½ x 11, $15.25**

• Home Wiring: Improvement, Extension, Repairs

How to repair electrical wiring in older homes, expand an existing electrical system in a house you're remodeling, and bring the electrical system up to modern standards. Shows how to use anticipated loads and demand factors to figure amperage and number of new circuits needed, and how to size and install wiring, conduit, switches, and auxiliary panels and fixtures. Explains how to test and troubleshoot fixtures, circuit wiring, and switches, and how to service or replace low voltage systems. **224 pages, 5½ x 8½, $15.00**

• Construction Estimating Reference Data

Provides the 300 most useful estimating reference tables. Labor requirements for nearly every type of construction, including: sitework, concrete work, masonry, steel, carpentry, thermal and moisture protection, doors and windows, finishes, mechanical and electrical. Each section details the work being estimated and gives the appropriate crew size and equipment needed. **368 pages, 8½ x 11, $30.00**

• Builder's Comprehensive Dictionary

Never let a construction term stump you again. Here you'll find almost 10,000 construction term definitions, over 1,000 detailed illustrations of tools, techniques, and systems, and a separate section of common legal, real estate, and management terms. **532 pages, 8½ x 11, $24.95**

• Estimating Framing Quantities

Gives you hundreds of time-saving estimating tips. Shows how to make thorough step-by-step estimates of all rough carpentry in residential and light commercial construction: ceilings, walls, floors, and roofs. Lots of illustrations showing lumber requirements, nail quantities, and practical estimating procedures. **285 pages, 5½ x 8½, $34.95**

• Residential Electrician's Handbook

Simple, clear instructions for wiring residences: understanding plans and specs, following the NEC, making simple load calculations, sizing wire and service equipment, installing branch and feeder circuits, and running wire. Explains how to estimate the cost of residential electrical systems, and speed up and simplify your estimates using composite unit prices. Includes forms and labor and material tables. **240 pages, 5½ x 8½, $16.75**

Craftsman Book Company
6058 Corte del Cedro
P. O. Box 6500
Carlsbad, CA 92018

In a hurry?
We accept phone orders charged to your MasterCard, Visa or American Express
Call 1-800-829-8123

Name ____________________
Address ____________________
Company ____________________
City/State/Zip ____________________
Total enclosed _______ In California at 7.25% tax
Use your ☐ Visa ☐ MasterCard or ☐ Am. Exp.
Card # ____________________
Expiration date ______________ Initials _______

10 Day Money Back GUARANTEE

- ☐ 26.50 Audio: Electrician's Exam Preparation Guide
- ☐ 65.00 Audio: Estimating Electrical Work
- ☐ 65.00 Audio: Estimating Remodeling
- ☐ 11.25 Blueprint Reading for the Building Trades
- ☐ 24.95 Builder's Comprehensive Dictionary
- ☐ 16.50 Building Cost Manual
- ☐ 28.00 Contractor's Guide to the Building Code Rev.
- ☐ 30.00 Construction Estimating Reference Data
- ☐ 18.00 Electrical Blueprint Reading Revised
- ☐ 28.50 Electrical Construction Estimator with **free** 5¼" high density *Electrical Estimate Writer* disk. If you need ☐ 5¼" or ☐ 3½" double density disks add$10 extra.
- ☐ 23.00 Electrician's Exam Preparation Guide
- ☐ 29.75 Estimating & Bidding for Builders & Remodelers
- ☐ 19.00 Estimating Electrical Construction
- ☐ 34.95 Estimating Framing Quantities
- ☐ 23.25 Fences & Retaining Walls
- ☐ 15.25 Finish Carpentry
- ☐ 15.00 Home Wiring: Improvement, Extension, Repairs
- ☐ 19.50 How to Succeed With Your Own Const. Busi.
- ☐ 31.50 National Construction Estimator with **free** 5¼" high density *Estimate Writer* disk. If you *need* ☐ 5¼" or ☐ 3½" double density disks add $10 extra.
- ☐ 32.25 National Plumbing & HVAC Estimator with **free** 5¼" high density *Plumbing & HVAC Estimate Writer* disk. If you need ☐ 5¼" or ☐ 3½" double density disks add $10 extra.
- ☐ 29.50 National Repair & Remodeling Estimator with **free** 5¼" high density *Repair & Remodeling Estimate Writer. If you need* ☐ 5¼" or ☐ 3½" double density disks add $10 extra.
- ☐ 19.75 Profits in Buying & Renovating Homes
- ☐ 11.50 Residential Electrical Design
- ☐ 16.75 Residential Electrician's Handbook
- ☐ 18.25 Residential Wiring
- ☐ 27.00 Spec Builder's Guide
- ☐ 19.75 Wood-Frame House Construction
- ☐ 26.75 Illustrated Guide to National Electrical Code
- ☐ Free Full Color Catalog

Craftsman Book Company
6058 Corte del Cedro
P. O. Box 6500
Carlsbad, CA 92018

In a hurry?
We accept phone orders charged to your MasterCard, Visa or American Express
Call 1-800-829-8123

Name ____________________
Address ____________________
Company ____________________
City/State/Zip ____________________
Total enclosed _______ In California at 7.25% tax
Use your ☐ Visa ☐ MasterCard or ☐ Am. Exp.
Card # ____________________
Expiration date ______________ Initials _______

10 Day Money Back GUARANTEE

- ☐ 26.50 Audio: Electrician's Exam Preparation Guide
- ☐ 65.00 Audio: Estimating Electrical Work
- ☐ 65.00 Audio: Estimating Remodeling
- ☐ 11.25 Blueprint Reading for the Building Trades
- ☐ 24.95 Builder's Comprehensive Dictionary
- ☐ 16.50 Building Cost Manual
- ☐ 28.00 Contractor's Guide to the Building Code Rev.
- ☐ 30.00 Construction Estimating Reference Data
- ☐ 18.00 Electrical Blueprint Reading Revised
- ☐ 28.50 Electrical Construction Estimator with **free** 5¼" high density *Electrical Estimate Writer* disk. If you need ☐ 5¼" or ☐ 3½" double density disks add$10 extra.
- ☐ 23.00 Electrician's Exam Preparation Guide
- ☐ 29.75 Estimating & Bidding for Builders & Remodelers
- ☐ 19.00 Estimating Electrical Construction
- ☐ 34.95 Estimating Framing Quantities
- ☐ 23.25 Fences & Retaining Walls
- ☐ 15.25 Finish Carpentry
- ☐ 15.00 Home Wiring: Improvement, Extension, Repairs
- ☐ 19.50 How to Succeed With Your Own Const. Busi.
- ☐ 31.50 National Construction Estimator with **free** 5¼" high density *Estimate Writer* disk. If you *need* ☐ 5¼" or ☐ 3½" double density disks add $10 extra.
- ☐ 32.25 National Plumbing & HVAC Estimator with **free** 5¼" high density *Plumbing & HVAC Estimate Writer* disk. If you need ☐ 5¼" or ☐ 3½" double density disks add $10 extra.
- ☐ 29.50 National Repair & Remodeling Estimator with **free** 5¼" high density *Repair & Remodeling Estimate Writer. If you need* ☐ 5¼" or ☐ 3½" double density disks add $10 extra.
- ☐ 19.75 Profits in Buying & Renovating Homes
- ☐ 11.50 Residential Electrical Design
- ☐ 16.75 Residential Electrician's Handbook
- ☐ 18.25 Residential Wiring
- ☐ 27.00 Spec Builder's Guide
- ☐ 19.75 Wood-Frame House Construction
- ☐ 26.75 Illustrated Guide to National Electrical Code
- ☐ Free Full Color Catalog

Mail This Card Today For a Free Full Color Catalog

Over 100 books, videos, and audios at your fingertips with information that can save you time and money. Here you'll find information on carpentry, contracting, estimating, remodeling, electrical work, and plumbing

All items come with an unconditional 10-day money-back guarantee. If they don't save you money, mail them back for a full refund.

Name ____________________
Company ____________________
Address ____________________
City/State/Zip ____________________

Craftsman Book Company / 6058 Corte del Cedro / P. O. Box 6500 / Carlsbad, CA 92018

NO POSTAGE
NECESSARY
IF MAILED
IN THE
UNITED STATES

BUSINESS REPLY MAIL

FIRST CLASS MAIL PERMIT NO. 271 CARLSBAD CA

POSTAGE WILL BE PAID BY ADDRESSEE

CRAFTSMAN BOOK COMPANY
PO BOX 6500
CARLSBAD CA 92018-9974

NO POSTAGE
NECESSARY
IF MAILED
IN THE
UNITED STATES

BUSINESS REPLY MAIL

FIRST CLASS MAIL PERMIT NO. 271 CARLSBAD CA

POSTAGE WILL BE PAID BY ADDRESSEE

CRAFTSMAN BOOK COMPANY
PO BOX 6500
CARLSBAD CA 92018-9974

NO POSTAGE
NECESSARY
IF MAILED
IN THE
UNITED STATES

BUSINESS REPLY MAIL

FIRST CLASS MAIL PERMIT NO. 271 CARLSBAD CA

POSTAGE WILL BE PAID BY ADDRESSEE

CRAFTSMAN BOOK COMPANY
PO BOX 6500
CARLSBAD CA 92018-9974